T0155469

Electronic and Photoelectron Spectroscopy

Fundamentals and Case Studies

Electronic and photoelectron spectroscopy can provide extraordinarily detailed information on the properties of molecules and are in widespread use in the physical and chemical sciences. Applications extend beyond spectroscopy into important areas such as chemical dynamics, kinetics, and atmospheric chemistry. This book provides the reader with a firm grounding in the basic principles and experimental techniques employed. The extensive use of case studies effectively illustrates how spectra are assigned and how information can be extracted, communicating the matter in a compelling and instructive manner.

Topics covered include laser-induced fluorescence, resonance-enhanced multiphoton ionization, cavity ringdown and ZEKE spectroscopy. The book is for advanced undergraduate and graduate students taking courses in spectroscopy and will also be of use to anyone encountering electronic or photoelectron spectroscopy during their research.

ANDREW ELLIS has research interests which encompass various aspects of electronic spectroscopy. He has taught numerous courses in physical chemistry and chemical physics and is currently a Senior Lecturer at the University of Leicester.

MIKLOS FEHER is Director of Computational Chemistry at Neurocrine Biosciences, San Diego, California. He has taught various invited lecture courses throughout the world and has published a textbook on quantum chemistry.

TIMOTHY WRIGHT received his doctorate in photoelectron spectroscopy at the University of Southampton in 1991. He is now Reader in the School of Chemistry, University of Nottingham.

Electronic and Photoelectron Spectroscopy

Fundamentals and Case Studies

ANDREW M. ELLIS

Department of Chemistry
University of Leicester, UK

MIKLOS FEHER

Neurocrine Biosciences
San Diego, USA

TIMOTHY G. WRIGHT

School of Chemistry
University of Nottingham, UK

CAMBRIDGE UNIVERSITY PRESS
Cambridge, New York, Melbourne, Madrid, Cape Town, Singapore,
São Paulo, Delhi, Dubai, Tokyo, Mexico City

Cambridge University Press
The Edinburgh Building, Cambridge CB2 8RU, UK

Published in the United States of America by Cambridge University Press, New York

www.cambridge.org
Information on this title: www.cambridge.org/9780521520638

First published 2005
First paperback edition 2010

A catalogue record for this publication is available from the British Library

Library of Congress Cataloguing in Publication Data
Ellis, Andrew M. (Andrew Michael), 1963–
Electronic and photoelectron spectroscopy : fundamentals and case studies / Andrew M. Ellis, Miklos Feher, Timothy G. Wright.
p. cm.
Includes bibliographical references and index.
ISBN 0 521 81737 4 (hardback : alk. paper)
1. Photoelectron spectroscopy – Study and teaching. 2. Electron spectroscopy – Study and teaching. I. Fehér, Miklós, 1960– II. Wright, Timothy G. (Timothy Grahame), 1965–
III. Title.
QC454.P48E44 2005
543′.62 – dc22 2004052544

ISBN 978-0-521-81737-0 Hardback
ISBN 978-0-521-52063-8 Paperback

Contents

Preface

Modern spectroscopic techniques such as laser-induced fluorescence, resonance-enhanced multiphoton ionization (REMPI), cavity ringdown, and ZEKE are important tools in the physical and chemical sciences. These, and other techniques in electronic and photoelectron spectroscopy, can provide extraordinarily detailed information on the properties of molecules in the gas phase and see widespread use in laboratories across the world. Applications extend beyond spectroscopy into important areas such as chemical dynamics, kinetics, and analysis of complicated chemical systems such as plasmas and the Earth's atmosphere. This book aims to provide the reader with a firm grounding in the basic principles and experimental techniques employed in modern electronic and photoelectron spectroscopy. It is aimed particularly at advanced undergraduate and graduate level students studying courses in spectroscopy. However, we hope it will also be more broadly useful for the many graduate students in physical chemistry, theoretical chemistry, and chemical physics who encounter electronic and/or photoelectron spectroscopy at some point during their research and who wish to find out more.

There are already many books available describing the principles, experimental techniques, and applications of spectroscopy. However, our aim has been to produce a book that tackles the subject in a rather different way from predecessors. Students at the advanced undergraduate and early graduate levels should be in a position to develop their knowledge and understanding of spectroscopy through contact with the research literature. This has the benefit of introducing the students to the cutting edge of modern spectroscopic work and can provide insight into the thought processes involved in spectral assignment and interpretation. However, the spectroscopic research literature can initially prove daunting even to the most committed and able of students because of the range of prior knowledge assumed, the brevity of explanations, and the extensive use of jargon.

We felt that there would be benefit in taking a number of focussed, and mostly modern, research studies and presenting them in a form that is palatable for the newcomer to advanced spectroscopy. We have called these mini-chapters Case Studies and they form the heart of this book. In essence we have taken original research findings, often directly from research papers, and describe selected aspects of them in a way which not only shows the original data and conclusions, but also tries to guide the reader step-by-step through the assignment and interpretation process. In other words, we have in many cases tried to put the reader in the shoes of the research team that first recorded the spectrum or spectra, and then tried to show them *how* the spectrum was assigned.

Jargon cannot be avoided entirely – indeed it is an essential part of the language of modern spectroscopy – but we have attempted to define any specialized jargon that does arise as we encounter it.

Of course some basic background knowledge is essential before encountering more advanced concepts, and so the first two parts describe some of the principles and experimental techniques employed in modern electronic and photoelectron spectroscopy. These two parts are not intended to be exhaustive, but rather contain the basic tools necessary for delving into the Case Studies. Some of the more advanced concepts met in spectroscopy, such as vibronic coupling, nuclear spin statistics, and Hund's coupling cases, are met only in certain specific Case Studies and can be entirely avoided by the reader if desired.

As much as possible, we have tried to make the majority of the Case Studies independent. This means that the reader can dip into only those that interest him/her. At the same time, this approach inevitably leads to some repetition of material but we consider this an acceptable price to pay for producing a book in this style.

We view the Case Studies as a useful bridge between traditional teaching and fully independent learning through the research level literature. We do not in any way claim to have covered all of the important topics in modern electronic spectroscopy, nor have we attempted to treat any particular topic in great depth. However, we believe that most of the material in electronic spectroscopy encountered in advanced undergraduate and early graduate level spectroscopy courses is covered within this book. Furthermore, we hope that the focus on research material will give the reader a flavour of the kind of work that currently takes place in the spectroscopic community and will encourage him/her to explore new avenues. Whether we have been successful or not is purely for the reader to judge.

Finally, the authors would like to take this opportunity to thank Cambridge University Press for showing great patience on the numerous occasions when the finishing date for the manuscript was postponed!

Journal abbreviations

Abbreviations are used for journal titles in the list of references at the end of each chapter. The full title of each journal is listed below.

Angew. Chemie Int. Edn.	*Angewandte Chemie, International Edition in English*
Ber. Bunsenges. Phys. Chem.	*Berichte der Bunsengesellschaft für Physikalische Chemie*
Chem. Phys.	*Chemical Physics*
Chem. Phys. Lett.	*Chemical Physics Letters*
Chem. Rev.	*Chemical Reviews*
Comput. Phys. Commun.	*Computer Physics Communications*
Found. Phys.	*Foundations of Physics*
Instrum. Sci. Technol.	*Instrumentation Science and Technology*
Int. Rev. Phys. Chem.	*International Reviews in Physical Chemistry*
J. Chem. Educ.	*Journal of Chemical Education*
J. Chem. Phys.	*Journal of Chemical Physics*
J. Chem. Soc.	*Journal of the Chemical Society*
J. Electron Spectrosc. Rel. Phenom.	*Journal of Electron Spectroscopy and Related Phenomena*
J. Mol. Spectrosc.	*Journal of Molecular Spectroscopy*
J. Opt. Soc. Am.	*Journal of the Optical Society of America*
J. Phys. Chem.	*Journal of Physical Chemistry*
Math. Comp.	*Mathematics of Computation*
Mol. Phys.	*Molecular Physics*
Philos. Trans. Roy. Soc.	*Philosophical Transactions of the Royal Society of London*
Phys. Rev.	*Physical Review*
Vib. Spectrosc.	*Vibrational Spectroscopy*
Z. Phys.	*Zeitschrift für Physik*
Z. Wiss. Photogr. Photophys. Photochem.	*Zeitschift für Wissenschaftliche Photographie, photophysik und photochemie*

Part I

Foundations of electronic and photoelectron spectroscopy

1 Introduction

1.1 The basics

It is convenient to view electrons in atoms and molecules as being in orbitals. This idea is ingrained in chemistry and physics students early on in their studies and it is a powerful concept that provides explanations for a wide variety of phenomena. It is important to stress from the very beginning that the concept of an orbital in any atom or molecule possessing more than one electron is an *approximation*. In other words, orbitals do not actually exist, although electrons in atoms and molecules often behave to a good approximation as if they were in orbitals.

An orbital describes the spatial distribution of a particular electron. For example, we expect that an electron in a $1s$ orbital in an atom will, on average, be much closer to the nucleus than an electron in a $2s$ orbital in the same atom. Qualitatively, we would picture the electron as being represented by a charge cloud with a much greater density near the nucleus for the $1s$ orbital than the $2s$ orbital. Similarly, we know that the electron in a $2p_z$ orbital does not have a spherically symmetric distribution, as does an s electron, but instead is distributed in a cylindrically symmetric fashion about the z axis with the charge cloud consisting of lobes pointing along both the $+z$ and $-z$ directions.

Within the constraints of the orbital approximation, electronic spectroscopy is the study of transitions of electrons from one orbital to another, induced by the emission or absorption of a quantum of electromagnetic radiation, i.e. a photon. Each orbital in an atom or molecule has a specific energy, E_n, and to induce a transition between these orbitals the photon must satisfy the resonance condition

$$E_2 - E_1 = h\nu = \frac{hc}{\lambda} \tag{1.1}$$

where ν and λ are the frequency and wavelength of the radiation, respectively, and h is the Planck constant (see Appendix A). Under normal circumstances, only one electron is involved in the promotion or demotion process, and therefore we say that we are dealing with *one-electron transitions*. Thus all other electrons remain in their original orbitals, although their energies may have changed as a result of the electronic transition.

In electronic emission spectroscopy, an electron drops to an orbital of lower energy with the concomitant emission of a photon. Owing to the quantization of orbital energies, only photons of certain discrete wavelengths are produced and an *emission spectrum* can therefore be obtained by measuring the emitted radiation intensity as a function of wavelength. In absorption, the reverse process operates and an *absorption spectrum* can be obtained by measuring the change in intensity of radiation, such as that produced by a continuum lamp, as a function of wavelength after passing it through a sample.

Photoelectron spectroscopy is essentially a special case of electronic absorption spectroscopy[1] in which the electron is given enough energy to take it beyond any of the bound orbitals: in other words, the electron is able to escape the binding forces of the atom or molecule and is said to have exceeded the *ionization limit*. The minimum energy required to do this is the *ionization energy* for an electron in that particular orbital. Photoionization differs from an absorption transition involving two bound orbitals in that there is more than one photon energy which can bring it about. In fact any photon with an energy high enough to promote an electron above the ionization limit can, in principle, bring about photoionization. Notice that this does not defy the resonance condition: the resonance condition equivalent to the requirement that energy be conserved and is still satisfied because the electron is able to take away any excess energy in the form of electron kinetic energy.

Since there are no discrete absorption wavelengths (only discrete absorption *onsets*), photoelectron spectroscopy is carried out in a very different manner from conventional electronic absorption spectroscopy. As the name implies, it is electron energies rather than photon energies which are measured. For an atom, part of the energy ($h\nu$) provided by the incoming photon is used to ionize the atom. The remainder is partitioned between the atomic cation and the electron kinetic energy and so, from the conservation of energy,

$$h\nu = IE_i + T_{\text{ion}} + T_{\text{e}} \tag{1.2}$$

where IE_i is the ionization energy of an electron in orbital i and T_{ion} and T_{e} are the cation and electron kinetic energies, respectively. Given that an electron is very much lighter than an atomic nucleus, conservation of momentum dictates that the ion recoil velocity will be very low and most of the kinetic energy will be taken away by the electron. As a result, T_{ion} can usually be neglected and a spectrum can therefore be obtained by fixing $h\nu$ and measuring the electron current as a function of electron kinetic energy. This is the basic idea of the traditional photoelectron spectroscopy experiment. In the case of atoms, peaks will appear at various electron energies in the spectrum corresponding to ionization of electrons from the various occupied orbitals. A peak at a given T_{e} can be converted to an orbital ionization energy using equation (1.2) provided the ionizing photon frequency ν is known.

Photoelectron spectroscopy is a good example of the tremendous changes that have taken place in spectroscopic techniques over the past two or three decades. Although conventional photoelectron spectroscopy as outlined above is still important and widely used, a relatively new method of electron spectroscopy, *zero electron kinetic energy (ZEKE)* spectroscopy, is now capable of extracting the same type of information but at much higher resolution. ZEKE spectroscopy is one of those techniques that has benefited from the introduction of the laser as a spectroscopic light source. There are many other laser-based spectroscopic techniques, some relatively simple and some which are very complicated. Most of the spectroscopic

[1] To minimize verbosity the term *electronic spectroscopy* will often be used to encompass both 'normal' electronic spectroscopy and photoelectron spectroscopy.

data presented in this book have been obtained using *laser spectroscopy* of one form or another, which should indicate its importance in the study of molecules in the gas phase.

However, it is not the aim of this book to describe the wide variety of methods that are available for electronic spectroscopy, although some experimental details are given in Part II. Rather the focus is primarily on the spectra themselves and in particular how they can be interpreted and what they reveal. The underlying principles needed to do this are common to a variety of different spectroscopic techniques, and in this part we develop the basic theoretical background.

1.2 Information obtained from electronic and photoelectron spectra

Before addressing some of the theoretical principles, we want to convince the reader that electronic spectroscopy is worthwhile doing. In particular, what information can be extracted from an electronic or photoelectron spectrum? This will be addressed in some detail when specific examples are met in Part III, but let us outline at this early stage some of the extraordinary range of information that can be deduced.

First and most obviously, information is obtained on orbital energies. In particular, the spectroscopic transition energy can be equated with the *difference* in energy between the two orbitals involved in an electronic transition (assuming that the orbital energies are unchanged as a result of the electron changing orbitals, which is only approximately true). Photoelectron spectroscopy is even more informative in this regard, since in the upper state the electron has no binding energy and can therefore be regarded as being in an orbital with zero potential energy. Consequently, ionization energies are a direct measure of orbital energies in the neutral atom or molecule, and can therefore be used to construct a molecular orbital diagram.

However, electronic spectroscopy is able to provide much more than just a measure of absolute orbital energies or orbital energy differences. Very often, particularly for molecules in the gas phase, vibrational and rotational structure can be resolved. Vibrational structure leads directly to vibrational frequencies. As will be seen later, not all vibrations need be active in electronic spectra. Excitation of some vibrations may be *forbidden* because of their symmetries. This may seem unfortunate, but in fact the absence of certain vibrational features can also have the benefit of providing qualitative, and sometimes even quantitative, information on the structure of the molecule, as will be shown later.

Rotational structure tends to be difficult to resolve in electronic spectra, except for small molecules, but when it is obtained it can be highly informative. Accurate equilibrium structures in both upper and lower states may be extracted from a rotational analysis. In addition, the exact details of the rotational structure are not only dependent on molecular structure, but also on the symmetries of the electronic states involved. Consequently, rotationally resolved spectra provide a reliable means of establishing electronic state symmetries. When spectra are of exceptionally high resolution there is even more information that can be extracted, although such ultra-high resolution spectra will not be considered in any detail in this book.

Finally, one should note that some of the laser-based methods of electronic spectroscopy are extremely sensitive, and are therefore able to detect very small quantities of a particular sample. This has many different uses, particularly in analytical chemistry. Furthermore, it is possible to detect and characterize molecules that are extremely unstable or reactive and therefore inevitably have a fleeting presence and/or very low concentrations. Species in this category would include free radicals and molecular ions, and we will show a number of examples in Part III.

2 Electronic structure

2.1 Orbitals: quantum mechanical background

In this and the subsequent chapter the reasoning behind the concept of orbitals in atoms and molecules is outlined. An appreciation of what an orbital is, and what its limitations are, is vital for an understanding of electronic spectroscopy. Some may find this section frustrating in that little justification is given for many of the statements made. However, the theoretical treatment of electronic structure is a complicated subject and is for the most part beyond the scope of this book, although some effort is made to summarize some of the technical issues in Appendix B. For a detailed account, including proof of the statements made below, the reader should consult some of the more advanced texts listed at the end of this chapter.

2.1.1 *Wave–particle duality and the Schrödinger equation*

An orbital defines the spatial distribution of an electron within an atom or molecule. It arises from the application of quantum mechanical ideas to atomic and molecular structure. Central to the wave mechanical view of quantum mechanics is the identification of a wavefunction, ψ, of a system, which is a solution of the Schrödinger equation

$$H\psi = E\psi \tag{2.1}$$

This simple-looking and very famous equation is deceptive, for it is more complicated than it first appears. H, the so-called Hamiltonian, is actually a mathematical *operator* composed of, among other things, second-order differential operators such as $\mathrm{d}^2/\mathrm{d}x^2$. On its own it is therefore an abstract mathematical quantity. The detailed form of the Hamiltonian appropriate for describing the electronic structure of atoms and molecules is given in Section 2.1.3. The Hamiltonian is an *energy* operator which, when it operates on the wavefunction on the left-hand side of equation (2.1), generates an energy, E, multiplied by the wavefunction on the right-hand side. The energy is said to be an *observable*, i.e. it is a physical property that can, in principle, be measured.

The Schrödinger equation provides the means for describing physical behaviour at the atomic and molecular level. Underlying this description is the implication that all matter possesses wave-like properties, and that this becomes particularly significant when dealing with sub-atomic particles, such as electrons, protons, and neutrons, and collections of these particles in atoms and molecules. The possession of both wave and particle properties is known as *wave–particle duality*. The wave characteristics are represented mathematically by the wavefunction, ψ, and it is important to have some feel for what it is that the wavefunction describes. Although oversimplified, it is useful to view the wavefunction as describing the amplitude of a matter wave, such as that associated with an electron, throughout space. If we persist in trying to think of the electron as a particle, then the alternative wave description clearly muddies any effort to specify the precise location of the electron at any instant in time. Instead, we can only specify the *probability* that the electron will be found at a particular place at a particular instant in time in an experimental measurement. This probabilistic interpretation was made quantitative by Born, who associated the square of the wavefunction, ψ^2, evaluated at some point in space, with the probability of the particle being at that point in space at any instant in time.[1] Since ψ is a continuous function and the particle must be located somewhere in space, we insist that

$$\int \psi^2 \, \mathrm{d}V = 1 \tag{2.2}$$

where, although no integration limits have been shown, the implication is that integration is over all accessible space (V is the volume). This is known as the *normalization* condition.

2.1.2 *The Born–Oppenheimer approximation*

The Schrödinger equation for molecules is complicated, since it must describe not only the motion of a collection of electrons, but also nuclear motion as well. However, there is an important simplification that can be made, which follows from the large mass difference between electrons and nuclei. Given that the mass of a proton is 1836 times larger than that of an electron, electrons in a molecule will generally move at far greater speeds than the nuclei. When the nuclei make small changes in their relative positions, such as during a molecular vibration, the electron cloud almost instantaneously adjusts to the new set of nuclear positions. This means that the electrons are almost completely unaffected by the speed with which the nuclei move. This statement is one version of a very important approximation known as the *Born–Oppenheimer approximation*.

The utility of the Born–Oppenheimer approximation is that it makes it possible to separate the total energy of a molecule into two terms, namely,

$$E_{\text{total}} = E_{\text{elec}} + E_{\text{nkin}} \tag{2.3}$$

where E_{elec} is the energy consisting of the potential energy due to all electrostatic interactions (see next section) plus the electron kinetic energies, and E_{nkin} is the kinetic energy due to nuclear motions (vibrations and rotations). Since the electronic structure is affected by the nuclear coordinates but not their rate of change, E_{nkin} can be ignored for the time being.

---

1 Solution of the Schrödinger equation can yield complex wavefunctions in some instances, i.e. ψ may have both real and imaginary parts. Since we only attach a physical interpretation to the square of the wavefunction, rather than the wavefunction itself, this causes no practical problems. It is simply necessary to ensure that the square of the wavefunction is a real quantity, and so for complex wavefunctions $\psi^*\psi$ must be used in place of ψ^2, where ψ^* is the complex conjugate of ψ.

2.1.3 The Schrödinger equation for many-electron atoms and molecules

If the Born–Oppenheimer approximation is invoked, the Hamiltonian in the Schrödinger equation (2.1), for a molecule with fixed nuclear positions has the general form

$$H = -\frac{\hbar^2}{2}\sum_i \frac{1}{m_i}\left(\frac{\partial^2}{\partial x_i^2}+\frac{\partial^2}{\partial y_i^2}+\frac{\partial^2}{\partial z_i^2}\right) - \sum_{i,A}\frac{Z_A e^2}{4\pi\varepsilon_0 R_{iA}} + \sum_i\sum_{j\neq i}\frac{e^2}{4\pi\varepsilon_0 r_{ij}} + \sum_A\sum_{B\neq A}\frac{Z_A Z_B e^2}{4\pi\varepsilon_0 R_{AB}} \tag{2.4}$$

In the above expression we make use of the general relationship from classical electrostatics that the electrostatic potential energy between two particles with charges q_i and q_j separated by distance r is given by $q_i q_j/4\pi\varepsilon_0 r$, where ε_0 is the permittivity of free space. In this specific case Z_A is used to designate the charge on nucleus A and e is the fundamental charge (an electron has charge $-e$). The quantity $\hbar$ is shorthand notation for $h/2\pi$.

Although it looks formidable, equation (2.4) has a simple interpretation. Four groups of operators can be identified inside the summations in (2.4). The first group is the total electron kinetic energy operator, which is the sum of kinetic energy operators for each electron. The second summation represents the electron–nuclear electrostatic interactions, where R_{iA} is the electron–nuclear distance, with the subscripts i and A labelling electrons and nuclei respectively. The third term, the first of the double summations, is the operator for electron–electron repulsion, while the fourth is for nuclear–nuclear repulsion. The Hamiltonian is therefore logical in the sense that it is a total energy operator constructed from the summation of kinetic energy operators for each individual electron and the operators describing all electron–nuclear, electron–electron, and nuclear–nuclear electrostatic interactions in the molecule. This is also illustrated in Figure 2.1 for a two-electron diatomic molecule. Had we not invoked the Born–Oppenheimer approximation, the molecular Schrödinger equation would also have to have included terms containing nuclear kinetic energy operators, which would clearly be an added complication.

Despite the simplification brought about by the Born–Oppenheimer approximation, the Schrödinger equation containing the Hamiltonian in (2.4) still cannot be solved exactly for any molecule containing more than one electron. The problem lies with the third term in the Hamiltonian, the electron–electron repulsions. If one were to imagine creating a molecule containing several electrons but these electrons interacted only with the nuclei, i.e. there were no electron–electron repulsions (clearly an imaginary situation!), then equation (2.4) would be rather easy to solve. In this limit the electronic wavefunction is a product of wavefunctions for each individual electron, i.e.

$$\psi = \phi_1(1)\phi_2(2)\ldots\phi_N(N) \tag{2.5}$$

where N is the number of electrons and $\phi_i(i)$ is the wavefunction of electron i. The product form of the wavefunction makes it possible to separate the full Schrödinger equation into a series of individual and independent Schrödinger equations, one for each electron, each of

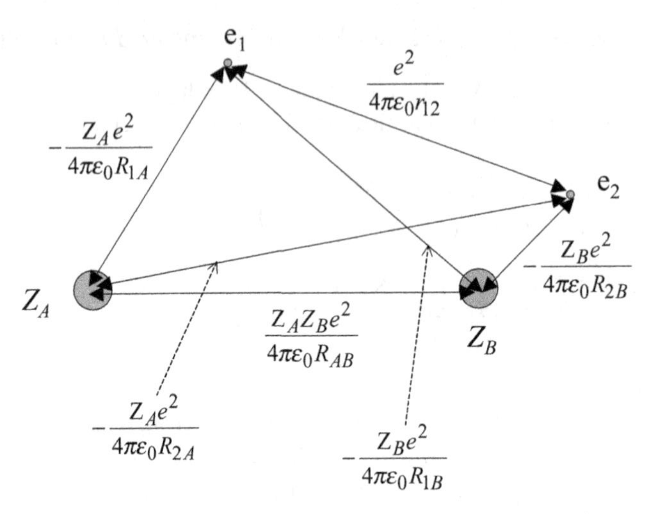

Figure 2.1 A schematic illustration of the electrostatic potential energies in a two-electron diatomic molecule at some particular instant in time.

which can then be solved exactly after some mathematical effort.[2] Each of the wavefunctions for the individual electrons describes the spatial probability distribution of that electron: in other words ϕ_1 is a function of the coordinates only of electron 1 and ϕ_1^2 at any point in space describes the probability that the electron is at that point in space. The individual wavefunctions ϕ_i are referred to as *orbitals*.

2.1.4 The orbital approximation

Electron–electron repulsion destroys the orbital picture given above. This arises because each electron–electron repulsion operator is a function of the coordinates of *two* electrons, and therefore the position of one electron affects all of the others. Consequently, the true electronic wavefunction in a molecule (or atom) containing two or more electrons is not a product of independent one-electron wavefunctions (orbitals).

This would seem to be unfortunate, since the ability to be able to describe electrons as being in separate orbitals offers a great simplification in the description of electronic structure. Fortunately, all is not lost since it is possible to retain the orbital concept if the following approach is adopted. We know that, strictly speaking, the total electronic wavefunction cannot be expressed exactly as a product of orbital wavefunctions. However, suppose that we in any case *choose* to express the total wavefunction as such an orbital product. This constraint allows the many-electron Schrödinger equation to be converted

[2] This process, known as separation of variables, is used in solving many quantum mechanical problems. For example, it is employed for relatively simple problems such as the quantum mechanics of a single particle in a two- or three-dimensional box, and at a more sophisticated level is used to obtain solutions of the Schrödinger equation for the hydrogen atom. Examples of its use can be found in textbooks on quantum mechanics, such as References [1] and [2].

into a new set of equations, known as the Hartree–Fock equations, which allow orbitals and their energies to be calculated. The way in which this is done is outlined in Appendix B.

The Hartree–Fock method allows for most of the electron–electron repulsion, but it treats it in an averaged fashion, i.e. it effectively takes each electron in turn and calculates the repulsive energy for this electron interacting with the time-averaged charge cloud of all the other electrons. In reality, the instantaneous electron–electron interactions tend to keep electrons further apart than is the case in the Hartree–Fock model. This inadequate treatment of *electron correlation* is the weakness of the Hartree–Fock method, and it is the price paid for clinging on to the concept of orbitals. Nevertheless, it can be used to make rather good calculations of atomic and molecular properties from first principles. Further details on these so-called *ab initio* calculations can be found in Appendix B.

References

1. *Quantum Chemistry*, I. N. Levine, New Jersey, Prentice Hall, 2000.
2. *Molecular Quantum Mechanics*, 3rd edn., P. W. Atkins and R. S. Friedman, Oxford, Oxford University Press, 1999.

3 Angular momentum in spectroscopy

The quantization of angular momentum is a recurring theme throughout spectroscopy. According to quantum mechanics only certain specific angular momenta are allowed for a rotating body. This applies to electrons orbiting nuclei (orbital angular momentum), electrons or nuclei 'spinning' about their own axes (spin angular momentum), and to molecules undergoing end-over-end rotation (rotational angular momentum). Furthermore, one type of angular momentum may influence another, i.e. the angular motions may *couple* together through electrical or magnetic interactions. In some cases this coupling may be very weak, while in others it may be very strong.

This chapter is restricted to consideration of a single body undergoing angular motion, such as an electron orbiting an atomic nucleus; the case of two coupled angular momenta is covered in Appendix C. In classical mechanics, the orbital angular momentum is represented by a vector, $\boldsymbol{l}$, pointing in a direction perpendicular to the plane of orbital motion and located at the centre-of-mass. This is illustrated in Figure 3.1. If a cartesian coordinate system of any arbitrary orientation and with the origin at the centre-of-mass is superimposed on this picture, then the angular momentum can be resolved into independent components along the three axes (l_x, l_y, l_z). If the z axis is now chosen such that it coincides with the vector $\boldsymbol{l}$, then clearly both l_x and l_y are zero and l_z becomes the same as $\boldsymbol{l}$. If only l_z is non-zero, then the rotation is *solely* in the xy plane, that is rotation is *about* the z axis. The larger the angular momentum is, the larger will be the magnitude of the vector $\boldsymbol{l}$ (or l_z).

In classical mechanics an orbiting or rotating body may have any angular momentum (and therefore any angular kinetic energy). However, quantum mechanics imposes restrictions. In particular, the following are found:

(i) The magnitude of the angular momentum can only take on certain specific values, i.e. it is quantized. The allowed values are $\hbar\sqrt{l(l+1)}$ where l is an angular momentum *quantum number* having the possible values 0, 1, 2, 3, 4, . . . , and $\hbar = h/2\pi$.

(ii) The angular momentum is also quantized *along one particular axis*, and the component of the angular momentum along this axis has the magnitude $m_l\hbar$ where m_l may have any one of the possible values $l, l-1, l-2, \ldots, -l+2, -l+1, -l$.

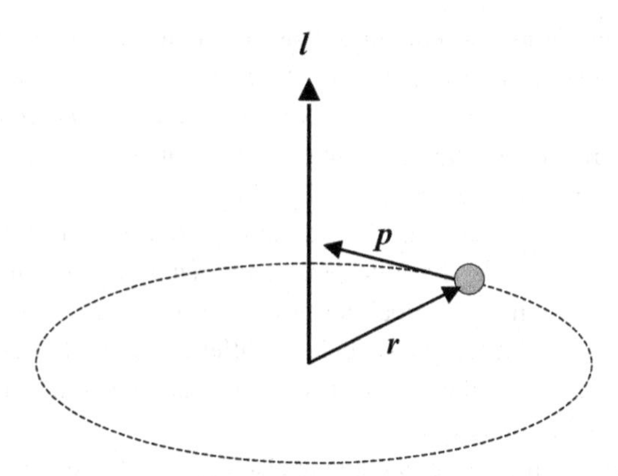

Figure 3.1 Vector representation of orbital angular momentum. Mathematically, the orbital angular momentum is given by the vector product $\boldsymbol{l} = \boldsymbol{r} \times \boldsymbol{p}$, where $\boldsymbol{r}$ is the position vector of the electron relative to the centre-of-mass and $\boldsymbol{p}$ is the instantaneous linear momentum. Notice that the angular momentum vector is perpendicular to the plane containing $\boldsymbol{r}$ and $\boldsymbol{p}$, although the actual orbital motion is in that plane.

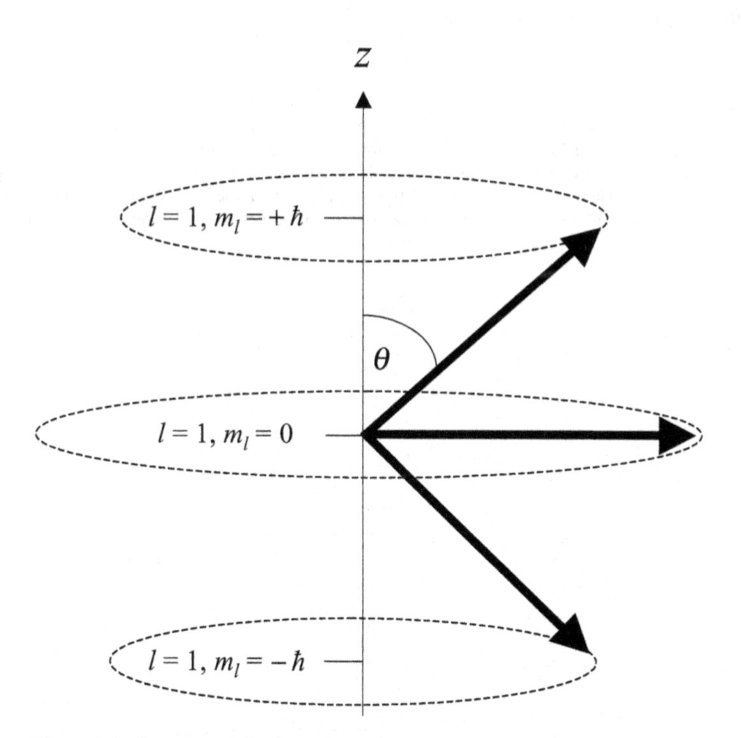

Figure 3.2 Space quantization of angular momentum for $l = 1$. All three possible angular momentum vectors have the same magnitude ($\sqrt{2}\hbar$) and precess about the z axis. However, they have different (constant) projections (m_l) on this axis. From simple trigonometry the angle θ is given by $\cos\theta = m_l/\sqrt{l(l+1)}$, which is 45° for $l = 1$.

These two points are illustrated in Figure 3.2. One finds that the angular momentum vector *precesses*[1] about the axis of quantization such that the *projected* value of the angular momentum along this axis, $m_l\hbar$, remains constant. This is known as *space quantization*. In the absence of external electric or magnetic fields, all of the possible values of m_l for a given l correspond to the same kinetic energy, i.e. they are $(2l+1)$-fold degenerate. Furthermore, although there is nominally an axis of quantization, in a free atom there is no way of knowing in which direction it lies! In other words, the concept is somewhat academic, and only becomes of practical consequence when an external perturbation specifically defines the axis of space quantization. For example, if an electric field of sufficient strength is applied in the laboratory, the field direction defines the axis of space quantization, and the angular momentum vector will then precess around this axis.

As will be seen in a later chapter, the above comments also apply to the quantization of molecular rotations, except for the fact that a different symbol, usually J, is employed to designate the angular momentum quantum number. For electron or nuclear spin angular momenta, the only twist in the tale is that half-integer values of the spin quantum number are also possible. For the particular case of an electron, the spin quantum number is given the symbol s, where $s=\frac{1}{2}$, and therefore the possible values of the corresponding projection quantum number, given the symbol m_s, are $\pm\frac{1}{2}$.

1 This idea of a precessing angular momentum vector may have been encountered elsewhere by readers, notably in the description of the principles of magnetic resonance spectroscopy. Magnetic resonance is concerned with spin angular momentum, either nuclear spin or electron spin. The earlier description of spin as arising from the spinning of these charged particles about their own axis is not strictly correct. Nevertheless, it is a useful picture to retain since it helps in envisaging the properties of the spin angular momentum vector. The precession that this will undergo is entirely analogous to the orbital case, and in both cases it is often referred to as the Larmor precession.

4 Classification of electronic states

The partitioning of electrons into molecular orbitals (MOs) provides a useful, albeit not exact, model of the electronic structure in a molecule. The MO picture makes it possible to understand what happens to the individual electrons in a molecule. Taking the electronic structure as a whole, a molecule has a certain set of quantized *electronic states* available. Electronic spectroscopy is the study of transitions between these electronic states induced by the absorption or emission of radiation. Within the MO model an electronic transition involves an electron moving from one MO to another, but the concept of quantized electronic states applies even if the MO model breaks down.

Different electronic states are distinguished by labelling schemes which, at first sight, can seem rather mysterious. However, understanding such labels is not a difficult task once a few examples have been encountered. We begin by considering the more familiar case of atoms, before moving on to molecules.

4.1 Atoms

If we accept the orbital approximation, then the starting point for establishing the electronic state of an atom is the distribution of the electrons amongst the orbitals. In other words the *electronic configuration* must be determined. Individual atomic orbitals are given quantum numbers to distinguish one from another, leading to labels such as $1s$, $3p$, $4f$, and so on. The number in each of these labels specifies the *principal quantum number*, which can run from 1 to infinity. The principal quantum number, n, defines the number of radial nodes in an orbital, of which there are $n - 1$. As the number of radial nodes increases, the orbital energy increases. The second label, the letter, specifies the orbital angular momentum quantum number, l, of an electron in the orbital. For $l = 0, 1, 2, \ldots$ the corresponding orbital symbols are $s, p, d, \ldots$ Most readers will be very familiar with this already and will also be aware of the fact that we can use three further labels for electrons in atoms, all of which were mentioned in the previous chapter, namely the orbital angular momentum projection quantum number m_l, the spin quantum number $s(= \frac{1}{2})$, and the spin projection quantum number m_s $(= \pm\frac{1}{2})$. Not all of these labels are necessarily meaningful in all circumstances, as will be seen shortly, but if it is assumed for the moment that they are, then each electron has a unique set of values of n, l, m_l, s, and m_s, and these quantum numbers are said to be *good* quantum numbers.

The $2l + 1$ possible values of m_l for a given value of l represent different orbitals of the same energy but with different orientations. It is this idea that gives rise to the concept of directional orbitals such as p_x or d_{yz}.

So far we have been considering quantum numbers associated with a single electron. However, an electronic state of a many-electron atom is the result of the contributions of all the electrons within it and therefore the composite system must be considered to generate a suitable label. The key factor in this process is consideration of the way in which the electronic orbital and spin angular momenta of individual electrons couple in a composite system. The theory of angular momentum necessary to do this is very well-established and is briefly covered in Appendix C. Here we concentrate on the results and, in particular, those points that will also be relevant when dealing with molecules. The model that we will employ, known as the Russell–Saunders approximation, tends to be a good one except for atoms of large atomic number (so-called heavy atoms).

The essence is as follows. As an electron orbits a nucleus, the rotating electric field it generates will interact with the rotating electric field generated by another electron. In other words, there will be a tendency for these electrons to precess in sympathy about a common axis and they will generate a total orbital angular momentum vector $\boldsymbol{L}$. We would like to know the total orbital angular momentum quantum number, L, which results from the coupling of the two individual vectors $\boldsymbol{l}_1$ and $\boldsymbol{l}_2$. The rules for this coupling dictate that for two electrons occupying orbitals with respective angular momentum quantum numbers l_1 and l_2, then L can have any one of the values $l_1 + l_2,\ l_1 + l_2 - 1,\ l_1 + l_2 - 2, \ldots, |l_1 - l_2|$.

Likewise, the electron spins can couple together in a manner entirely analogous to the orbital angular momentum case. Here, the interaction is not electrostatic, as in the orbital angular momentum case, but instead is magnetic, since spin is a magnetic effect. Since $s = \frac{1}{2}$, the total spin quantum number S can only be $1\,(= s_1 + s_2)$ or $0\,(= s_1 - s_2)$ for the two-electron case. This coupling procedure is explained in terms of a simple vector model in Appendix C.

The final part of the Russell–Saunders approximation is to assume that the interactions between the orbital and spin angular momenta will be relatively weak compared to the orbital–orbital and spin–spin interactions. This does not mean that orbital–spin interactions, referred to as *spin–orbit coupling*, can be ignored, as we will see in several examples later on. However, it does require that they are modest in magnitude, otherwise the Russell–Saunders approximation will fail.

We now have a recipe for determining the possible values of L and S given knowledge of the electronic configuration of an atom, since the comments made above can be readily extended to three or more electrons. At first sight it might seem a formidable task to calculate all of the allowed values of L and S for an atom with many electrons. However, there is an important simplification that greatly reduces the amount of work. This arises when sub-shells are completely full, such as ns^2, np^6, or nd^{10}. In full sub-shells the individual electron orbital and spin angular momenta completely cancel each other out yielding a zero net contribution to the total orbital angular momentum and spin angular momentum of the atom. *The angular momenta of electrons in filled sub-shells can therefore be ignored in determining the overall electronic state.*

Turning to the process of labelling an electronic state, in the Russell–Saunders scheme a particular state is designated as $^{2S+1}L_J$ where L and S have already been defined. The quantity $2S+1$ is referred to as the spin multiplicity, since it specifies the degeneracy of the spin part of the electronic state. The subscript J refers to the total electronic (orbital + spin) angular momentum quantum number. This is important when dealing with atoms where both L and S are non-zero. In such circumstances the orbital and spin angular momenta may couple together. This spin–orbit coupling produces spin–orbit states, each with a different value of J, which have different energies. The possible values of J are $L+S,\ L+S-1, \ldots, |L-S|$,[1] and in electronic spectra these give rise to spin–orbit splittings of bands. It increases in magnitude as the atomic number increases, and therefore, while it may be modest for light atoms, it can become very large for heavy atoms.[2]

To bring this section to a close we consider an example, neon. A neon atom has the electronic configuration $1s^2 2s^2 2p^6$. All of the sub-shells are full and so both L and S (and therefore J) are zero. Thus the electronic state arising from this configuration would be labelled $^1S_0$ in the Russell–Saunders scheme. Note that $L = 0$ but instead of using the numerical value of L the letters S, P, D and F are used to label states with $L = 0$, 1, 2 and 3 by direct analogy with the angular momentum labels used for individual atomic orbitals. Since there is only one possible value of J for the ground state of neon, it is common to omit this from the electronic state label. There are many other configurations of neon which will give rise to other (higher energy) $^1S_0$ states, e.g. $1s^2 2p^6 3s^2$. Thus the $^{2S+1}L_J$ label will not uniquely specify a particular electronic state in an atom. There is no additional quantum number that we can use to distinguish between states with the same Russell–Saunders label, and consequently it is useful to specify the configuration from which a particular state arises in order to distinguish it from another having the same Russell–Saunders label.

Now suppose that an electron is removed from the $2p$ sub-shell of neon; it is clear that L must change by one unit (since $l = 1$ for a p orbital) and therefore $L = 1$ in the resulting cation. Similarly, now $S = \frac{1}{2}$. J can now have more than one value, specifically $\frac{1}{2}$ or $\frac{3}{2}$. Thus two states arise which have similar but different energies, a $^2P_{1/2}$ and a $^2P_{3/2}$ state, with the latter happening to have the lower energy (the $^2P_{1/2} - {}^2P_{3/2}$ splitting is 782 cm^{-1}). These two spin–orbit states can be viewed as arising from antiparallel or parallel orientations of the total orbital and spin angular momenta.

4.2 Molecules

The classification of electronic states of molecules builds on the methodology employed for atoms. To appreciate this, one should note that the labels S, P, D, etc., for the total orbital

---

[1] Note the similarity to the rules used for determining L and S from the orbital and spin angular momenta of the individual electrons. This similarity is not accidental.

[2] The Russell–Saunders coupling model is a poor approximation when spin–orbit coupling is large, as is often the case for heavy atoms. In such circumstances, L and S are not good quantum numbers because the spin–orbit coupling mixes the orbital and spin angular momenta in such a manner that they can no longer be independently specified. In this event, alternative coupling schemes, such as jj-coupling, are an improvement. For further details see Appendix C and/or References [1–3].

angular momentum in an atom are actually symmetry labels for the electronic orbital angular momentum wavefunctions. It is assumed that readers are already familiar with the idea of symmetry through studies of *point group* symmetry in molecules, and that they have a working knowledge of the use of *character tables*. If this is not the case then a summary of key points can be found in Appendix D, along with a listing of the more commonly used character tables. The label used to indicate the symmetry of some molecular property, such as an orbital or a vibration, is given by a particular *irreducible representation* in the character table of the appropriate point group. The ability to be able to identify the symmetry of a particular property in a molecule is of great importance in spectroscopy. For example, from knowledge of the symmetries of the initial and final states in a potential spectroscopic transition, group theoretical arguments can quickly be used to determine whether that transition is allowed or forbidden without having to do any lengthy calculations. This will be employed on many occasions in this book.

The concept that the orbital angular momentum labels for electronic states in atoms might also be symmetry labels may seem strange, since atoms do not have any interesting point group symmetry. Nevertheless they are indeed symmetry labels arising from the so-called *full three-dimensional rotation group*; more specificially they are irreducible representations of this group (which in fact has an infinite number of irreducible representations). The three-dimensional rotation group is applicable to systems where unimpeded rotation in three dimensions is possible, and this is clearly the case for the orbital motion of an electron (or collection of electrons) around a single atomic nucleus. We will not dwell on the atomic case, but instead will focus on molecules. Electrons in molecules have more restricted motion due to the presence of more than one nucleus, and it is point group irreducible representations that specify symmetries.

To see this, we first consider a simple example, molecular hydrogen, in some detail, and then move on to consider the electronic structure of a more complicated molecule.

4.2.1 *Low-lying molecular orbitals of H_2*

In the simplest molecular orbital picture the ground (lowest) electronic state of H_2 is formed by bringing together two H 1*s* orbitals. If the two atomic orbitals have the same phase then a bonding MO results, whereas opposite phases give rise to an antibonding MO. These possibilities are indicated pictorially in Figure 4.1.

Our concern is with the overall electronic state. To identify this, the symmetries of the occupied MOs must first be established. For H_2 this is a straightforward task. Initially, the point group of the molecule must be determined, which for H_2 is $D_{\infty h}$. Next we consult the $D_{\infty h}$ character table, which is shown in Table 4.1, and determine how the sole occupied MO is affected by the symmetry operations of the $D_{\infty h}$ point group.

The symmetry operations of the point group, which are defined in Appendix D, are shown along the top row of the character table. The table looks formidable, but in fact its interpretation is for the most part straightforward. None of the symmetry operations have any distinguishable effect on the lowest energy bonding MO of H_2, as may be seen by referring to Figure 4.1 and applying each of the symmetry operations in turn. Consequently, this MO transforms as the totally symmetric irreducible representation, which is always the uppermost one in the character table. It is conventional to use lower case symbols for

Table 4.1 *Character table for $D_{\infty h}$ point group*

$D_{\infty h}$	E	$2C_\infty^\phi$	...	$\infty\sigma_v$	i	$2S_\infty^\phi$	...	∞C_2		
Σ_g^+	1	1	...	1	1	1	...	1		x^2+y^2, z^2
Σ_g^-	1	1	...	−1	1	1	...	−1	R_z	
Π_g	2	$2\cos\phi$	...	0	2	$-2\cos\phi$	...	0	(R_x, R_y)	(xz, yz)
Δ_g	2	$2\cos 2\phi$	...	0	2	$2\cos 2\phi$	...	0		(x^2-y^2, xy)
...	...	...	...	...	...	...	...	...		
Σ_u^+	1	1	...	1	−1	−1	...	−1	z	
Σ_u^-	1	1	...	−1	−1	−1	...	1		
Π_u	2	$2\cos\phi$	...	0	−2	$2\cos\phi$	...	0	(x, y)	
Δ_u	2	$2\cos 2\phi$	...	0	−2	$-2\cos 2\phi$	...	0		
...	...	...	...	...	...	...	...	...		

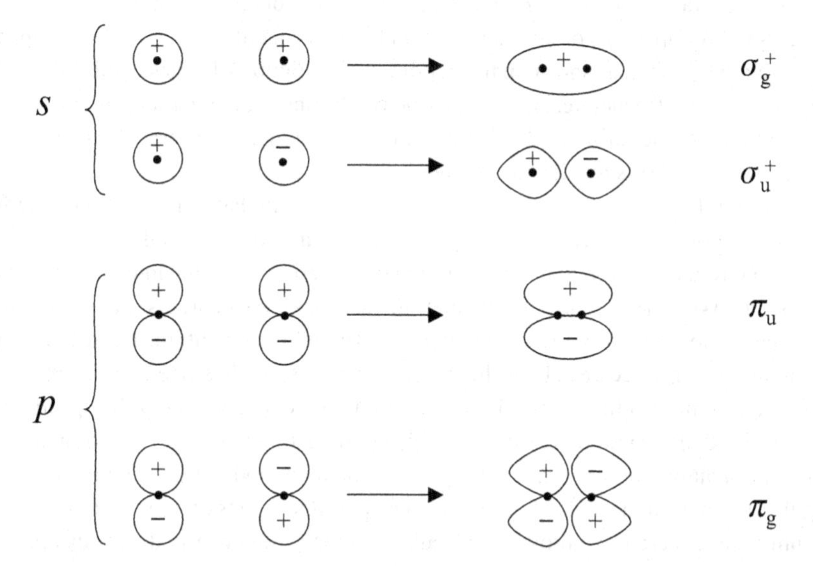

Figure 4.1 Formation of σ and π bonding and antibonding MOs by overlap of s and p AOs, respectively, in a homonuclear diatomic molecule. The + and − signs refer to the phases of the orbitals or orbital lobes. Other orbital combinations are possible but not shown, e.g. σ orbitals formed by combination of an s orbital on one atom with a $p\sigma$ or $d\sigma$ orbital on the other.

labelling the symmetries of MOs, and so the lowest MO in H_2 is designated as σ_g^+. In other words, we would say that this MO has σ_g^+ symmetry.

The antibonding MO is different, however, owing to the change in orbital phase across the nodal plane. Rotation of the molecule about the internuclear axis or reflection in a plane passing through the internuclear axis has no distinguishable effect on the MO, i.e. the characters for both the C_∞^ϕ and σ_v operations are unity. However, the effect of the S_∞^ϕ and C_2 operations is to change the sign of the orbital wavefunction, since the positive and negative lobes exchange places. Consequently, the characters for these operations are −1. Armed with these facts, inspection of the $D_{\infty h}$ character table reveals that the antibonding MO has σ_u^+ symmetry.

4.2.2 *Higher energy molecular orbitals of H_2*

The σ_g^+ and σ_u^+ molecular orbitals mentioned above are not the only orbitals of H_2 possessing these symmetries. Overlap of higher energy *s* atomic orbitals will also generate both σ_g^+ and σ_u^+ molecular orbitals. Clearly symmetry alone is not a unique label for a molecular orbital, in the same way that the *s*, *p*, *d*, and *f* orbital angular momentum labels are not sufficient to label specific atomic orbitals. A numbering scheme is therefore added, akin to the principal quantum numbering of atomic orbitals, to specify a particular MO. The numbering for orbitals of a particular symmetry is 1, 2, 3, . . . , the 1 specifying that it is the lowest energy orbital of this particular symmetry, 2 indicates it is the second lowest energy orbital of this symmetry, and so on. Thus the two molecular orbitals arising primarily from overlap of 1*s* orbitals in H_2 are designated $1\sigma_g^+$ and $1\sigma_u^+$, while the corresponding orbitals arising primarily from 2*s* overlap are designated $2\sigma_g^+$ and $2\sigma_u^+$.[3]

The focus so far has been on σ molecular orbitals in H_2, but other symmetries are possible. For example overlap of $2p_x$ or $2p_y$ atomic orbitals can form either π_g or π_u MOs depending on their relative phases, as shown in the bottom half of Figure 4.1. These molecular orbitals are clearly doubly degenerate, since a rotation of 90° about the internuclear axis produces equivalent but distinct orbitals. A character of 2 for the identity operation (*E*) in the $D_{\infty h}$ character table confirms the double degeneracy.

The orbital degeneracy has an important consequence in that it makes it possible for an electron in a molecule to have orbital angular momentum. In a π orbital, unlike a σ orbital, it is possible for an electron to undergo unimpeded rotation about the internuclear axis (but not about an axis perpendicular to the internuclear axis). As in atoms, the orbital motion in a diatomic molecule must give rise to quantized angular momentum. The orbital angular momentum vector precesses about the internuclear axis, as illustrated in Figure 4.2. The overall angular momentum is poorly defined in a molecule and cannot be specified by a meaningful (good) quantum number. The only good quantum number is the orbital angular momentum quantum number λ, which specifies the magnitude of the angular momentum along the internuclear axis (see Figure 4.2). The possible values of λ are 0, 1, 2, etc. These quantum numbers are linked to the molecular symmetry and one finds, for example, that the σ, π, δ, and ϕ irreducible representations listed in the $D_{\infty h}$ character table correspond to molecular orbitals with $\lambda = 0, 1, 2$, and 3, respectively.

4.2.3 *Electronic states of H_2*

We now have sufficient information to establish the electronic states of H_2. There are two main steps in this task, determining (i) the spatial symmetry and (ii) the net spin.

For H_2, the lowest energy electronic configuration is $(1\sigma_g^+)^2$, that is both electrons are in the most strongly bonding MO. To determine the symmetry of the electronic state arising

[3] It is not only *s* atomic orbitals that can contribute to σ MOs. For example p_z orbitals, where *z* lies along the internuclear axis, have σ symmetries and, depending on their relative phases, can therefore contribute to both σ bonding and antibonding MOs. This is an example of a more general situation, namely that every MO should be regarded as an admixture of various AOs of the correct symmetry. However, it is often the case that one type of AO makes a dominant contribution.

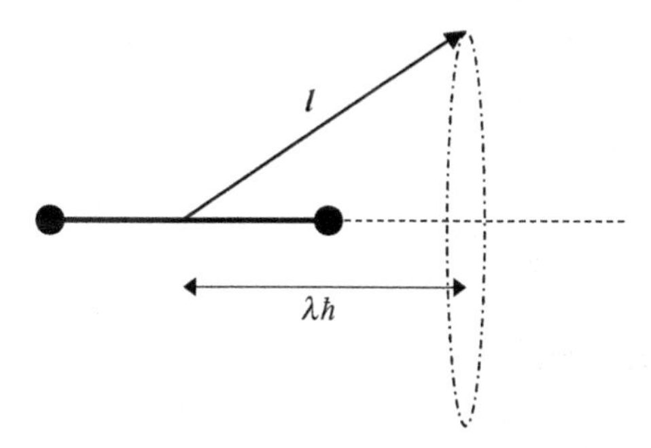

Figure 4.2 Diagram showing the orbital angular momentum vector, ***l***, for an electron in a diatomic molecule. This vector precesses about the internuclear axis maintaining a constant projection λ. The orbital angular momentum quantum number, ***l***, will be poorly defined in such circumstances and is not used, but λ would be a useful (good) quantum number.

from this configuration, group theory is employed. According to the orbital approximation, the total electronic wavefunction is a product of the orbital wavefunctions of each individual electron. It must therefore be possible to obtain the spatial symmetry of the overall electronic wavefunction by multiplying the irreducible representations for the individual electrons. In general this gives a reducible representation that can be reduced by standard group theoretical methods. This is rarely necessary, however, since most books containing character tables also include so-called *direct product* tables that do virtually all of the work for us. Selected direct product tables are provided in Appendix D. The direct product table for the $D_{\infty h}$ point group gives

$$\sigma_g^+ \otimes \sigma_g^+ = \sigma_g^+$$

where $\otimes$ is used instead of $\times$ to indicate that this is not multiplication in the normal sense of multiplying numbers (see Appendix D for more details). The spatial symmetry of the electronic state is therefore Σ_g^+. Notice that upper case Greek characters are used when referring to an electronic state, while lower case characters are used for MOs. To complete the task, the spin multiplicity, $2S + 1$, is required. Since the two electrons are paired in the $1\sigma_g^+$ orbital, $S = 0$ and therefore the electronic state is written as ${}^1\Sigma_g^+$, where the spin multiplicity appears as a pre-superscript (cf. Russell–Saunders notation for atoms given earlier).

Now consider some possible excited states of H_2. For example, the excited electronic configuration $(1\sigma_g^+)^1(1\sigma_u^+)^1$ can give rise to two excited states with the same spatial symmetry but different spin multiplicities. The spatial symmetry can be obtained from the direct product $\sigma_g^+ \otimes \sigma_u^+ = \sigma_u^+$, while both singlet and triplet spin multiplicities are now possible since the two electrons are in different orbitals. Thus both ${}^1\Sigma_u^+$ and ${}^3\Sigma_u^+$ states can arise.

As a final example for H_2, consider the states that would arise out of the excited configuration $(1\sigma_g^+)^1(1\pi_u)^1$. Once again, both singlet and triplet spin multiplicities are possible and the direct product of spatial symmetries $\sigma_g^+ \otimes \pi_u = \pi_u$; hence the two possible states are ${}^1\Pi_u$ and ${}^3\Pi_u$. Note, however, that this is not quite the end of the story, since spin–orbit coupling is possible in the ${}^3\Pi_u$ state. In the same way that an electron in a π MO has an orbital angular momentum about the internuclear axis corresponding to $\lambda = 1$, so a

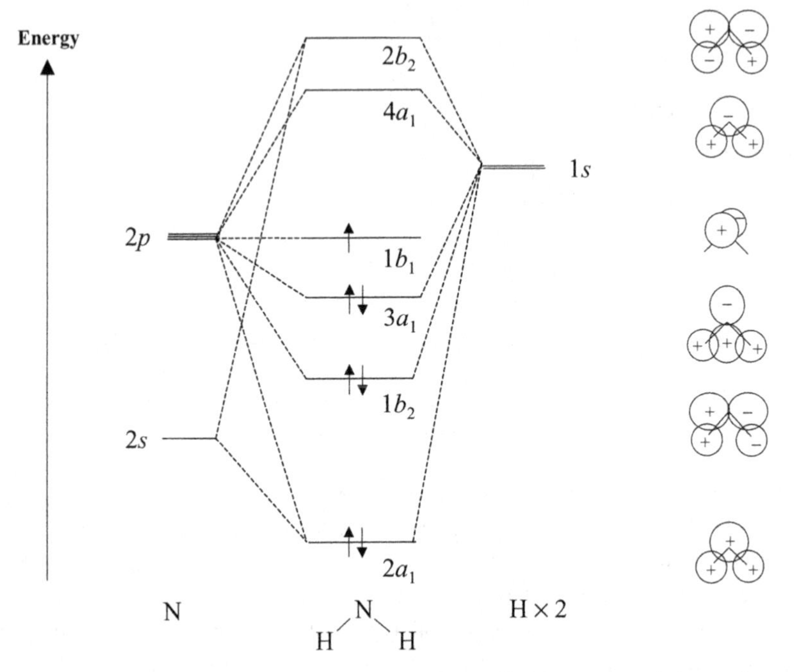

Figure 4.3 Valence MO diagram for the NH_2 free radical in its ground electronic state. The doubly occupied $1a_1$ MO is essentially a N $1s$ atomic orbital and is not shown. A pictorial indication of the AO contributions to each MO is given to the right of the figure.

molecule in a Π electronic state has a net electronic orbital angular momentum quantum number (Λ) of the same magnitude. This orbital motion of the electron generates a magnetic field that can also couple the spin angular momentum to the internuclear axis. When this spin–orbit coupling occurs, and it will only occur for states that are not singlets and for $\Lambda \neq 0$, the projection of the spin angular momentum, given the symbol Σ (not to be confused with the Σ used to label electronic states with $\Lambda = 0$), is also a good quantum number. In general the possible values of Σ are

$$\Sigma = S, S-1, S-2, \ldots, -S$$

For the $^3\Pi_u$ state, the possible values of Σ are 0, ± 1. The combined effect of the coupling of Λ and Σ is denoted by the quantum number Ω ($= |\Lambda + \Sigma|$). The allowed values of Ω are therefore 2, 1 or 0 and, just as was seen earlier for atoms, the resulting spin–orbit states will have different energies. To label a specific spin–orbit component the value of Ω is added to the state label as a subscript. Consequently, the $^3\Pi_u$ state will therefore split into the spin–orbit states $^3\Pi_{u(2)}$, $^3\Pi_{u(1)}$, and $^3\Pi_{u(0)}$, where the value of Ω is given in parentheses. In practice the spin–orbit splitting is very small for H_2, as would be expected given the comments made in Section 4.1, but for molecules containing much heavier atoms the spin–orbit splitting can be quite large (ranging from tens to even thousands of cm^{-1}).

4.2.4 The amidogen free radical, NH_2

This molecule is a non-linear molecule with C_{2v} point group symmetry. We must therefore use irreducible representations from the C_{2v} character table to describe the symmetries of the molecular orbitals. A molecular orbital diagram for NH_2 is shown in Figure 4.3. This can be arrived at using qualitative bonding arguments and is confirmed by sophisticated *ab initio* calculations (see Appendix B) and by experiment. The orbital occupancy shown in Figure 4.3 corresponds to the electronic configuration

$$(1a_1)^2(2a_1)^2(1b_2)^2(3a_1)^2(1b_1)^1$$

At first sight it may look like a complicated task to ascertain the spatial symmetry of the electronic state(s) arising from the above configuration, since there are many electrons to deal with. However, a quick inspection of the direct product table for the C_{2v} point group reveals that filled orbitals always contribute a totally symmetric spatial symmetry, i.e. they have a_1 symmetry regardless of whether the orbital itself is totally symmetric or not (this is analogous to the atomic case, where all filled sub-shells make no contribution to the net angular momentum of an atom). Filled orbitals also make no contribution to the spin multiplicity, since the spins of the paired electrons cancel each other out. These conclusions apply to all point groups, and greatly simplify the process of determining the spatial symmetries of electronic states in molecules. The only orbital in NH_2 that therefore needs to be considered in order to determine the overall electronic state spatial symmetry is the $1b_1$ orbital, which contains a single unpaired electron. Since this electron is in an orbital with b_1 symmetry, we can quickly conclude that there is only one state arising from the above configuration, a $^2B_1$ state.

Consider what would happen if NH_2 was now ionized by removing an electron, say, from the $1b_2$ orbital. There would now be two half-filled orbitals, and we need to consider both of these (but only these two) when determing the spin multiplicity and spatial symmetry. Since $S = 0$ or 1, the spin multiplicity is either 1 or 3, giving a singlet or triplet state. The spatial symmetry is obtained by determining the direct product $b_1 \otimes b_2$ (the order of the multiplication is immaterial), which gives a_2. There are therefore two possible electronic states, $^3A_2$ and $^1A_2$, with different energies.

References

1. *Quantum Chemistry*, I. N. Levine, New Jersey, Prentice Hall, 2000.
2. *Molecular Quantum Mechanics*, 3rd edn., P. W. Atkins and R. S. Friedman, Oxford, Oxford University Press, 1999.
3. *Elementary Atomic Structure*, G. K. Woodgate, Oxford, Oxford University Press, 1983.
4. *Molecular Symmetry and Spectroscopy*, P. R. Bunker and P. Jensen, Ottawa, NRC Research Press, 1998.

5 Molecular vibrations

So far molecules have been treated as if they contained nuclei fixed in space. However, molecules can of course move through space (translation), they can rotate, and internuclear distances can be altered by vibrations. Translational motion is uninteresting from the point of view of spectroscopy since it is essentially unquantized motion. Vibrations and rotations are, however, very important to spectroscopists and so each will be considered in some detail, beginning with molecular vibrations.

5.1 Diatomic molecules

5.1.1 *The classical harmonic oscillator*

There is an internuclear separation in a diatomic molecule for which the sum of the electrostatic potential energies and the electron kinetic energies, the quantity labelled E_{elec} in Section 2.1.2, is a minimum. This internuclear separation corresponds to the *equilibrium bond length*. If the internuclear separation is now altered from the equilibrium position, whether by stretching or compressing the bond, there will now be an opposing force, known as a *restoring* force, trying to pull the system back to equilibrium. The obvious analogy here is with a spring.

Experiment has shown that the restoring force, F, for a spring is directly proportional to the displacement, x, from equilibrium ($x = 0$), providing the displacement is small. In other words,

$$F = -kx \tag{5.1}$$

where the constant of proportionality, k, is known as the force constant. The force constant is a measure of the stiffness of the spring to distortion, with much greater energy being required to distort a spring a certain distance when k is large compared with when k is small. The minus sign in equation (5.1) arises because the restoring force acts in a direction opposite to that of the displacement. Equation (5.1) is a statement of Hooke's law, and any oscillating system satisfying Hooke's law is said to be a *harmonic oscillator*.

The potential energy, V, stored in a distorted spring can be readily calculated by making use of the following well-known relationship from classical mechanics:

$$F = -\frac{\mathrm{d}V}{\mathrm{d}x} \tag{5.2}$$

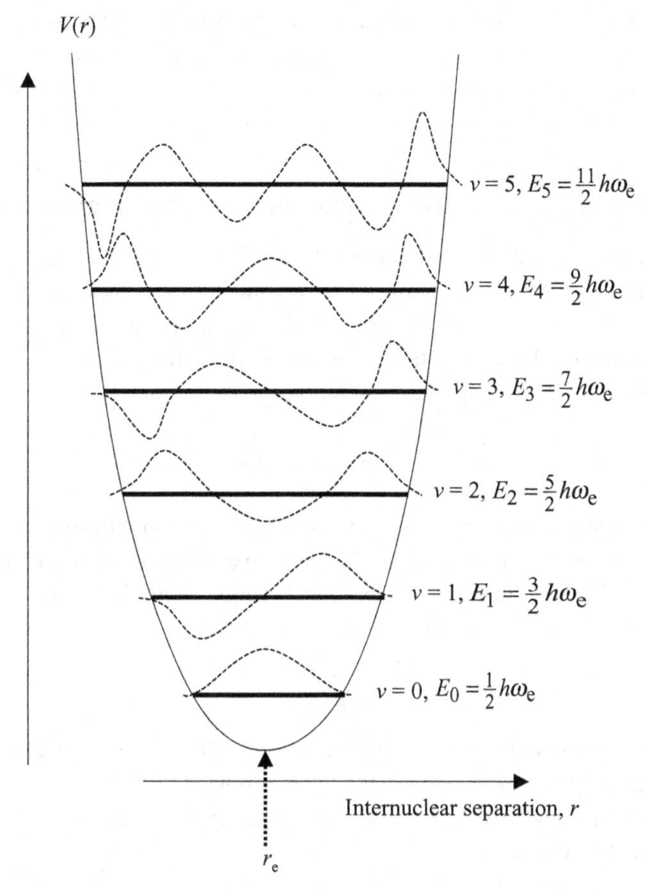

Figure 5.1 Plot showing the parabolic potential energy curve of a diatomic simple harmonic oscillator. Superimposed on the curve are the first few energy levels expected for a quantized harmonic oscillator, each being labelled by a unique value of the vibrational quantum number v. Also shown (dashed lines) are the corresponding vibrational wavefunctions.

Substituting for F in equation (5.1) using the expression in (5.2) followed by integration gives

$$V = \tfrac{1}{2}kx^2 \tag{5.3}$$

If it is assumed that a diatomic molecule is also subject to Hooke's law, then the potential energy due to distortion will be given by equation (5.3) where $x = r - r_e$, r is the internuclear separation, and r_e is the equilibrium bond length. A plot of V versus r is shown in Figure 5.1. The potential energy curve is parabolic, being symmetrical about equilibrium (where $r = r_e$). If one imagines the bond being stretched to a certain displacement and then released, the stored potential energy is progressively converted to kinetic energy as the bond shortens until at r_e all of the energy is kinetic. The system then passes through the equilibrium position and gradually converts the kinetic energy back to potential energy as the bond is compressed. Once the kinetic energy has all been converted into potential energy, the molecule is at its inner turning point and then reverses its motion by progressively stretching the bond again. The total amount of energy in this vibrational motion, potential + kinetic, is constant and is referred to as the *vibrational energy*.

5.1.2 *The quantum mechanical harmonic oscillator: vibrational energy levels*

The above discussion is a classical view of vibrational motion. In the classical world a vibrating diatomic molecule may have any vibrational energy. However, once quantum mechanics is taken into account this is no longer the case. To determine the quantum mechanical energies in the harmonic oscillator limit, the potential energy expression in (5.3) is substituted into the Schrödinger equation to obtain the following:

$$\left(-\frac{\hbar^2}{2m}\frac{\mathrm{d}^2}{\mathrm{d}x^2}+\frac{1}{2}kx^2\right)\psi = E\psi \tag{5.4}$$

Equation (5.4) is only valid when one of the atoms is essentially of infinite mass, in which case it is only the lighter atom, of mass m, which moves in a vibration. In reality no atom is infinitely heavy and therefore to describe vibrational motion about the *centre-of-mass*,[1] m is replaced in (5.4) with the *reduced mass* μ where

$$\mu = \frac{m_A m_B}{m_A + m_B} \tag{5.5}$$

and m_A and m_B are the masses of the two atoms. Equation (5.4) can be solved, although it is a rather involved process; we focus here solely on the results.

It is found, not surprisingly, that the energy is now quantized. The allowed vibrational energies are given by the expression

$$E_v = h\omega_e\left(v + \tfrac{1}{2}\right) \tag{5.6}$$

where ω_e is the *harmonic vibrational frequency* (in Hz),[2] and v is the *vibrational quantum number*, which can have the values 0, 1, 2, 3, . . . The vibrational frequency depends on both the bond force constant and the reduced mass in the following fashion:

$$\omega_e = \frac{1}{2\pi}\sqrt{\frac{k}{\mu}} \tag{5.7}$$

Equation (5.6) shows that the quantized harmonic oscillator consists of a series of equally spaced energy levels, the separation between adjacent levels being $h\omega_e$. This is illustrated in Figure 5.1. According to equation (5.7), ω_e will increase as the bond force constant

[1] By specifying atomic displacements relative to the centre-of-mass, no overall translational energy of the molecule is included in a calculation of the vibrational energy. In a centre-of-mass system, if the two atoms have different masses, a displacement x involves the lighter atom moving further than the heavier atom (which moves in the opposite direction) such that the centre-of-mass remains stationary.

[2] The subscript on ω_e indicates that vibration is, rather obviously, about the equilibrium position. It is customary to retain it when referring to diatomic molecules but for polyatomic molecules the e subscript will be omitted and instead the subscript will be a number designating a particular vibrational mode (see Section 5.2.1).

increases. Stronger bonds tend to be stiffer bonds and therefore vibrational frequencies normally increase with increasing bond strength. The reduced mass acts in the opposite direction, with an increase in μ leading to a decrease in $h\omega_e$. It should also be noted that even in the lowest energy level, corresponding to $v = 0$, the vibrational energy is non-zero. This residual energy is known as the *zero point energy*, and plays a celebrated role in quantum mechanics [1].

It is more usual to employ wavenumber units than energies when dealing with vibrational transitions in spectroscopy. Equation (5.6) can be re-written as a vibrational *term value*, $G(v)$, given by

$$G(v) = \omega_e\left(v + \tfrac{1}{2}\right) \tag{5.8}$$

where ω_e is now interpreted as the harmonic vibrational wavenumber, usually expressed in cm^{-1}, and therefore $G(v)$ has the same units. Term values will be used extensively throughout this book.

5.1.3 The quantum mechanical harmonic oscillator: vibrational wavefunctions

Full solution of the Schrödinger equation (5.4) also yields the vibrational wavefunctions in addition to energies. These have the mathematical form

$$\psi_v = N_v H_v(\eta) \exp(-\eta^2/2) \tag{5.9}$$

where N_v is a normalization constant and $\eta = 2\pi(\mu c\omega_e/h)^{1/2}(r - r_e)$. Note that η is directly proportional to the displacement $(r - r_e)$. The quantity $H_v(\eta)$ is a Hermite polynomial in the coordinate η, and the first few Hermite polynomials are

$$\begin{aligned} H_0(\eta) &= 1 \\ H_1(\eta) &= 2\eta \\ H_2(\eta) &= 4\eta^2 - 2 \\ H_3(\eta) &= 8\eta^3 - 12\eta \\ H_4(\eta) &= 16\eta^4 - 48\eta^2 + 12 \end{aligned} \tag{5.10}$$

We can therefore write down the wavefunctions for the first few vibrational levels of a diatomic molecule as follows:

$$\begin{aligned} \psi_0 &= N_0 \exp(-\eta^2/2) \\ \psi_1 &= 2N_1\eta \exp(-\eta^2/2) \\ \psi_2 &= N_2(4\eta^2 - 2) \exp(-\eta^2/2) \\ \psi_3 &= N_3(8\eta^3 - 12\eta) \exp(-\eta^2/2) \\ \psi_4 &= N_4(16\eta^4 - 48\eta^2 + 12) \exp(-\eta^2/2) \end{aligned} \tag{5.11}$$

Although the above functions may look quite cumbersome, they have a simple form when plotted, as shown in Figure 5.1. A number of important conclusions arise. First, the probability of the molecule being at any particular internuclear separation is given by ψ_v^2 at the particular value of $r - r_e$. Thus for $v = 0$ quantum mechanics predicts that the most probable internuclear separation is r_e. This is counterintuitive when one thinks of a spring, since a vibrating spring will spend most of its time in the region of the two turning points and is moving at its fastest at the equilibrium position! Note also that there is some 'leakage' of the wavefunction outside of the harmonic oscillator potential well, a phenomenon that is impossible in the classical case. This 'leakage' grows in importance as the vibrational energy is increased.

For levels $v = 1$ and higher, nodes appear in the wavefunction and indeed the number of nodes is equal to the vibrational quantum number. As v increases, the wavefunction progressively heads towards behaviour that is anticipated classically, i.e. the most probable internuclear separations shift towards the turning points, with the probability of finding the molecule at the equilibrium separation becoming rather small.

5.1.4 *The anharmonic oscillator*

The justification for treating a vibrating diatomic molecule like a quantized vibrating spring is that the harmonic oscillator model works rather well. Spectroscopic measurements demonstrate that the separation between adjacent pairs of vibrational levels is indeed approximately constant, as equation (5.6) predicts. However, if we look more closely at experimental data, and if we think more clearly about the implications of the potential well shown in Figure 5.1 we conclude that the harmonic oscillator model is only an approximation and there are circumstances where its failure can be very serious.

The most obvious deficiency is that no allowance is made for the fact that any bond, or for that matter any spring, will eventually break when sufficiently stretched. The harmonic oscillator potential energy curve is infinitely deep, which would lead to the nonsensical conclusion that a chemical bond is infinitely strong.

Qualitatively, the potential energy curve of a real diatomic molecule would be expected to have the same shape as that shown in Figure 5.2. To determine the pattern of vibrational energy levels for such an oscillator, which is now referred to as *anharmonic* because of the asymmetry of the potential energy curve, a mathematical form for the potential energy is needed, which can be substituted into the Schrödinger equation (5.4) in place of the harmonic potential (5.3). A number of different mathematical functions give rise to a curve of similar shape to that shown in Figure 5.2 but the most widely used is the *Morse potential function*, which has the form

$$V = D_e\{1 - \exp[-a(r - r_e)]\}^2 \tag{5.12}$$

where D_e is the *dissociation energy* of the molecule, measured from the bottom of the potential well. The quantity a is a constant that varies from one molecule to another (and one electronic state to another), as does D_e.

Solution of the Schrödinger equation with the potential in equation (5.12) gives

$$E_v = h\omega_e\left(v + \tfrac{1}{2}\right) - h\omega_e x_e\left(v + \tfrac{1}{2}\right)^2 \tag{5.13}$$

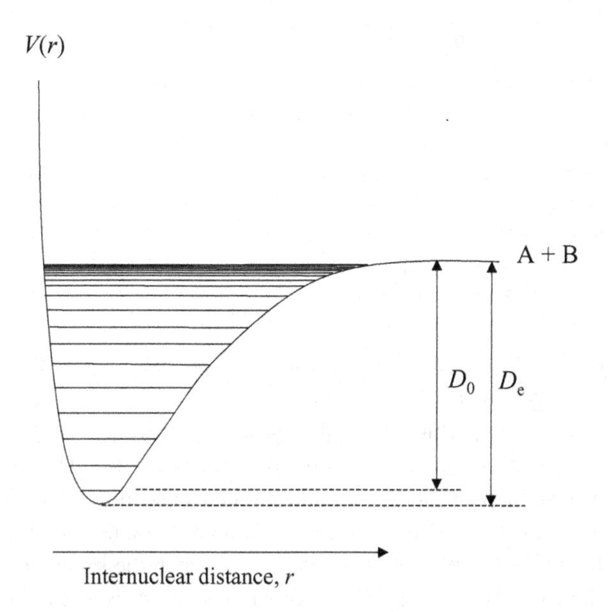

Figure 5.2 Morse potential energy curve for a diatomic molecule AB with quantized vibrational energy levels superimposed. Two different definitions of dissociation energies are shown, dissociation energy D_e measured from the bottom of the potential well, and D_0, measured from the zero point level. Since all molecules must have at least the zero point vibrational energy, D_0 is the more useful quantity.

where ω_e is in Hz or, expressed as a term value,

$$G(v) = \omega_e\left(v + \tfrac{1}{2}\right) - \omega_e x_e\left(v + \tfrac{1}{2}\right)^2 \tag{5.14}$$

with ω_e in wavenumbers. Notice that the first term on the right-hand side of (5.13) is identical to the harmonic oscillator energy expression (5.6). The second term differs in two ways. First, it depends on the square of $v + \frac{1}{2}$, and second it contains the dimensionless quantity x_e, which is known as the *anharmonicity constant*. For almost all diatomics the anharmonicity constant is small, typically <0.01, and therefore if v is small the second term in (5.13) and (5.14), the anharmonic correction, is also small. The harmonic oscillator approximation is therefore a good one for vibrational energy levels near the bottom of the potential well. However, the anharmonic correction quickly grows in importance as v increases, due to the quadratic dependence on $v + \frac{1}{2}$. Furthermore, the fact that the anharmonic correction is subtracted from the harmonic term means that adjacent vibrational levels get closer together as the vibrational ladder is climbed, and in the limit that dissociation is reached the energy levels form a continuum. This convergence of energy levels is illustrated in Figure 5.2.

5.1.5 *Vibrations in different electronic states*

A Morse potential of the type shown in Figure 5.2 and equation (5.12) is normally a good approximation to the vibrational potential energy of a real diatomic molecule. However,

Table 5.1 *Spectroscopic constants for the ground and first excited electronic states of* CO

Parameter	$X^2\Sigma^+$	$A^2\Pi$
r_e/Å	1.1281	1.2351
ω_e/cm^{-1}	2170.21	1515.61
$\omega_e x_e/\text{cm}^{-1}$	13.46	17.25
D_e/cm^{-1}	90 230	25 160

let us think more closely about the factors that determine the precise form of the Morse potential for a particular electronic state of a molecule.

As an example consider CO, which has the ground electronic configuration $1\sigma^2 2\sigma^2 3\sigma^2 4\sigma^2 5\sigma^2 1\pi^4$ and therefore has a $^1\Sigma^+$ ground electronic state.[3] The first four σ orbitals are bonding/antibonding pairs, and so have the effect of cancelling each other out in a bonding sense. However, the 5σ and 1π orbitals are bonding orbitals and since both are full the molecule is held together by a triple bond in very much the same way as N_2 (which is isoelectronic with CO). The carbon and oxygen atoms will therefore be strongly bound together in the $^1\Sigma^+$ ground electronic state, and indeed the dissociation energy D_e is 1074 kJ mol^{-1} (90 230 cm^{-1}), which is very large. In addition, a strong bond would be expected to yield a relatively short equilibrium bond length and a relatively high vibrational frequency (since the bond force constant will be large).

Now suppose that an electron is excited from the 5σ MO to the vacant 2π MO. Providing the spin of this electron maintains the same orientation, the excited state will be a $^1\Pi$ state. The 2π MO is strongly antibonding, so the dissociation energy should decrease significantly. Concomittantly, the vibrational frequency should also decrease and the equilibrium bond length should increase. This is precisely what is found experimentally, as illustrated by the data for CO collected in Table 5.1.

In the general case, different shaped potential energy curves are expected for different electronic states. The minimum of each of these curves represents the pure electronic energy of the state. A diagram showing the potential energy curves for the two states of CO that we have just considered is shown in Figure 5.3. This figure is rather simple, but if every potential energy curve of the known electronic states of CO were shown on this diagram it would look very complicated, particularly at high energies. This point is illustrated by Figure 5.4, which shows some of the potential energy curves of PbH, a free radical. Figure 5.4 clearly shows that a molecule in some particular electronic state need not dissociate to the ground state atoms. If this were not true, then it would be impossible to have a bound state with an electronic energy above that of the two ground state atoms, which would be contrary to experimental observations. The factors that determine which electronic states of the atoms correlate with which molecular electronic state is beyond the scope of this book (see [2]).

3 Strictly speaking the σ MOs of CO should be labelled σ^+ to distinguish them from σ^- symmetry. However, since all of the σ MOs have σ^+ symmetry it is common to drop the superscript.

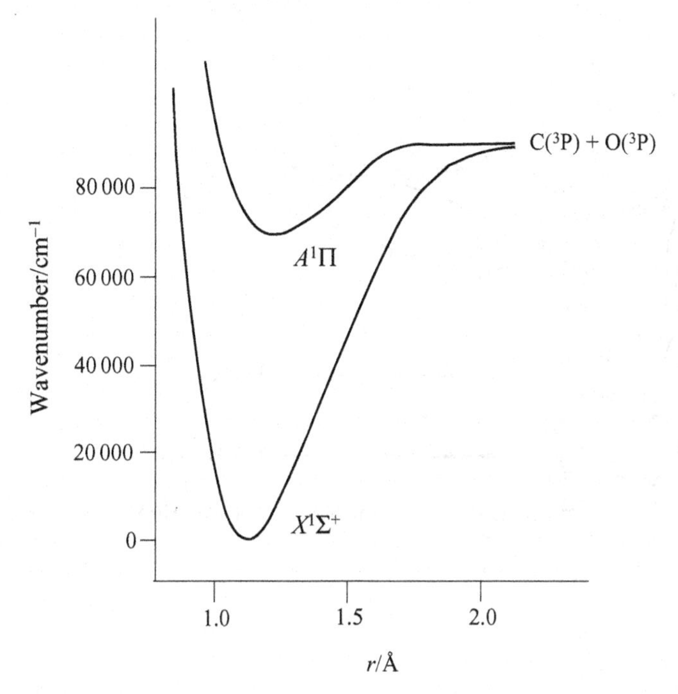

Figure 5.3 Potential energy curves for the $X^1\Sigma^+$ and the $A^1\Pi$ states of CO. Clearly the former is far more strongly bound than the latter. Both of these states correlate with the lowest dissociation asymptote, formation of ground state C and O atoms.

5.2 Polyatomic molecules

5.2.1 *Normal vibrations*

The vibrating diatomic molecule is relatively easy to describe since it has only one bond, and therefore only one vibrational mode. The situation is clearly more complicated for polyatomic molecules, since there is more than one bond that may be stretched/compressed,[4] and there are also bond angles that can be changed by vibrations. At first sight a pessimist might conclude that it would be difficult, if not impossible, to solve the quantum mechanics of polyatomic vibrations. However, this is not the case, although there are indeed additional complications.

It is helpful to focus on small molecules to bring out the key features applicable to more complicated molecules. In fact we will consider three triatomic molecules, CO_2, OCS, and H_2O, as an illustration.

[4] It is obvious that a vibrating bond will undergo both stretching and compression as it oscillates about the equilibrium position. However, it is pedantic to keep referring to it as a stretching/compressing motion, and from now on it will just be called a *stretch*.

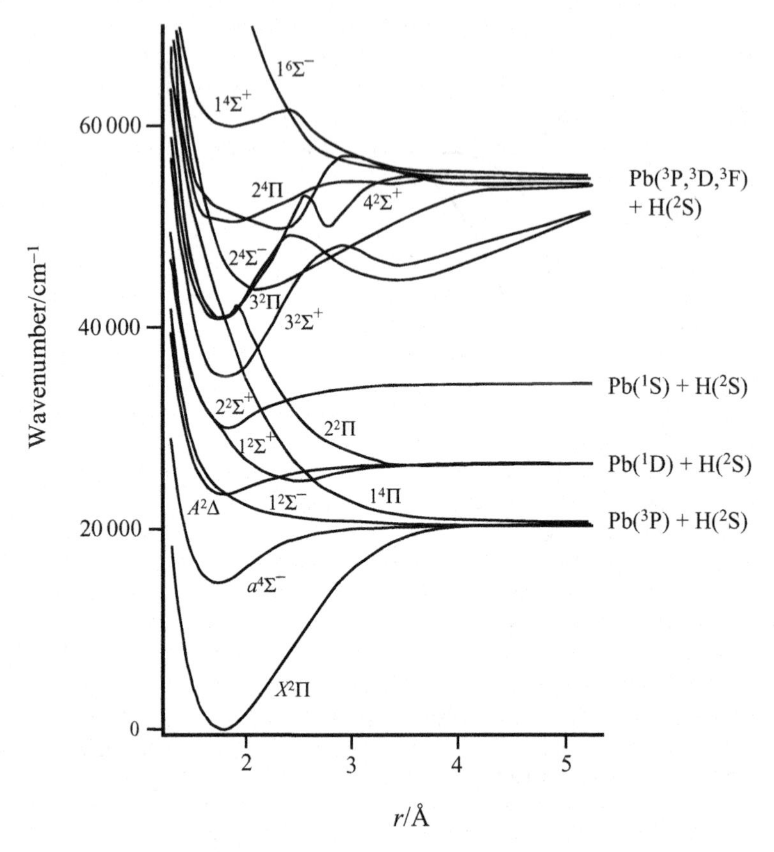

Figure 5.4 Potential energy curves for PbH obtained from sophisticated *ab initio* calculations. This figure is adapted from work reported by A. B. Alekseyev and co-workers (*Mol. Phys.* **88** (1996) 591). Notice that not all potential curves are Morse-like. In this figure several *repulsive* curves can be seen which are unbound at all internuclear separations; an example is the $1^4\Pi$ state, which dissociates to ground state atoms. In addition, some curves have double wells caused by mixing of character with other potential energy curves of the same symmetry.

CO_2

CO_2 is a linear molecule. When it vibrates, both carbon–oxygen bonds will stretch and compress, and in addition the molecule may undergo bending vibrational motion in which the bond angle oscillates about the equilibrium angle, $\theta_e = 180°$. To keep things as simple as possible, the bending motion will be ignored to begin with and the focus will be solely on the bond stretches. We have seen that for diatomic molecules the vibrational potential energy is approximately quadratic in the distortion coordinate, $r - r_e$. It is therefore reasonable to suppose that the same type of potential energy relationship holds for each bond in a polyatomic molecule. The vibrational potential energy would therefore be

$$V = \tfrac{1}{2}k_{CO}r_1^2 + \tfrac{1}{2}k_{CO}r_2^2 \tag{5.15}$$

where r_1 and r_2 are the *displacements* of the two CO bonds from their equilibrium positions and k_{CO} is the force constant for a C=O bond. This potential energy function could be inserted into the Schrödinger equation (5.4), but it would be incomplete because the kinetic energy part would also need modifying. The kinetic energy term can also be written in terms of the displacements r_1 and r_2, and one obtains the slightly lengthy expression

$$T = \frac{m_O(m_O + m_C)}{2M}\left(\dot{r}_1^2 + \dot{r}_2^2\right) + \frac{m_O^2}{M}\dot{r}_1\dot{r}_2 \tag{5.16}$$

where $M = m_C + 2m_O$ and $\dot{r}_1$ and $\dot{r}_2$ are the first derivatives of r_1 and r_2 with respect to time.

The key point to note is that the final term on the right-hand side of (5.16) is a cross-term in the coordinates r_1 and r_2. If this cross-term was absent the vibrational Schrödinger equation could be solved by the method of separating variables, which would involve transforming the Schrödinger equation into two separate equations, one involving only r_1 and the other only r_2. These equations would each be equivalent to the Schrödinger equation for a diatomic harmonic oscillator, and so the solutions would possess the same general form. However, the presence of the cross-term in the kinetic energy operator (5.16) prevents such a separation, in much the same way that the interelectronic repulsion terms in the electronic Schrödinger equation prevent a separation in that case (see Section 2.1.4).

Fortunately, there is a way to remove the cross-term. This involves switching to a different coordinate system, the *normal coordinate* system. It is possible to show that for *all* molecules, providing we make the assumption of simple harmonic oscillation in each internal coordinate (a bond length or bond angle), a set of coordinates can be chosen which give no cross-terms in either the kinetic or potential energy operators; these are the normal coordinates for the molecule. Methods are available for working out the form of these coordinates (see References [3, 4]), but here we concentrate on the results. For CO_2 there are two stretching normal coordinates, designated Q_1 and Q_3, which are as follows:

$$\begin{aligned} Q_1 &= \sqrt{\frac{m_O}{2}}(x_1 - x_3) \\ Q_3 &= \sqrt{\frac{1}{2M}}(m_O x_1 - 2m_C x_2 + m_O x_3) \end{aligned} \tag{5.17}$$

The quantities x_1, x_2, and x_3 are the displacements of the individual atoms (1, O; 2, C; 3, O), and M is the total mass of the molecule ($= m_C + 2m_O$). If the kinetic and potential energy operators are recast as functions of the two normal coordinates, no cross-terms arise and a diatomic-like vibrational Hamiltonian is obtained for each normal coordinate Q_i, i.e.

$$H_i = -\frac{\hbar^2}{2\mu}\frac{\partial^2}{\partial Q_i^2} + \frac{1}{2}k_i Q_i^2 \tag{5.18}$$

Solution of the Schrödinger equation for each normal coordinate is achieved in exactly the same manner as for diatomics. However, it is important to recognize that the vibrational coordinate may involve displacements of more than two atoms. We now have, therefore, a rather simple picture of a vibrating CO_2 molecule. Providing we restrict it to linear geometries, it has two *normal modes* of vibration, each having a vibrational term value

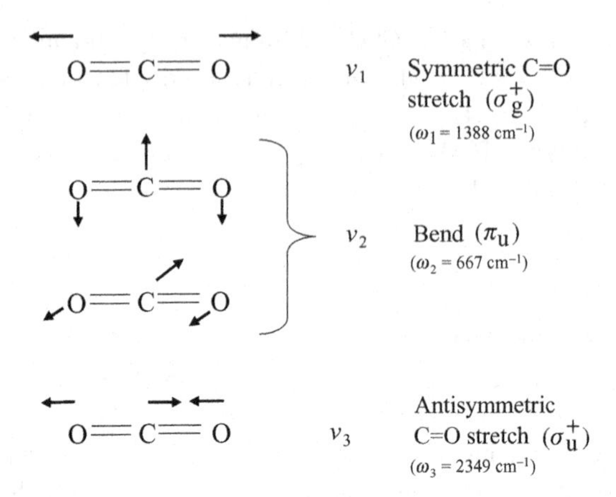

Figure 5.5 Schematic illustration of the vibrational normal modes of CO_2. Also included in the figure are the symmetries and the harmonic frequencies of each mode. Lower case labels are commonly used to show the symmetries of individual vibrations.

given by the diatomic-like harmonic oscillator expression

$$G(v_i) = \omega_i\left(v_i + \tfrac{1}{2}\right) \tag{5.19}$$

where i identifies the particular normal mode, v_i is the vibrational quantum number and ω_i is the corresponding harmonic vibrational wavenumber of this mode. Each mode has a diatomic-like wavefunction of the form

$$\psi_{iv} = N_v H_v\left(2\pi\sqrt{\frac{c\omega_i}{h}}Q_i\right)\exp\left(-\frac{\pi^2 c\omega_i}{h}Q_i^2\right) \tag{5.20}$$

and the overall vibrational wavefunction is a product of the wavefunctions of the individual modes. It is important to recognize that the vibrational quantum numbers for each mode are independent quantities and so any combination of values is possible.

Let us now try to visualize what is happening for CO_2 in the light of the above results, and then extend this picture to other molecules. Two independent normal vibrational modes have been identified. One of these, designated ν_1 and having a normal coordinate Q_1, involves in-phase stretching and compressing of the two C=O bonds, as can be seen from the form of the normal coordinate in (5.17), which is also shown pictorially in Figure 5.5. The centre-of-mass must be stationary during a vibration, otherwise the motion will be a mixture of vibration and overall molecular translation and we are not interested in the latter. The centre-of-mass does not move during vibration ν_1, since any displacement of one O atom is exactly compensated for by motion of the other O atom in the opposite direction (the C atom is at the centre-of-mass and it therefore does not move during this vibration). This mode is called the *symmetric stretch* because it maintains the equilibrium point group symmetry of the molecule at all stages of the vibrational motion.

In contrast, the normal mode ν_3 corresponds to stretching of one C=O bond and compression of the other, and is therefore referred to as the *antisymmetric stretch*.[5] The actual nuclear motion involves displacement of both O atoms in the same direction, and therefore the C atom must move by a sufficient amount in the opposite direction to keep the centre-of-mass unmoved. Inspection of the mathematical form of Q_3 in (5.17) shows that the C atom does indeed move in the opposite direction to the O atoms. If C were replaced by a heavier atom, a smaller displacement, x_2, would be needed to maintain a stationary centre-of-mass. Once again, this behaviour is reflected in the mathematical form of Q_3 since the mass of the central atom is the multiplier of displacement x_2.

If CO_2 is no longer restricted to being linear, the possibility of bending motion now arises. In fact there are two bending vibrations in two mutually perpendicular planes, as shown in Figure 5.5. We can extend the arguments given above for the stretching modes to the bending vibrations, and it is possible to define normal coordinates and therefore to obtain relationships identical to those shown in equations (5.19) and (5.20). Note however that, apart from a 90° rotation of the molecule about the central axis, the two bending vibrations are equivalent. Consequently, they are degenerate and the pair form the *degenerate bending mode*, ν_2, of CO_2.

Finally, it is worth emphasizing the simplicity that the normal coordinate picture provides. If it were possible to view the overall vibrational motion of a polyatomic molecule, even one as simple as a triatomic, it would appear very complicated. By using normal modes, this complicated motion can be treated as a *superposition* of normal vibrations, in each of which the atoms are displaced at the same frequency and phase. The normal vibrations are much simpler to visualize, as well as providing the mathematical simplifications in the quantum mechanics mentioned earlier.

OCS

Like CO_2, OCS is linear at equilibrium. Its vibrations have much in common with CO_2 in that three normal modes can be identified, two of them stretches and one a doubly degenerate bend. An important difference, however, is that the two bonds are no longer equivalent, and so the two stretching vibrations cannot be described respectively as symmetric and antisymmetric stretches. In fact the two stretching vibrations now have identical symmetries. Owing to the different strengths of the two bonds and, more importantly in this case, the substantial difference in masses of the O and S atoms, the stretching vibrations show a degree of bond localization and can be thought of as separate C=O and C=S stretches. This is only an approximation, but a comparison of the harmonic vibrational frequencies of

[5] There is a convention for labelling vibrational modes. For triatomic molecules, these are labelled ν_1, ν_2, and ν_3 (not to be confused with the vibrational quantum numbers). For historical reasons, the two stretching modes are always designated by ν_1 and ν_3 and the bending mode by ν_2. The convention for all other polyatomics requires the vibrations to be grouped according to their symmetries, starting from the highest symmetry and descending to progressively lower symmetries. If there are two or more vibrations of the highest possible symmetry, these are labelled $\nu_1, \nu_2, \ldots, \nu_n$ in order of descending harmonic frequency. One then moves to the next highest symmetry and again the mode labels are ordered in terms of descending frequency, and so on.

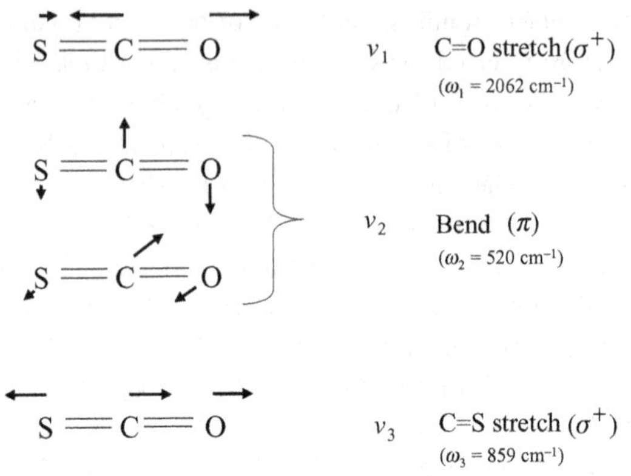

S=C=O ν_1 C=O stretch (σ^+) (ω_1 = 2062 cm^{-1})

S=C=O ν_2 Bend (π) (ω_2 = 520 cm^{-1})

S=C=O ν_3 C=S stretch (σ^+) (ω_3 = 859 cm^{-1})

Figure 5.6 Schematic illustration of the normal modes of OCS.

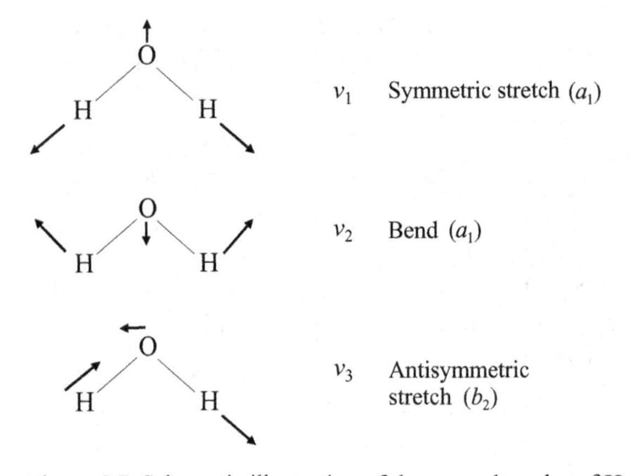

Figure 5.7 Schematic illustration of the normal modes of H_2O.

OCS, shown in Figure 5.6, with those of CO_2 shown in Figure 5.5, is consistent with this idea.

H_2O

Unlike the previous two examples, H_2O is bent at equilibrium. The two O—H bonds are equivalent, so by analogy with CO_2 the two stretching normal modes can be divided into a symmetric stretch (in-phase stretch of the two O—H bonds) and an antisymmetric stretch (antiphase stretching of the two O—H bonds). As for the bending motion, there is now only one way in which we can alter the bond angle and so the bending vibration is non-degenerate. All three normal coordinates of H_2O are shown in Figure 5.7.

H_2O illustrates the fact that non-linear molecules possess $3N - 6$ degrees of vibrational freedom, whereas linear molecules have $3N - 5$. The loss of one vibrational degree of freedom for a non-linear molecule is compensated by the gain of an additional rotational degree of freedom.

5.2.2 *Symmetries of vibrational coordinates and wavefunctions*

Just as the MOs of a molecule must transform as one of the irreducible representations of the molecular point group, so must the normal vibrations. This turns out to be extremely useful for not only does it make it possible to classify the modes according to their symmetries, but more importantly it provides vital information when it comes to establishing selection rules for transitions between vibrational levels in spectroscopy.[6] It is assumed that most readers will already be familiar with the use of symmetry for describing molecular vibrations; those who are not should consult an appropriate textbook (see, for example, Reference [5]). The symmetries of the normal modes of the three triatomics considered in the previous section are shown in Figures 5.5–5.7.

This section is concerned with establishing the groundwork necessary for the discussion of selection rules later on in this chapter, in which symmetry plays a central part. In particular we will need to consider the symmetries of the *vibrational wavefunctions*. This is a topic that often causes considerable difficulty because the symmetries of the normal coordinates and the vibrational wavefunctions are often confused with each other. The symmetry of the normal coordinate is determined by the motion of the atoms during a vibration. For example, if we consider the pictorial representation of the vibrations of H_2O in Figure 5.7 it is obvious, even without employing formal group theory, that the symmetric stretch and the bending mode are *totally symmetric* vibrations. In other words, the molecule maintains the same point group *symmetry* throughout both of these vibrations (even though the structure necessarily changes). However, the antisymmetric stretch involves the compression of one O—H bond and the stretching of another, and so at all stages of this vibration (except when it passes through equilibrium) the molecule is distorted to a lower symmetry (in fact C_s point group symmetry). The antisymmetric stretch should therefore more correctly be referred to as a *non-totally symmetric vibration*. The distinction between totally symmetric and non-totally symmetric vibrations is one that will be made use of frequently in this book. Of course, if one knows the irreducible representations for the vibrations, then finding out which are totally symmetric is trivial, since the totally symmetric irreducible representation is always the uppermost one listed in the corresponding character table.

A vital point to recognize is that *the symmetry of the vibrational wavefunction is not necessarily the same as the symmetry of the normal coordinate*. To see this, look at the general form of the vibrational wavefunction in equation (5.20). The normal coordinate for the ith vibrational mode is given the symbol Q_i in this equation, and it appears twice. Considering the exponential part first, this contains Q_i^2. If Q_i is a non-totally symmetric normal coordinate, then there will be *at least* one symmetry operation of the point group that will change the sign of Q_i. However, this will not change the sign of Q_i^2 and so the exponential term will be invariant to any symmetry operation of the point group whether or not Q_i is totally symmetric, i.e. the exponential term is always totally symmetric.

---

[6] Our concern in this book is with vibrational changes accompanying electronic transitions, but the use of symmetry in establishing vibrational selection rules is also extremely important in infrared and Raman spectroscopies.

Consequently, if the vibrational wavefunction is to be anything other than totally symmetric, it is the Hermite polynomial in equation (5.20) that brings this about. Look back at the form of the first few Hermite polynomials given in (5.10), where you should substitute Q_i for the diatomic normal coordinate η. When the vibrational quantum number v_i is even (including zero), then only even powers of Q_i appear in the Hermite polynomial. However, when v_i is odd, only odd powers of Q_i appear in the Hermite polynomial. Thus for the very same reason that we concluded that the exponential term was totally symmetric, the Hermite polynomial is also totally symmetric with respect to all symmetry operations of the point group if v_i is even. However, if v_i is odd, then the symmetry of the wavefunction is the same as the symmetry of the normal coordinate Q_i.

We may therefore conclude the following. *If the normal coordinate is totally symmetric, then the corresponding vibrational wavefunction is totally symmetric for all values of* v_i. *However, if the normal coordinate is non-totally symmetric, then the vibrational wavefunction will be totally symmetric when* v_i *is even and non-totally symmetric when* v_i *is odd.* We will make considerable use of these important results throughout the remainder of this book.

5.2.3 *Anharmonicity in polyatomic vibrations*

The expression of polyatomic vibrational motion in terms of a set of independent normal modes is only exact if harmonic motion is assumed for each vibration. In practice, anharmonicity occurs in vibrations of polyatomics just as it does for diatomics. This means that the energy level formula given in equation (5.19) must be modified to include the effects of anharmonicity, but the modification is somewhat more complicated than the diatomic case since there is more than one anharmonicity constant associated with each vibration. Specifically, one finds that the vibrational term value is given by[7]

$$G = \sum_i \omega_i \left(v_i + \tfrac{1}{2}\right) + \sum_i \sum_j x_{ij} \left(v_i + \tfrac{1}{2}\right)\left(v_j + \tfrac{1}{2}\right) \tag{5.21}$$

where the x_{ij} are anharmonicity constants. The so-called diagonal anharmonicity constant, x_{ii}, has a similar interpretation to the anharmonicity constant used for diatomics. Usually the x_{ii} are small and negative, thus causing the vibrational levels to get closer together as the vibrational ladder is climbed. However, it is worth noting that unlike the diatomic case, the diagonal anharmonicity constants can also sometimes be positive. The so-called off-diagonal anharmonicity constants, x_{ij} where $i \neq j$, arise from the mixing of normal modes. This mixing, caused by a breakdown of the harmonic oscillator approximation, is normally small for the lowest vibrational levels and can often be ignored. It is also limited to vibrations possessing the same symmetry. However, there are certain special cases where the mixing can be very large, even for low vibrational quantum numbers. *Fermi resonance* is such a

[7] For those molecules with degenerate vibrations, $(v_i + \frac{1}{2})$ should be replaced with $(v_i + d_i/2)$ where d_i is the degeneracy of mode i. There is an additional complication with degenerate modes in that they can possess vibrational angular momentum. This results in further modification to equation (5.21); readers who wish to find out more should consult References [3] and [4].

case, which occurs when two vibrational levels of the same symmetry are accidentally very close together in energy (perhaps just a few cm^{-1} apart in the harmonic oscillator limit); extensive coupling caused by anharmonicity is then possible and the normal coordinate picture is not valid.

References

1. J. Mehra and H. Rechenberg, *Found. Phys.* **29** (1999) 91–132.
2. *Molecular Spectra and Molecular Structure. I. Spectra of Diatomic Molecules*, Chapter 6, G. Herzberg, Malabar, Florida, Krieger Publishing, 1989.
3. *Molecular Vibrations*, E. Bright Wilson, J. C. Decius, and P. C. Cross, New York, Dover Publications, 1980.
4. *Introduction to the Theory of Molecular Vibrations and Vibrational Spectroscopy*, L. A. Woodward, Oxford, Clarendon Press, 1972.
5. *Molecular Symmetry and Group Theory*, R. L. Carter, New York, John Wiley and Sons, 1998.

6 Molecular rotations

The observation of rotational fine structure in electronic spectra has proven to be an invaluable source of information on molecular properties, particularly in quantifying molecular structures. We will take the same approach as for molecular vibrations, describing the quantized rotational motion of diatomic molecules before moving on to consider the more complicated case of polyatomic molecules.

6.1 Diatomic molecules

6.1.1 *The rigid rotor*

The energy levels of a rotating diatomic are particularly simple to describe if there is no other form of angular momentum in the molecule (e.g. electronic orbital or spin) with which the rotational angular momentum can interact. A rotating diatomic molecule can have two independent components of rotational angular momentum, which are shown in Figure 6.1 as rotation about two mutually perpendicular axes, designated x and y. The origin of these axes is at the centre-of-mass of the molecule. The corresponding angular momenta are represented by the vectors $\boldsymbol{R}_x$ and $\boldsymbol{R}_y$. Note that $\boldsymbol{R}_z$ is zero since the z axis contains the nuclei.[1] In classical mechanics the rotational energy would be

$$E = \frac{R_x^2}{2I_x} + \frac{R_y^2}{2I_y} \tag{6.1}$$

where I_x and I_y are the *moments of inertia* about the x and y axes. Put somewhat crudely, if we imagine that the molecule was initially not rotating, then the moment of inertia is related to the force that would need to be applied to make the molecule rotate at a certain speed. Everyday experience with much larger objects is sufficient to deduce that this force will be smaller if the atoms are lighter and/or if they are closer together. For a diatomic the moments of inertia about the x and y axes are the same and are given by

$$I = \mu r^2 \tag{6.2}$$

[1] This neglects the mass of the off-axis electrons, but this is a good approximation since the nuclei contain virtually all of the molecule's mass and can be viewed as point masses (i.e. have no significant size) located along the internuclear axis.

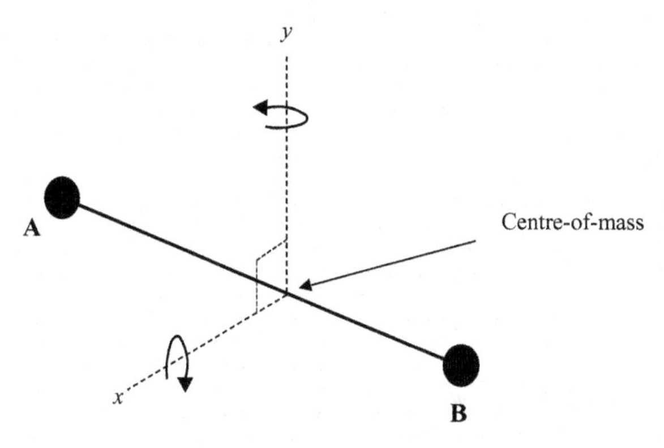

Figure 6.1 Diagram showing the two rotational degrees of freedom of a diatomic molecule.

where μ is the reduced mass (see equation (5.5)), r is the internuclear separation, and $I = I_x = I_y$.

In quantum mechanics the possible values of the angular momentum are restricted, as already seen for the case of an orbiting electron in an atom (Chapter 3). Mathematically, quantization can be introduced by recognizing that the square of the total rotational angular momentum, R^2, has the allowed values $(h/2\pi)^2 J(J+1)$ where J is the *rotational quantum number* ($J = 0, 1, 2, 3$, etc.).[2] Since (6.1) can be written as

$$E = \frac{R^2}{2I} \tag{6.3}$$

the quantum mechanical result is obtained by substituting in the eigenvalue of R^2, i.e.

$$E = \frac{h^2}{8\pi^2 I} J(J+1) = BJ(J+1) \tag{6.4}$$

The quantity B is referred to as the *rotational constant*. The pattern of energy levels that results is shown in Figure 6.2. Since the energies of rotational levels are a quadratic function of the rotational quantum number they are not equally spaced, in contrast to the vibrational energy levels of the harmonic oscillator. Equation (6.4) assumes that the molecule is a rigid rotor, i.e. the bond length is unaffected by the speed of rotation. This turns out to be a good approximation in most cases, although *centrifugal distortion* does become significant at high J.

Finally, note the link between the rotational constant and the equilibrium bond length, r_e, specifically

$$B = \frac{h^2}{8\pi^2 \mu r_e^2} \tag{6.5}$$

This very important result shows that the measurement of the rotational constant allows the bond length of a diatomic molecule to be calculated.

[2] Actually J is reserved for labelling the total angular momentum quantum number excluding nuclear spin. Thus the comments made in this section, and the rotational energy formula (6.4), must be modified when dealing with anything other than $^1\Sigma$ states of molecules. Such molecules will be encountered in some of the Case Studies.

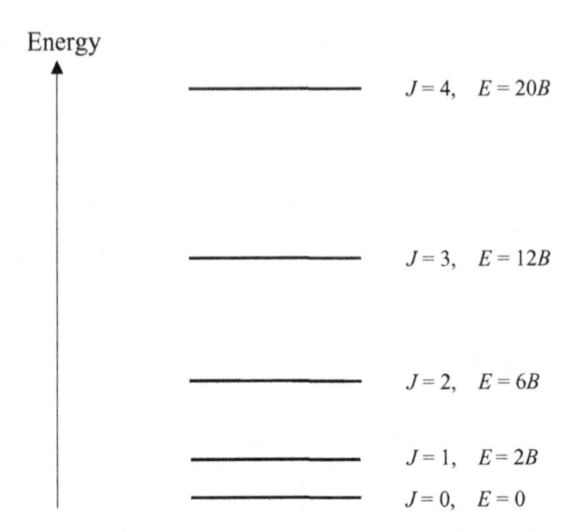

Figure 6.2 Rotational energy levels of a rigid diatomic molecule.

6.1.2 *Space quantization*

It is instructive to explore further the analogy between molecular rotation and the orbital motion of a single electron. We know that the orbital angular momentum of a single electron is characterized by the quantum number l, which is analogous to J in the case of molecular rotation. However, we also know that there is a second quantum number arising in the case of orbital motion, the projection quantum number m_l, which defines the possible orientations of the orbital angular momentum vector relative to some arbitrary axis; this is known as space quantization. The $2l + 1$ possible values of m_l give rise to a $2l + 1$ degeneracy for each value of l.

By analogy, space quantization would also be expected for molecular rotation. The projection quantum number in this case is denoted by the symbol M_J, and can have $2J + 1$ values ranging from $-J$ to $+J$. The projection quantum number is significant in two respects. First, the degree of degeneracy of a rotational level is J-dependent and increases with J. For a sample composed of diatomic molecules, this can therefore give rise to the situation where there are more molecules in excited rotational levels than in the lowest ($J = 0$), since the number of molecules in level J, N_J, is given by the Boltzmann distribution

$$N_J \propto (2J + 1)\exp[-BJ(J + 1)/kT] \tag{6.6}$$

at thermal equilibrium. The degeneracy increases with J while the exponential term decreases with J; consequently, a maximum in N_J at some non-zero value of J is possible.

Although the direction of space quantization is arbitrary for a freely rotating molecule, this is no longer the case if the molecule is placed in an electric or magnetic field. If the molecule has a permanent electric dipole moment, the rotational angular momentum vector will be forced to precess about an axis parallel to the field direction providing the field is sufficiently strong, thus defining the direction of space quantization. Furthermore, the $2M_J + 1$ degeneracy is removed, a phenomenon known as the Stark effect. The degeneracy will also be removed in a magnetic field if the molecule has a magnetic moment, e.g. if it possesses unpaired electrons. This is known as the Zeeman effect. For both the Zeeman and Stark effects the splitting will increase as the field strength increases and may therefore not be noticeable if only weak fields are present.

6.2 Polyatomic molecules

6.2.1 *Classical limit*

It turns out to be very useful to classify polyatomic molecules into various groups according to their moments of inertia. As will be seen shortly, this makes it possible to ascertain whether standard formulae will apply for their rotational energy levels, or whether a more in-depth analysis is required.

By analogy with the diatomic case considered earlier, let us treat a rotating polyatomic molecule classically as a rigid rotating body. We will also retain the good approximation that all of the mass of a molecule is contained within the nuclei and that each nucleus is a point mass. In the general case the moments of inertia can be defined about any three mutually perpendicular axes passing through the centre-of-mass of the molecule. If we make an arbitrary choice of axes, then the rotational kinetic energy about any one of these axes contains three terms. For example, for the x axis

$$E_x = \tfrac{1}{2} I_{xx}\omega_x^2 + I_{xy}\omega_x\omega_y + I_{xz}\omega_x\omega_z \tag{6.7}$$

The quantity I represents the moment of inertia, which has already been met for diatomic molecules, but notice the addition of double subscripts in equation (6.7). The double subscript indicates that the moment of inertia is not a vector with three components along the x, y, and z axes, but in fact has cross-terms yielding a total of nine components altogether. The moment of inertia is an example of a second-rank tensor quantity.[3] This was not an important issue to consider for diatomic molecules, but it is important for non-linear polyatomic molecules. The diagonal moment, I_{xx}, is given by

$$I_{xx} = \sum_i m_i r_{ix}^2 \tag{6.8}$$

where m_i is the mass of nucleus i and r_{ix} is the distance of the ith nucleus from the x axis (measured along a line perpendicular to the x axis, as shown in Figure 6.3). The off-diagonal moments of inertia are given by

$$I_{xy} = -\sum_i m_i r_{ix} r_{iy}$$

$$I_{xz} = -\sum_i m_i r_{ix} r_{iz} \tag{6.9}$$

[3] A vector is a first-rank tensor, and it can be represented in a neat fashion by a column matrix with three rows. A second-rank tensor is most clearly expressed when written as a 3 × 3 matrix. It turns out that this matrix is always symmetrical, e.g. $I_{xy} = I_{yx}$, and so there are actually only six independent components at most. There are other physical properties that are also represented by second-rank tensors, e.g. polarizability.

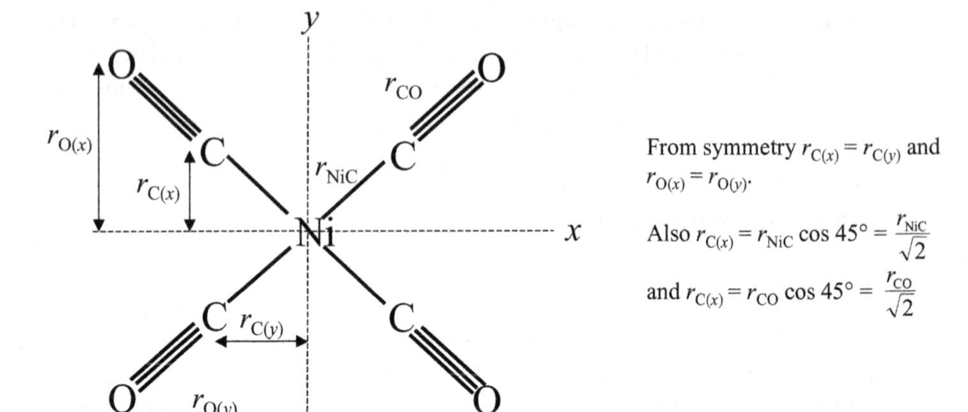

Moments of inertia

$$I_{xx} = 4m_{\mathrm{C}} r_{\mathrm{C}(x)}^2 + 4m_{\mathrm{O}} r_{\mathrm{O}(x)}^2 = 2m_{\mathrm{C}} r_{\mathrm{NiC}}^2 + 2m_{\mathrm{O}}(r_{\mathrm{NiC}} + r_{\mathrm{CO}})^2; \quad I_{yy} = I_{xx}; \quad I_{xy} = I_{yx} = 0;$$

$$I_{zz} = 4m_{\mathrm{C}} r_{\mathrm{NiC}}^2 + 4m_{\mathrm{O}}(r_{\mathrm{NiC}} + r_{\mathrm{CO}})^2; \quad I_{xz} = I_{zx} = I_{yz} = I_{zy} = 0$$

Figure 6.3 Moments of inertia for the square planar molecule $Ni(CO)_4$. The inertial axes x, y, and z have their origin at the centre-of-mass of the molecule, which is the Ni nucleus. Notice that the off-diagonal contributions to the moments of inertia are zero for the chosen axis system, which can be seen by careful application of equations (6.9). According to the classification given later, $Ni(CO)_4$ is an oblate symmetric top.

The appearance of off-diagonal moments of inertia is a complication we would like to avoid. Fortunately, it turns out that it is *always* possible to find a set of inertial axes where the off-diagonal moments of inertia are zero. These axes are called the *principal axes*, and they are conventionally labelled as a, b, and c to distinguish them from any arbitrary set of cartesian axes x, y, and z. In the principal axis system, the overall rotational kinetic energy is given by

$$E = \tfrac{1}{2} I_a \omega_a^2 + \tfrac{1}{2} I_b \omega_b^2 + \tfrac{1}{2} I_c \omega_c^2 \tag{6.10}$$

in the classical limit.

There is an important convention used in labelling the principal axes. This convention stipulates that, once the axes have been identified (see next section), they are labelled according to the requirement that $I_a \leq I_b \leq I_c$.

6.2.2 *Classification of polyatomic rotors*

It was stated in the previous section that a set of principal axes always exists for any molecule, but how are they identified? Symmetry, should the molecule possess any, is of great help here.

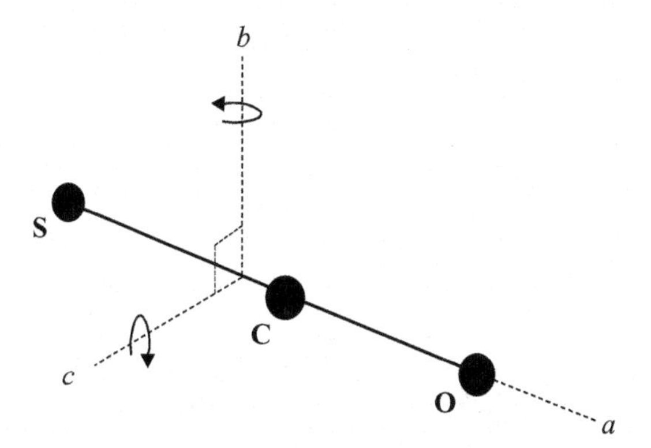

Figure 6.4 Diagram showing the principal inertial axes for the linear OCS molecule. Notice that the centre-of-mass lies between the C and S atoms because of the larger mass of S compared with O. The moments of inertia are $I_a = 0$ and $I_b = I_c$ for a linear molecule.

Consider as an initial example a linear polyatomic molecule. Since all the nuclei in such a molecule lie along a single axis it is fairly obvious that this axis coincides with one of the principal axes. Confirmation is provided by inspection of the form of the off-diagonal elements of the inertial tensor of the type shown in (6.9). The other two principal axes must be perpendicular to the internuclear axis, but beyond that the choice is arbitrary because of the cylindrical symmetry of the molecule.

The principal axes of a linear polyatomic are illustrated in Figure 6.4. The allocation of a, b, and c to the axes shown in Figure 6.4 is made on the basis of the rule that $I_a \leq I_b \leq I_c$. I_a is in fact zero, whereas $I_b = I_c$.

There are other classifications based on the relative magnitudes of the principal moments of inertia. If all three moments of inertia are equal then the molecule is said to be a *spherical top*. This occurs for molecules with very high point group symmetries, such as T_d and O_h. Thus molecules such as CH_4 and SF_6 belong in this category (see Figure 6.5(a)). Because of their high symmetry, any choice of mutually perpendicular axes passing through the centre-of-mass will be principal axes.

If only two principal moments of inertia are equal, then the molecule is classified as a *symmetric top*. A linear polyatomic (or diatomic) molecule is a special case of a symmetric top where one moment of inertia (I_a) is zero. For non-linear molecules symmetric tops can only occur when the molecule has a C_3 or higher axis of rotational symmetry. Some examples of symmetric top molecules are shown in Figure 6.5 (b and c). There are two subdivisions of the symmetric top: those for which $I_a < I_b = I_c$, which are referred to as *prolate symmetric tops*, and those for which $I_a = I_b < I_c$, which are known as *oblate symmetric tops*. It is usually straightforward to ascertain if a molecule is a symmetric top by inspection. Distinguishing between prolate and oblate symmetric tops is also often straightforward, but there are exceptions to this statement. For example, if ammonia were to adopt a planar equilibrium geometry, as in fact it does in at least one of its excited electronic states, then it is clearly an oblate symmetric top. At the other extreme, if the ammonia molecule became non-planar with an extremely small H—N—H bond angle in some electronic state, then it would be a prolate symmetric top. Clearly at some intermediate H—N—H bond angle there

(a) **Spherical tops**

(b) **Prolate symmetric tops**

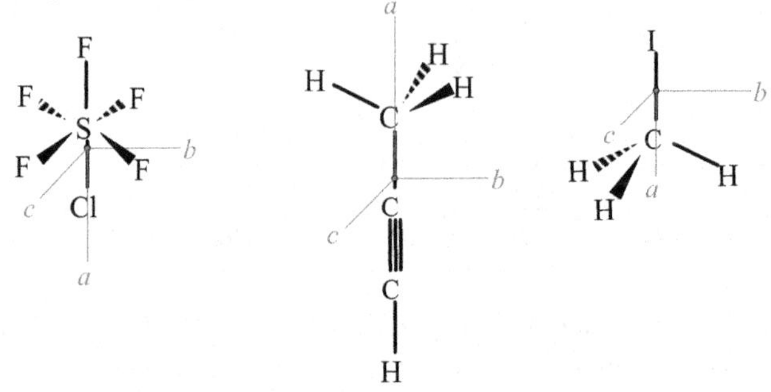

(c) **Oblate symmetric tops**

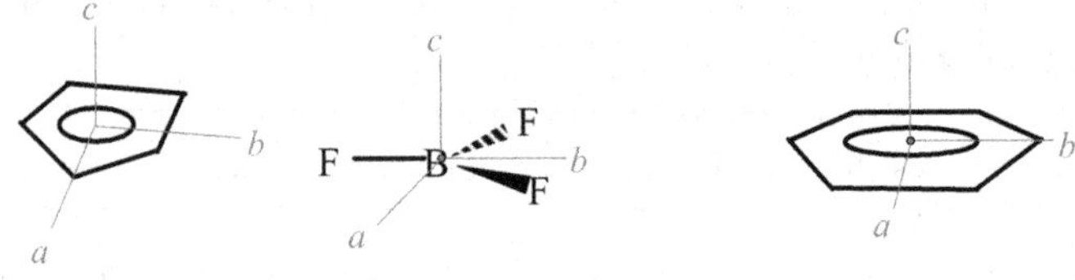

(d) **Asymmetric tops**

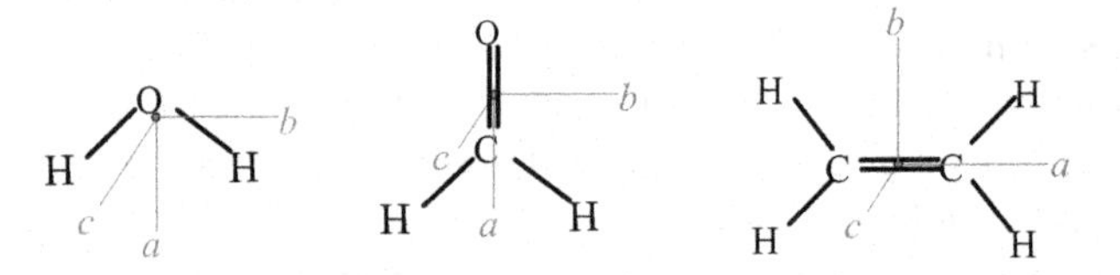

Figure 6.5 Examples of spherical tops, prolate and oblate symmetric tops, and asymmetric tops molecules. In each case the approximate position of the centre-of-mass is shown by the location of the origin of the inertial axes.

will be a transition from a prolate to an oblate symmetric top, and at angles close to this point of transition it is not obvious, without performing calculations, which case applies.

Last, but by no means least, we come to the class to which most molecules belong, namely the *asymmetric tops*. For these molecules, the three principal moments of inertia are all different. Some examples are shown in Figure 6.5(d). For those molecules with a reasonable amount of symmetry, such as H_2O, the principal axes are easy to locate; for H_2O one of the axes (the a axis) must coincide with the C_2 symmetry axis. However, for molecules with less symmetry the principal axes can only be established by calculation using either a known or assumed geometry.

Table 6.1 *Convention for labelling principal inertial axes in polyatomic molecules*

Rotor	Inertial relationship
Spherical top	$I_a = I_b = I_c$
Linear molecule	$I_a = 0; I_b = I_c$
Prolate symmetric top	$I_a < I_b = I_c$
Oblate symmetric top	$I_a = I_b < I_c$
Asymmetric top	$I_a < I_b < I_c$

Table 6.1 summarizes the inertial classification system for polyatomic molecules. Figure 6.3 illustrates how moments of inertia are related to structural parameters for an example of an oblate symmetric top, $Ni(CO)_4$.

6.2.3 Rotational energy levels of linear polyatomic molecules

Given that there are only two non-zero and equal principal moments of inertia for a linear polyatomic molecule, just as for a diatomic, the rotational energy level formula (6.4) applies.

However, an important difference lies in the relationship between the rotational constant B and the atomic masses and bond lengths. For a diatomic molecule B is dependent on only one bond length, whereas in a linear polyatomic molecule B depends on all of the bond lengths. Formulae can be derived to link B to the bond lengths (and the atomic masses) but we omit them here. It suffices to say that, unlike a diatomic molecule, the precise structure of a linear polyatomic molecule cannot be established from measurement of a single rotational constant,[4] except in those cases where all bond lengths are assumed (or are known) to be equal.

6.2.4 Rotational energy levels of symmetric tops

Formulae for the rotational energy levels of symmetric tops can be derived using an extension of the approach employed for diatomics in Section 6.1.1. In a non-linear molecule the rotational kinetic energy has three independent components and is given by

$$E = \frac{R_a^2}{2I_a} + \frac{R_b^2}{2I_b} + \frac{R_c^2}{2I_c} \tag{6.11}$$

in a principal axis system. Consider a prolate symmetric top, for which $I_b = I_c$. In this case equation (6.11) becomes

$$E = \frac{R_a^2}{2I_a} + \frac{1}{2I_b}(R_b^2 + R_c^2) \tag{6.12}$$

[4] There are two ways around this difficulty. For a linear triatomic, one way would be to assume a reasonable value for one of the bond lengths and then use the rotational constant to determine the other. An alternative and better approach is to measure the rotational constant for more than one isotopomer, i.e. use isotopic substitution of one atom and measure a new rotational constant. The two rotational constants this provides are sufficient to determine the two equilibrium bond lengths.

The total angular momentum is the sum of the squares of the individual components, so therefore

$$R_b^2 + R_c^2 = R^2 - R_a^2 \tag{6.13}$$

Substituting (6.13) into (6.12) gives

$$E = \frac{R^2}{2I_b} + R_a^2\left(\frac{1}{2I_a} - \frac{1}{2I_b}\right) \tag{6.14}$$

To get the quantum mechanical result, we must now treat the total rotational angular momentum, R, and its component, R_a, as operators and replace them with their eigenvalues. Indeed, equation (6.11) has been converted into the form shown in (6.14) to make this possible. Recall that the total angular momentum is characterized by a quantum number, J, such that the total rotational angular momentum is $(h/2\pi)[J(J+1)]^{1/2}$. The operator R^2 can therefore be replaced by the square of the total rotational angular momentum, $(h^2/4\pi^2)[J(J+1)]$. There is also a quantized component of angular momentum along some arbitrary axis in the general case. However, in a symmetric top this axis of quantization is no longer arbitrary, but rather corresponds to the axis of highest rotational symmetry. For a prolate symmetric top the axis of highest symmetry is the a axis, and therefore the component of rotational angular momentum along this axis is quantized. The corresponding quantum number is given the symbol K and the angular momentum along this axis is $Kh/2\pi$. Thus the operator R_a^2 can therefore be replaced by $K^2h^2/4\pi^2$.

The final step in replacing equation (6.14) with something more useful is to recognize that $h^2/8\pi^2 I_a$ and $h^2/8\pi^2 I_b$ are rotational constants, which we label as A and B. Thus the final result for the rotational energy levels of a rigid prolate symmetric top is

$$E = BJ(J+1) + (A - B)K^2 \tag{6.15}$$

The rotational energy level arrangement obtained from application of this formula is shown in Figure 6.6(a). Note that because $Kh/2\pi$ is the projection of the total rotational angular momentum along the a axis, then $K \leq J$. For each value of J there is a stack of levels corresponding to the possible values of K, which range from 0 to J in integer steps. In fact each of the K levels (except $K = 0$) is doubly degenerate, a consequence of the equivalence of clockwise or anticlockwise rotation about the top axis. Clearly the rotational structure for symmetric tops is more complicated than for linear molecules given that there are now two quantum numbers and two different rotational constants.

Oblate symmetric tops can be dealt with in a similar manner to prolate symmetric tops. The main difference is that now the c axis is the highest symmetry axis and so this is the axis of quantization. After several steps we obtain

$$E = BJ(J+1) + (C - B)K^2 \tag{6.16}$$

This looks similar to the prolate formula (6.15), but an important difference is that $C - B$ is negative while $A - B$ is positive (since $A \geq B \geq C$). Consequently, whereas the gap between adjacent rotational levels increases as K increases for a given J in a prolate symmetric top, it

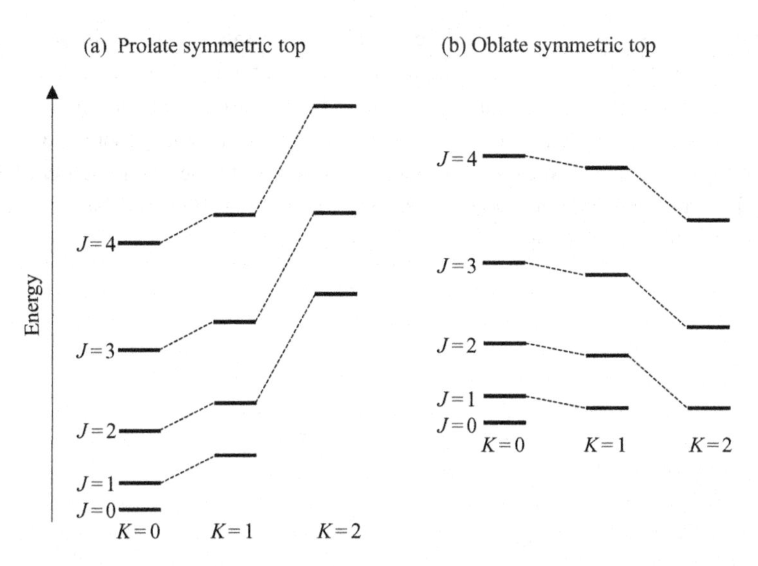

Figure 6.6 Rotational energy level diagrams for (a) prolate and (b) oblate symmetric tops.

decreases for oblate tops. This can be seen by comparison of the two energy level diagrams in Figure 6.6.

6.2.5 Rotational energy levels of spherical tops

In a spherical top $I_a = I_b = I_c$ and therefore the rotational operator is just

$$H_{\text{rot}} = \frac{R^2}{2I_b} \tag{6.17}$$

Clearly a rotational energy level formula identical in appearance to the linear molecule case is obtained, i.e.

$$E = BJ(J+1) \tag{6.18}$$

6.2.6 Rotational energy levels of asymmetric tops

Asymmetric tops are characterized by the fact that all three principal moments of inertia, and therefore all three rotational constants, are unequal. As a result, no factorization of the rotational Hamiltonian (6.11) is possible and the mathematical problem is far more challenging to solve. It is possible to derive formulae for specific rotational energy levels of asymmetric tops but a general formula cannot be obtained. The more usual way of predicting rotational energies of asymmetric tops is to use numerical solution of the rotational problem on a computer, a topic beyond the scope of this book but touched upon briefly in Appendix H.

It is worth bearing in mind, however, that many molecules quite closely approximate the prolate or oblate symmetric top limits. For example, both water and formaldehyde are asymmetric tops but in both cases the B and C constants are quite similar. Thus the prolate symmetric top formula (6.15) should be a reasonably good approximation of their rotational energy levels, and this is indeed found to be the case from experiment. However, as one would also expect, there are differences; in particular the K degeneracy observed in symmetric tops is removed in asymmetric tops, giving rise to the phenomenon of K-type doubling. In strongly asymmetric tops the K-type doubling is so severe that a comparison with symmetric tops is meaningless.

7 Transition probabilities

Depending on the resolution, a spectrum may consist of well-resolved discrete peaks, each of which is attributable to a single specific transition, or it may consist of broader bands that are actually composed of several unresolved transitions. In either case, the intensities will depend on a number of factors. The sensitivity of the spectrometer is crucial. So too is the concentration of the absorbing or emitting species. However, our interest in the remainder of this chapter is with the intrinsic transition probability, i.e. the part that is determined solely by the specific properties of the molecule. The key to understanding this is the concept of the *transition moment*.

7.1 Transition moments

Consider two pairs of energy levels, one pair in molecule A and one pair in a completely different molecule B. Assume for the sake of simplicity that the energy separation between the pair of levels is exactly (and fortuitously) the same for both molecules. Suppose that a sample of A is illuminated by a stream of monochromatic photons with the correct energy to excite A from its lower to its upper energy level. There will be a certain probability that a molecule is excited per unit time. Now suppose sample A is replaced with B, keeping the concentration and all other experimental conditions unchanged. In general the probability of photon absorption per unit time for B would be different from A, perhaps by a very large amount. The conclusion we must draw is that there is some factor dependent on the specific details of the energy levels which determines whether A or B has the higher transition probability. This factor is known as the *transition moment*.

For radiation to be absorbed, there must be an interaction between the radiation and the molecule. This is not the only condition,[1] but it is clearly of fundamental importance. Both the electric and magnetic fields of electromagnetic radiation may interact with any electric or magnetic fields present in a molecule. For the types of spectroscopy that we will consider it is the electrical rather than magnetic interaction that is normally important, although an exception to this will be met in the Case Studies. Molecules may have non-zero

[1] A photon also possesses quantized angular momentum, a strange thought given that photons have zero rest mass as, but one which has nevertheless been proven by experiment. Since angular momentum must be conserved in all processes, there is also a momentum restriction that limits the possible spectroscopic transitions [1].

electric fields for a number of reasons, such as the presence of a permanent electric dipole moment, or because a particular vibration induces an oscillating dipole moment, or because the instantaneous motion of one or more electrons produces a transient electric field.

It is possible to go beyond this simple picture and perform a quantum mechanical analysis of the probability that absorption will take place due to the coupling of the electric fields from the radiation and the molecule. The derivation is complex, but the result is simple and of great significance. The intrinsic transition probability is given by $|\boldsymbol{M}_{21}|^2$, where $\boldsymbol{M}_{21}$ is the transition dipole moment for a transition from energy level 1 up to level 2. The transition moment, which is labelled in bold typescript to indicate that it is a vector quantity, is

$$\boldsymbol{M}_{21} = \int \Psi_2 \boldsymbol{\mu} \Psi_1 \, d\tau \tag{7.1}$$

where Ψ_1 and Ψ_2 are the wavefunctions of the lower and upper states, respectively, and $d\tau$ includes all relevant coordinates (i.e. spatial and spin). The vector quantity $\boldsymbol{\mu}$ is the electric dipole moment *operator*. For a system of n particles, each of charge Q_n, the dipole moment operator is given by

$$\boldsymbol{\mu} = \sum_n Q_n \boldsymbol{x}_n \tag{7.2}$$

where $\boldsymbol{x}_n$ is the position vector of the nth charged particle. It is useful, as will be seen later, to split the summation in (7.2) into two terms, one involving the electrons and the other the nuclei, such that $\boldsymbol{\mu} = \boldsymbol{\mu}_e + \boldsymbol{\mu}_n$.[2]

There are two important points to note at this stage. The first is that the electric dipole moment operator is *not* the same as the electric dipole moment of a molecule. In quantum mechanics the electric dipole moment of a molecule in some state with wavefunction Ψ_i is given by

$$\boldsymbol{\mu}_{\text{edm}} = \int \Psi_i \boldsymbol{\mu} \Psi_i \, d\tau \tag{7.3}$$

The wavefunction of only one state appears in (7.3), as opposed to two in the transition dipole moment expression in (7.1). The difference is a crucial one, for we can interpret the transition dipole moment as quantifying an *instantaneous* change in dipole moment brought about by the movement of electrical charge during the transition from the state with wavefunction Ψ_1 to the state with wavefunction Ψ_2. Consequently, a permanent electric dipole moment is not required for electronic transitions to take place. The transition from Ψ_1 to Ψ_2 normally involves only a single electron moving from one MO to another.

It is important to recognize that equation (7.1) is only an approximation, albeit usually a very good one, known as the *electric dipole approximation*. Transitions governed by the transition moment in (7.1) are said to be electric dipole transitions and they are by far the most important for the topics covered in this book. However, the reader should be aware

[2] In a cartesian coordinate system $\boldsymbol{M}_{21}$ and $\boldsymbol{\mu}$ will have components in the x, y, and z directions. The transition probability is a scalar quantity and is given by $|\boldsymbol{M}_{21}|^2 = (M_{21(x)})^2 + (M_{21(y)})^2 + (M_{21(z)})^2$.

that transitions may also be induced by the magnetic part of the radiation, giving rise to magnetic dipole transitions (responsible for NMR and ESR spectroscopy),[3] or can arise from higher order electrical effects, notably electric quadrupole transitions.

7.1.1 Absorption and emission

The discussion in the previous section referred specifically to the absorption of radiation, but much of what was said could also be applied to emission. The absorption of radiation is a *stimulated* process, with the incident photon stimulating the molecule into action. This may seem obvious, but it is a point of great significance given that emission can occur in *two* ways, *stimulated* and *spontaneous* emission.

Stimulated emission is the reverse of absorption. If a molecule is in some upper energy level, E_2, it can be induced to fall to a lower level, E_1, by emission of a photon if another photon of energy $E_2 - E_1$ is incident upon it. The new photon produced will share the same frequency, phase and direction as the stimulating photon. In other words, the process is a *coherent* one. The transition moment for stimulated emission is equal to that of stimulated absorption, i.e. $\boldsymbol{M}_{21} = \boldsymbol{M}_{12}$.

Although their transition moments are the same, the probabilities of stimulated emission and absorption will not normally be the same in practice because of differences in populations of the upper and lower energy levels. The rate of absorption or stimulated emission can be treated quantitatively by using a rate equation approach directly analogous to that employed in chemical kinetics. The rate of absorption will be directly proportional to both the number of molecules in the lower state, N_1, and to the density of incident radiation, $\rho(\nu)$, at the resonance frequency ν, and so can be expressed as

$$-\frac{dN_1}{dt} = \frac{dN_2}{dt} = BN_1\rho(\nu) \tag{7.4}$$

The proportionality constant, B (not to be confused with rotational constants), which is analogous to a second-order rate constant in kinetics, is known as the *Einstein B coefficient* and is dependent on the transition moment in the following manner:

$$B = \frac{8\pi^2}{12\varepsilon_0 h^2}|\boldsymbol{M}_{21}|^2 \tag{7.5}$$

The rate of stimulated emission is

$$\frac{dN_1}{dt} = -\frac{dN_2}{dt} = BN_2\rho(\nu) \tag{7.6}$$

and so combining (7.4) and (7.6) gives

$$\frac{dN_1}{dt} = B(N_2 - N_1)\rho(\nu) \tag{7.7}$$

[3] Electric dipole transitions dominate in electronic spectroscopy, as well as in IR and microwave spectroscopy. However, because they involve the 'flipping' of magnetic spins, it is *magnetic dipole transitions* which are responsible for ESR and NMR spectra.

For a system at thermal equilibrium, the population ratio N_2/N_1 is given by the *Boltzmann distribution*,

$$\frac{N_2}{N_1} = \frac{g_2}{g_1}\exp\left(-\frac{(E_2 - E_1)}{kT}\right) \tag{7.8}$$

where g_1 and g_2 are the degeneracies of the two levels. If E_2 and E_1 are electronic energy levels then in general $E_2 - E_1 \gg kT$ and so $N_2 \ll N_1$. Thus the right-hand side of (7.7) will be negative and so a net depletion of the population of level 1 occurs; in other words absorption, rather than stimulated emission, will dominate. If, on the other hand, $N_2 > N_1$, then stimulated emission will dominate. This unusual situation is termed a *population inversion* and is an essential requirement for the operation of lasers [2, 3].

Emission of a photon can also occur spontaneously, i.e. in the absence of a stimulating photon. In view of earlier comments this might be thought to be impossible because there is nothing obvious to 'kick-start' (stimulate) the emission process. The explanation for this apparent discrepancy can be extracted from a branch of quantum physics known as quantum electrodynamics. The full story is very involved but a brief explanation is as follows. Suppose a near-perfect vacuum was maintained inside a container such that only one molecule remained within it. Existence would seem to be dull for this molecule as it would encounter nothing but the walls as it bounced around in the chamber. However, according to quantum electrodynamics nothing could be further from the truth. Inside (and outside) the chamber there are rapid *zero-point fluctuations* in which photons burst into existence and then quickly disappear. This strange process is capable of providing the necessary stimulation and so spontaneous emission can be viewed as stimulated emission brought about by the momentary presence of photons produced by zero-point fluctuations.

Spontaneous emission is easily incorporated into the rate equation model. It results only in the depopulation of the upper state, and unlike absorption or emission is unaffected by the applied radiation density. Consequently, it is akin to a first-order chemical reaction with rate given by AN_2. The quantity A, which is analogous to a first-order rate constant, is known as the *Einstein A coefficient* and is given by

$$A = \left(\frac{8\pi h\nu^3}{c^3}\right)B \tag{7.9}$$

Modification of the rate equation (7.7) to include spontaneous emission yields

$$\frac{\mathrm{d}N_1}{\mathrm{d}t} = AN_2 + B(N_2 - N_1)\rho(\nu) \tag{7.10}$$

and so spontaneous emission competes with stimulated emission in depopulating the upper state. In fact in most circumstances one finds that spontaneous emission is far more important than stimulated emission.

Since A is directly proportional to B, the spontaneous emission probability depends on the magnitude of the transition moment. However, notice also that the spontaneous emission probability depends on ν^3. As a result, spontaneous emission rapidly increases in importance as the emitted radiation frequency increases. Drawing on the analogy with first-order chemical kinetics, or for that matter the first-order spontaneous decay of radioactive

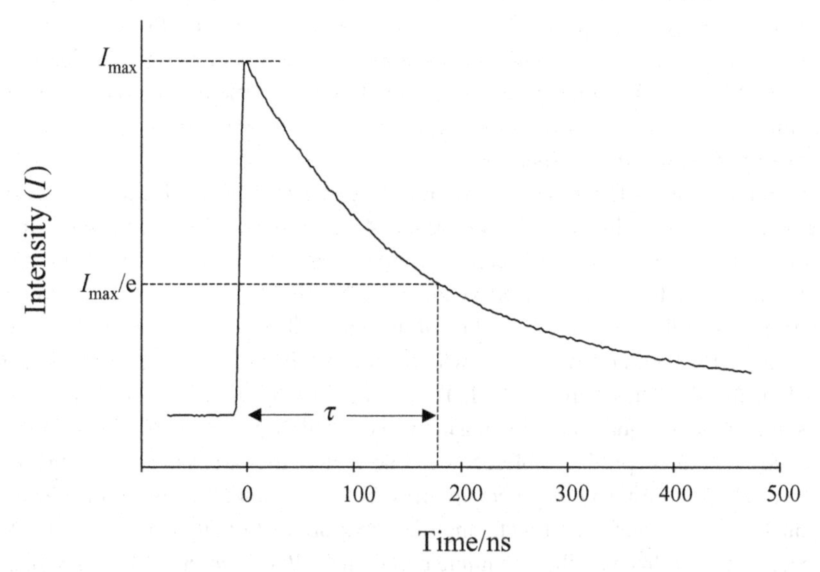

Figure 7.1 Typical radiative decay curve for an ensemble of molecules excited to some specific upper state by a short pulse of light. The radiative lifetime τ is defined as the time taken for the emission intensity to fall to 1/e (=1/2.718) of its original value. In this particular example the radiative lifetime is ~180 ns, a fairly typical value for an excited electronic state connected to a lower electronic state by an allowed transition.

nuclei, the decay of an ensemble of excited molecules by spontaneous emission is an exponential process given by the decay curve shown in Figure 7.1. This figure assumes that all molecules are excited simultaneously, e.g. by a pulse from a laser. The decay curve has the functional form e^{-At} or alternatively $e^{-t/\tau}$, where t is the time. The quantity τ, known as the spontaneous emission lifetime, or *radiative lifetime*, of the excited state, is the time taken for the spontaneous emission intensity to fall to a factor of $1/e$ of its original value. Since $\tau = 1/A$, the radiative lifetime will be short when A is large.

The frequency dependence of A is crucial in determining values of τ. In the visible and ultraviolet regions of the spectrum, excited state lifetimes in the range 10–1000 ns are the norm. In the infrared, lifetimes may be tens of microseconds or even milliseconds, while in the microwave and millimetre wave regions the lifetimes can run into seconds. With such long radiative lifetimes in long wavelength regions, the probability of spontaneous emission is very low and indeed other means of depopulating excited states, such as collisional processes, may become dominant. This is the reason why infrared, and particularly microwave, spectra are normally obtained as absorption rather than emission spectra.

7.1.2 Concept of selection rules

If all three components of the transition dipole moment are zero then the absorption and emission probabilities are zero. When this occurs the transition is said to be *forbidden*. To prove that a particular transition is forbidden, the absolute value of the transition moment could be determined by substituting the upper and lower state wavefunctions into (7.1) and evaluating the integral. However, it is rare that accurate wavefunctions are known, and in any case it is normally quite unnecessary to go to such trouble to determine whether or not the transition moment vanishes. Instead, a knowledge of the *symmetries* of the upper and lower state wavefunctions will suffice.

The importance of symmetry in establishing spectroscopic selection rules cannot be overstated. We have already seen in Chapters 4 and 5 that the symmetry of electronic and vibrational wavefunctions can be conveniently classified in terms of point group symmetry. Furthermore, we have seen that the symmetry of the product of two wavefunctions can be determined by taking the direct product of the irreducible representations of the individual wavefunctions. The integrand in (7.1) will transform as the reducible representation obtained by taking the direct triple product $\Gamma(\Psi_2) \otimes \Gamma(\boldsymbol{\mu}) \otimes \Gamma(\Psi_1)$, where Γ is shorthand notation for the symmetry of the quantity following in brackets. This triple product is easily evaluated and reduced using direct product tables. Notice that each of the cartesian components of $\boldsymbol{\mu}$ must be considered in turn, so this procedure must be carried out three times. Each triple direct product can be evaluated by first taking the direct product of any pair, and then taking the direct product of this with the remaining component. *If the final result does not include the totally symmetric irreducible representation of the point group, then the transition moment must be zero.* The reason for this conclusion is that a non-totally symmetric integrand will have two regions of space, one in which the integrand has a certain phase, and another of equal volume where the phase is reversed, because of its antisymmetry with respect to at least one of the point group symmetry operations. When integration is performed along the relevant coordinate the opposite phases of these two regions will cancel and so the integral vanishes.

Arguments along these lines can be used to establish transition *selection rules*. The Born–Oppenheimer approximation conveniently allows the selection rules to be sub-divided into electronic, vibrational, and rotational selection rules. The remainder of this chapter deals mainly with electronic and vibrational selection rules. Rotational selection rules can also be deduced by using symmetry arguments, but in general their derivations are difficult and are not included here. Rotational selection rules are briefly returned to in the final section of this chapter.

7.2 Factorization of the transition moment

When a molecule undergoes an electronic transition, its vibrational and rotational state may also change. In any one overall state the Born–Oppenheimer approximation allows the total wavefunction to be factorized into electronic, vibrational, and rotational parts, namely

$$\Psi(\boldsymbol{r}, \boldsymbol{R}) = \psi_e(\boldsymbol{r}, \boldsymbol{R}_e).\psi_v(\boldsymbol{R}).\psi_r(\boldsymbol{R}) \tag{7.11}$$

where $\boldsymbol{r}$ and $\boldsymbol{R}$ are generic symbols representing all electronic and nuclear coordinates, respectively. It is assumed in the above that the electronic wavefunction, ψ_e, is well approximated at all points during a vibration by the wavefunction at the equilibrium nuclear coordinates ($\boldsymbol{R}_e$), an approximation justified by the small amplitude of most vibrations in low-lying vibrational levels. The rotational wavefunction, ψ_r, is a function only of nuclear coordinates, since electron masses are very small by comparison. In fact for a fixed nuclear configuration the rotational wavefunction depends only on the *orientation* of the molecule relative to some arbitrarily defined set of laboratory axes. This is important information for determining rotational selection rules but we will not consider the matter any further here (see, for example, Reference [4] for further information).

If (7.11), with the rotational part removed, is substituted into (7.1), and the dipole moment operator is expressed as the sum of nuclear and electronic parts, then the transition moment becomes

$$\boldsymbol{M} = \iint \psi_e'(\boldsymbol{r}, \boldsymbol{R}_e).\psi_v'(\boldsymbol{R})(\boldsymbol{\mu}_e + \boldsymbol{\mu}_n).\psi_e''(\boldsymbol{r}, \boldsymbol{R}_e).\psi_v''(\boldsymbol{R})\,\mathrm{d}\boldsymbol{r}\,\mathrm{d}\boldsymbol{R} \tag{7.12}$$

The subscripts 1 and 2 have been omitted to avoid clashing with the wavefunction subscripts, and instead we use $'$ and $''$ to designate upper and lower states, respectively. Equation (7.12) can be separated into the sum of two parts, one involving $\boldsymbol{\mu}_n$ and the other $\boldsymbol{\mu}_e$. The former turns out to be zero[4] leaving

$$\boldsymbol{M} = \int \psi_e'(\boldsymbol{r}, \boldsymbol{R}_e).\boldsymbol{\mu}_e.\psi_e''(\boldsymbol{r}, \boldsymbol{R}_e)\,\mathrm{d}\boldsymbol{r} \int \psi_v'(\boldsymbol{R}).\psi_v''(\boldsymbol{R})\,\mathrm{d}\boldsymbol{R} \tag{7.13}$$

The above expression is extremely important because the first integral on the right-hand side is the basis for electronic selection rules, while the second determines the accompanying vibrational selection rules. We now consider each in turn.

7.2.1 *Electronic selection rules*

The application of group theory to the first integral in equation (7.13) allows the electronic selection rules to be predicted for any molecule. An example will serve to illustrate this, with more being found later in some of the Case Studies.

For linear molecules one of the electronic selection rules is $\Delta\Lambda = 0, \pm 1$, where Λ is the quantum number for the projection of the total electronic orbital angular momentum onto the internuclear axis. We will not prove this *per se*, but instead will show that it is consistent with (7.13) using simple group theoretical arguments. According to the $\Delta\Lambda$ selection rule a $\Sigma^+ \leftrightarrow \Pi$ transition in a molecule with $C_{\infty v}$ symmetry is allowed. To show that this is true, we take the direct product $\Sigma^+ \otimes \Pi$, which from direct product tables gives the Π irreducible representation. According to the $C_{\infty v}$ character table, the x and y components of the dipole operator also collectively have Π symmetry. The transition must therefore be allowed since the direct product of any irreducible representation with itself must always include the totally symmetric representation (and hence the electronic transition moment can be non-zero). Using the same sort of arguments it is easily shown that, for example, a $\Sigma^+ \leftrightarrow \Delta$ transition is forbidden.

---

4. The term involving $\boldsymbol{\mu}_n$ is zero because, on separating the variables, a product of two integrals is obtained, one of which is $\int \psi_e'(\boldsymbol{r}, \boldsymbol{R}_e).\psi_e''(\boldsymbol{r}, \boldsymbol{R}_e)\mathrm{d}\boldsymbol{r}$. Different electronic state wavefunctions must be orthogonal to each other, hence this overlap integral is zero.

There are other selection rules that are just as easy to deduce using group theory. For example, we can establish that $\Sigma^+ \leftrightarrow \Sigma^+$ and $\Sigma^- \leftrightarrow \Sigma^-$ transitions are allowed, whereas $\Sigma^+ \leftrightarrow \Sigma^-$ transitions are forbidden. Similarly, for electronic states in molecules having a centre of symmetry, a g or u subscript is added to indicate the symmetry with respect to the inversion operation, i (the g and u derive from the German words gerade and ungerade, meaning even and odd, respectively). All three components of the transition dipole moment operator in any molecule with a centre of symmetry have u inversion symmetry (in point groups where this symmetry operation is meaningful) and therefore only g $\leftrightarrow$ u transitions are allowed.

The arguments above refer to the spatial requirements for an allowed transition but the electron spin must also be considered when deciding whether a transition is allowed or not. The electronic transition moment as written in (7.13) does not explicitly include spin as a coordinate in the electronic wavefunctions, although it ought to be there. However, provided spin–orbit coupling is not large, the electron spins will be unaltered by electric dipole transitions since spin is a purely magnetic effect. Consequently, no change in spin multiplicity should occur for an electric dipole transition, i.e. the selection rule is $\Delta S = 0$. This is a good selection rule for molecules that contain relatively light atoms, but begins to weaken as spin–orbit coupling increases, as is often the case for molecules containing heavy atoms. A classic example of this breakdown is I_2, with strong singlet–triplet bands being well known in its electronic spectrum [5].

7.2.2 *Vibrational propensities for diatomic molecules*

The second integral in (7.13) is an overlap integral for the vibrational wavefunctions in the upper and lower electronic states. This determines the vibrational contribution to the transition probability. More precisely, the square of the overlap integral, which is known as the *Franck–Condon factor* (FCF), i.e.

$$\mathrm{FCF} = \left(\int \psi'_{\mathrm{v}} \psi''_{\mathrm{v}} \, \mathrm{d}R \right)^2 \tag{7.14}$$

determines the vibrational contribution to the transition probability. For a diatomic molecule, there is only one internal coordinate, the internuclear separation R, and so the general nuclear position coordinates symbolized by $\boldsymbol{R}$ in (7.13) are replaced by R in (7.14).

Equation (7.13) is a mathematical statement of the *Franck–Condon principle*. According to the Franck–Condon principle, an electron in an electronic transition moves from one orbital to another so rapidly that the nuclear positions are virtually the same immediately before and after the transition. In other words, the time taken for the electron promotion (or demotion) is very short compared with a vibrational period. This is consistent with the idea of separating the electronic and vibrational degrees of freedom as in the Born–Oppenheimer approximation.

Suppose that a diatomic molecule has very similar potential energy curves in two different electronic states, as illustrated by the lowest two curves in Figure 7.2. By similar we mean that not only do these two curves have the same depth and similar slopes at all points along the curves, but that they also have nearly identical equilibrium bond lengths. If a transition

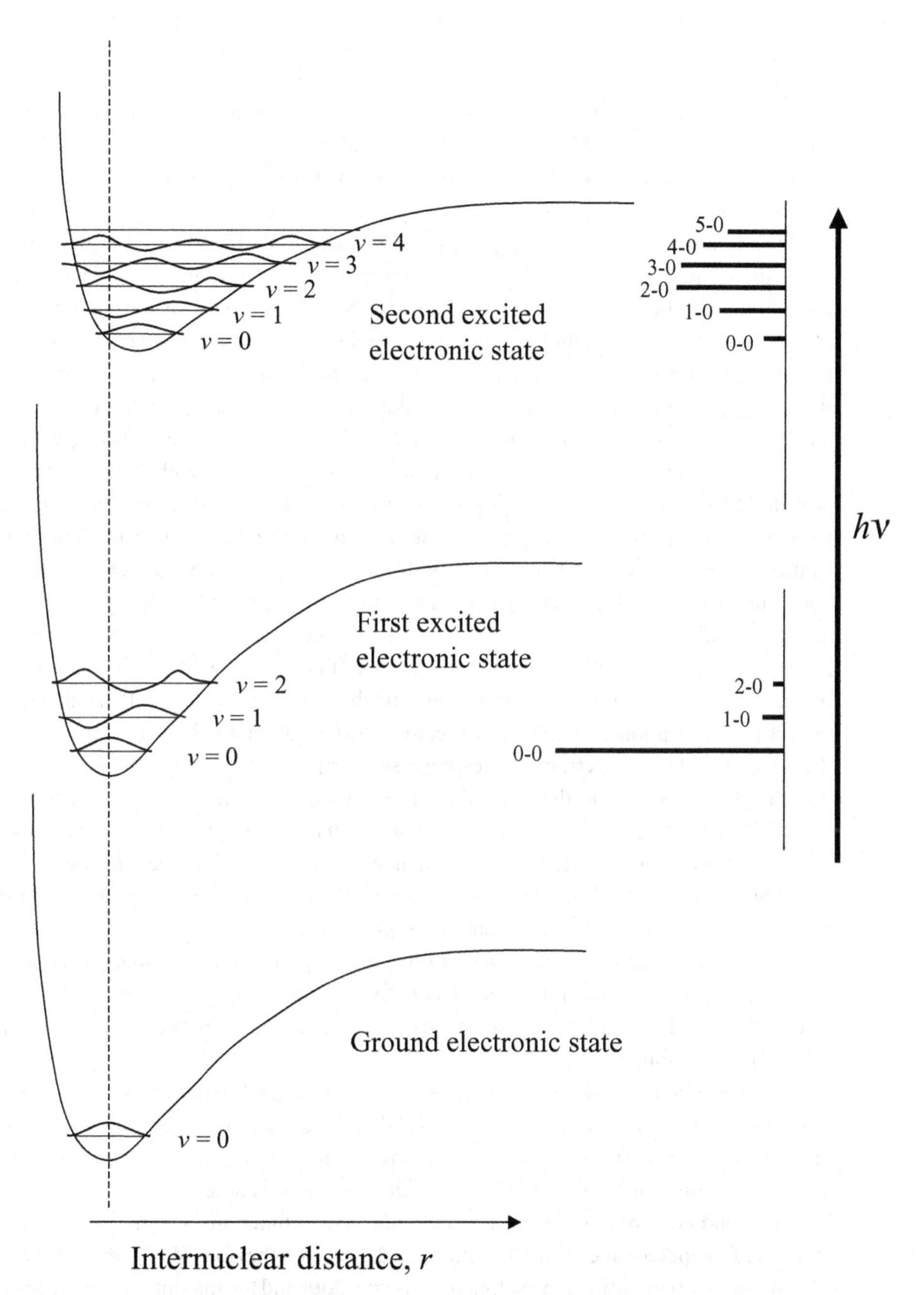

Figure 7.2 Diagram illustrating the source of vibrational structure in electronic absorption spectra. Two scenarios are illustrated. For the two lowest potential energy curves a spectrum (see stick diagram on right) dominated by the $v'' = 0 \rightarrow v' = 0$ transition is observed. This is the case I scenario described in the text. In contrast a transition from the ground electronic state to the second excited state yields a long vibrational progression (case II scenario). Notice that the intervals between vibrational bands are a direct measure of the separations between vibrational levels in the upper electronic state if all transitions take place from the ground vibrational level in the ground electronic state.

takes place between these two electronic states, the Franck–Condon factors (FCFs) could be used to determine the relative probabilities of transitions to different vibrational levels. To do this quantitatively the vibrational wavefunctions in the two electronic states must be known so that the FCF in (7.14) can be calculated. Although this can be done, we will focus on qualitative arguments that lead to some very important general conclusions.

The similarity of the potential energy curves means that the vibrational wavefunctions (and vibrational energies) will be very similar in the two electronic states. The vibrational contribution to the transition moment can then be assessed by considering the degree of overlap between the vibrational wavefunctions in the upper and lower electronic states. Consider absorption from $v'' = 0$. As can be seen by consulting Figure 7.2, the overlap of the $v'' = 0$ vibrational wavefunction with that of $v' = 0$ for the middle potential energy curve is excellent. To see this, imagine sliding the lower curve vertically upwards until the $v'' = 0$ vibrational level lies directly on top of the $v' = 0$ level. If the same process is followed for transitions to higher v', such as $v' = 1$ or 2, then one finds once again that the overlap is good. However, there are now both positive and negative contributions to the overlap and when integration is performed these approximately cancel, yielding a very small FCF. The FCF get rapidly smaller as v' increases and so the $\Delta v = 0$ transition dominates. If higher vibrational levels in the ground electronic state were populated, as might be the case at high temperatures, we would find a similar situation, namely that the overlap for $\Delta v = 0$ transitions will be very much larger than for $\Delta v \neq 0$. Thus one would expect the absorption (or emission) spectrum to be dominated by $\Delta v = 0$ transitions if the upper and lower electronic states have similar potential energy curves. We will call this case I behaviour, and the stick diagram in Figure 7.2 illustrates its consequences for the vibrational structure in a spectrum. Case I behaviour is typically observed when the electronic transition involves movement of an electron whose character changes little from one orbital to another, as would be the situation if the transition was from one non-bonding molecular orbital to another non-bonding orbital.

It is worth emphasizing, before we continue further along this track, that no selection rule has been established here; in fact on the contrary we have found that there is no vibrational selection rule! However, the arguments just used do reveal the *propensity* for a change of vibrational quantum numbers.

Consider now two potential energy curves that are very different, as would be obtained when an electron jumps between two orbitals of very different bonding character. The uppermost curve in Figure 7.2 represents a molecule in an excited electronic state having a longer equilibrium bond length and smaller dissociation energy than the same molecule in the ground electronic state. Typical vibrational wavefunctions are shown superimposed on each of the potential curves. In contrast to the case I behaviour above, the overlap of the $v'' = 0$ and $v' = 0$ vibrational wavefunctions is now poor and the maximum overlap shifts to a transition involving a substantial change in v. However, notice also, as made especially clear in the accompanying stick diagram of an absorption spectrum in Figure 7.2, that several different Δv transitions have comparable Franck–Condon factors and so the absorption (or emission) spectrum now consists of a long vibrational progression. This is typical of what we will call case II behaviour. Indeed if the potential curves are sufficiently dissimilar it may not be possible to observe the so-called electronic origin transition, $v' = 0 \leftarrow v'' = 0$, because the corresponding FCF is too small. Case II behaviour will occur when the electron involved in the transition shows a major change in bonding character, e.g. a non-bonding $\rightarrow$ antibonding transition.

7.2.3 Vibrational selection rules and propensities for polyatomic molecules

In the harmonic oscillator limit we have seen that the vibrational motion of polyatomic molecules can be reduced to a superposition of vibrations in $3N - 6$ normal modes (or $3N - 5$ for a linear molecule), as described in Section 5.2.1. The total vibrational wavefunction is then a product of the individual normal mode wavefunctions, i.e.

$$\psi_{\text{vib}} = \prod_{3N-6} \psi_1 \psi_2 \psi_3 \ldots \psi_{3N-6} \tag{7.15}$$

where the wavefunction ψ_i of each normal mode is given in equation (5.20) (be careful not to confuse the subscripts in (7.15), which label the particular normal mode, with the vibrational quantum number of a specific mode).

If we substitute equation (7.15) into (7.11), and carry out the same factorization process as employed in Section 7.2, a similar result to that shown in equation (7.14) is obtained. The only difference is that, instead of a single vibrational overlap integral, a product of overlap integrals, *one for each normal mode*, results. This remarkable outcome, which is brought about by the independence of the various normal coordinates, greatly simplifies the interpretation of vibrational structure in the electronic spectra of polyatomic molecules.

However, while useful analogies with diatomic Franck–Condon factors can be made, there are also some important and quite subtle differences. To bring these to the fore, we must focus on the symmetries of the vibrational wavefunctions. For a diatomic molecule, the vibrational wavefunction is always totally symmetric with respect to all symmetry operations of the point group regardless of electronic state or vibrational quantum number. It is for this reason that the integrand in the Franck–Condon factor is always totally symmetric for a diatomic and there are therefore no vibrational selection rules in its electronic spectroscopy. However, as explained in Section 4.2.2, polyatomic vibrational wavefunctions can be totally symmetric or non-totally symmetric depending on the symmetry of the normal coordinate and the vibrational quantum number. For a normal mode with a totally symmetric normal coordinate, the vibrational wavefunction is totally symmetric for all v. However, a vibration with a non-totally symmetric normal coordinate has a vibrational wavefunction that alternates from being totally symmetric to non-totally symmetric as v changes from even to odd.

Group theory makes it possible to quickly assess the impact this has on vibrational structure. Suppose that the vibrational wavefunction for a particular mode is totally symmetric in both upper and lower electronic states. In this case, the corresponding Franck–Condon factor will have a totally symmetric integrand since the direct product of something that is totally symmetric with something else that is totally symmetric must give a result that is totally symmetric. On the other hand, if either the upper *or* lower state vibrational wavefunction is non-totally symmetric, then the integrand will be non-totally symmetric.[5]

At this point it will be helpful to consider a specific example, CO_2. This has the three normal modes shown schematically in Figure 5.5. The symmetric C—O stretch ν_1, has a totally symmetric vibrational wavefunction for all values of the vibrational quantum number. However, the two non-totally symmetric modes, the degenerate bend, ν_2, and the antisymmetric stretch, ν_3, are different. For these modes, the vibrational wavefunction has a totally symmetric component for v even $(0, 2, 4, \ldots)$ but must be non-totally symmetric for v odd $(1, 3, 5, \ldots)$. We can now work out the symmetries of the integrands in the Franck–Condon factors and hence selection rules for all possible upper and lower state vibrational quantum numbers in an electronic transition.

For ν_1, any value of $\Delta\nu_1$ is possible, although as for diatomic molecules there will be a propensity for certain values. Suppose the equilibrium C—O bond lengths are substantially larger or smaller in the upper electronic state than in the lower. This will be the equivalent of the case II scenario in diatomics, and a long vibrational progression in ν_1 would be expected for this electronic transition. On the other hand, if the equilibrium bond lengths are virtually the same in the two electronic states, then this corresponds to the case I limit and $\Delta\nu_1 = 0$ transitions will dominate, i.e. no significant vibrational progression will be observed.

For modes ν_2 and ν_3, because they involve non-totally symmetric normal coordinates, only even quantum number changes $\Delta v = 0, \pm 2, \pm 4$, etc., are allowed. In fact a little more thought will show that $\Delta v = 0$ transitions will dominate for these modes. For example, unless one bond becomes longer than the other in the excited electronic state, then there is no change in equilibrium structure in the direction of normal coordinate ν_3. This is equivalent to a case I Franck–Condon situation applying for this mode. Similarly, if the molecule is linear in both electronic states then there is no propensity for $\Delta\nu_2 \neq 0$ transitions.

This illustrates a general and important point that will be met in many examples later, namely that the vibrational structure in electronic spectra is normally dominated by modes with totally symmetric normal coordinates. Furthermore, the propensity for formation of a progression in a particular mode will depend on whether there is a change in equilibrium structure in the direction of that coordinate. If there is a substantial structural change in the direction of only one coordinate, then only this mode will show any significant activity in the spectrum. Thus one may have, and often finds, very simple vibrational structure arising in the spectrum of a relatively complicated molecule.

---

5 A useful analogy is to liken the direct product of representations with products of the numbers $+1$ and -1, where $+1$ represents totally symmetric and -1 represents a non-totally symmetric representation. We can therefore instantly see that the direct product of two totally symmetric representations will give a totally symmetric result since $(+1) \times (+1) = +1$. The direct product of two (identical) non-totally symmetric representations will also give a totally symmetric result, since $(-1) \times (-1) = +1$. On the other hand, the direct product of a totally symmetric and non-totally symmetric representation (or vice versa) will give a non-totally symmetric representation, since $(+1) \times (-1) = -1$.

7.2.4 *Rotational selection rules*

When viewed from a classical perspective, photons possess some strange properties. They have no mass but an advanced theoretical treatment shows that they possess angular momentum.[6] This is an important conclusion because it impacts on the selection rules for spectroscopic transitions. In particular, one of the fundamental tenets of mechanics is that angular momentum must be conserved. Consequently, whenever a photon is absorbed or emitted the overall angular momentum of the system must be maintained. Many of the key rotational selection rules can be justified on these grounds [1]. The basic premise is that each photon possesses one unit of quantized angular momentum. As a result, the quantized angular momentum of a molecule cannot change by more than one unit during photon absorption or emission. A more sophisticated analysis bringing together the transition moment and the symmetry properties of the rotational wavefunctions leads to additional selection rules. Proof of these selection rules for the various types of electronic transitions and various molecular symmetries is beyond the scope of this book. However, the results for a few simple cases are summarized below.

Consider a single-photon electronic transition in a diatomic molecule. If the upper and lower electronic states are both $^1\Sigma$ states, the rotational selection rule turns out to be $\Delta J = \pm 1$, which is easily justified on the basis of the comments above. Transitions where $\Delta J = +1$ are said to be R branch transitions, while those for which $\Delta J = -1$ are known as P branch transitions. The convention in labelling specific transitions is to follow the P or R designation with the rotational quantum of the *lower state* in parentheses, e.g. $R(3)$ refers to the transition from $J = 3$ in the lower electronic state to $J = 4$ in the upper electronic state. Transition energies can easily be determined by combining the rotational selection rule with equation (6.4). Designating the energy of the electronic + vibrational transition as ΔE_{ev}, the general R branch transition $R(J)$ should appear at

$$\begin{aligned}\Delta E &= \Delta E_{\text{ev}} + B'J'(J'+1) - B''J''(J''+1)\\ &= \Delta E_{\text{ev}} + (B' - B'')J^2 + (3B' - B'')J + 2B' \qquad (7.16)\end{aligned}$$

using the notation $J'' = J$, $J' = J + 1$. If the rotational constants in the upper and lower electronic states are approximately the same, which will be the case if the bond length is largely unchanged by the electronic transition, then (7.16) approximates to $\Delta E = \Delta E_{\text{ev}} + 2B(J+1)$, i.e. a series of lines in the R branch with adjacent members approximately $2B$ apart is obtained. Similarly, it is easy to derive an analogous formula for P branch transitions and one finds once again that adjacent members in the P branch are approximately $2B$ apart when $B' \approx B'' = B$. In practice, substantial differences between B' and B'' are common in electronic transitions (but not in infrared transitions). The effect that this has on rotational structure is encountered in several examples in Case Studies later on in this book.

When one of the electronic states possesses net orbital angular momentum, $\Delta J = 0$ transitions are possible. These transitions are called Q branch transitions, a transition from

[6] Strictly speaking photons only have no mass when at rest, which they never are. According to special relativity mass and energy are interconvertible so from a practical point of view photons do possess mass.

a specific J level being referred to as $Q(J)$. Q branches are impossible for ${}^1\Sigma - {}^1\Sigma$ electronic transitions because the absorption or emission of a photon *must* change the angular momentum of the molecule. If one of the electronic states has angular momentum there is now a mechanism by which the angular momentum of the photon can be compensated for within the molecule without changing the rotational state.

Observation of rotational structure in spectra is useful because it provides structural information on the molecule via the rotational constant(s). However, notice also that the type of rotational structure depends on the symmetries of the electronic states. In electronic spectroscopy the assignment of electronic states is frequently made through analysis of the rotational structure. An inverse approach is adopted whereby the observed rotational structure is first analysed and used to determine the rotational selection rules in operation. A comparison with the selection rules expected for certain specific types of electronic transitions then leads to the assignment.

The rotational selection rules for closed-shell non-linear polyatomic molecules are more involved than for the diatomic case. The quantum number J in the general case is reserved for the total angular momentum[7] of a molecule and for a single photon transition the change in J is still limited to a maximum of ± 1 (because of conservation of angular momentum). However, Q branch transitions are now possible regardless of the symmetries of the electronic states. For symmetric tops, the rotational quantum number K must also be considered. If the electronic transition moment is polarized along the inertial axis on which K is quantized, then the selection rule is $\Delta K = 0$. Otherwise, the selection rule is $\Delta K = \pm 1$. Further information on the rotational selection rules for electronic transitions in closed-shell molecules, including asymmetric tops, can be found in Reference [6].

Finally, we note that when a molecule possesses a non-zero net electron spin, as would be the case for free radicals, there are additional factors to be considered when analysing the rotational structure. Case Studies 22, 24, and 28 provide specific examples of this behaviour.

References

1. A. M. Ellis, *J. Chem. Educ.* **76** (1999) 1291.
2. *Principles of Lasers*, O. Svelto, New York, Plenum Publishing Corporation, 1998.
3. *Laser Fundamentals*, W. T. Silvast, Cambridge, Cambridge University Press, 1996.
4. *Molecular Spectroscopy*, Chapter 11, J. D. Graybeal, New York, McGraw Hill, 1988.
5. J. I. Steinfeld, R. N. Zare, L. Jones, M. Lesk, and W. Klemperer, *J. Chem. Phys.* **42** (1965) 25.
6. *Molecular Spectra and Molecular Structure. III. Electronic Spectra and Electronic Structure of Polyatomic Molecules*, G. Herzberg, Malabar, Florida, Krieger Publishing, 1991.

[7] Excluding nuclear spin. If nuclear spin is included, the total angular momentum quantum number is given the symbol F.

Part II
Experimental techniques

Modern electronic spectroscopy is a broad and constantly expanding field. A detailed description of the experimental techniques available for this one area of spectroscopy could fill several books of this size. This part is therefore restricted to giving an introduction to some of the underlying principles of experimental spectroscopy, together with brief descriptions of some of the more widely used and easily understood methods employed in electronic spectroscopy.

8 The sample

This book is concerned with the spectroscopy of molecules, primarily in the gas phase. Broadly speaking, there are two types of gas source that are commonly used in laboratory spectroscopy. One is a *thermal* source, by which we mean that the ensemble of molecules is close to or at thermal equilibrium with the surroundings. An alternative, and non-equilibrium, source is the *supersonic jet*. Both are discussed below. Individual molecules can also be investigated in the condensed phase by trapping them in rigid, unreactive solids. This *matrix isolation* technique will also be briefly described.

8.1 Thermal sources

A simple gas cell may suffice for many spectroscopic measurements. This is a leak-tight container that retains the gas sample and allows light to enter and leave. It may be little more than a glass or fused silica container, with windows at either end and one or more valves for gas filling and evacuation. The cell can be filled on a vacuum line after first pumping it free of air (if necessary). If the sample under investigation is a stable and relatively unreactive gas at room temperature, this is a trivial matter.

If the sample is a liquid or solid with a low vapour pressure at room temperature, then the cell may need to be warmed with a heating jacket to achieve a sufficiently high vapour pressure. Residual air, together with volatile impurities that may be trapped in the condensed sample, can be removed using one or more freeze–pump–thaw cycles. This relies on the desired species being less volatile than impurities. As the name implies, a freeze–pump–thaw cycle begins with the cell being cooled to a temperature at which the sample is frozen and hence has a negligible vapour pressure, perhaps using a dry ice or liquid nitrogen bath. It is then pumped on for a short time to remove undesirable volatile species (but not the frozen sample) before closing the vacuum tap and warming the cell up to the desired operating temperature. A repeat of this process will help to improve the sample purity.

If the aim is to study highly reactive molecules, such as free radicals or molecular ions, some means of generating these molecules from a suitable precursor will be required. For these more exacting experiments it is frequently necessary to replenish the sample by using a constant flow of gas through the cell. Free radicals can be made by a number of methods, the most common being ultraviolet photolysis or electrical discharge. Electrical discharges through gases are also excellent sources of molecular ions. High temperature pyrolysis or vaporization may also be used to generate reactive or unusual molecules, and this can be done inside a gas cell with careful design.

The production of highly reactive molecules is encountered again in Section 8.2.3.

8.2 Supersonic jets

8.2.1 General principles

For a typical molecular gas at room temperature, many rotational energy levels will have significant populations. Furthermore, while the population of vibrational levels other than the zero point level is likely to be small, this may not be true as the temperature is raised significantly above room temperature. Thus the spectrum of a molecule at room or higher temperatures may consist of transitions out of many different energy levels. If the spectral resolution is relatively low, this will result in broadened bands consisting of unresolved rotational structure and perhaps even unresolved vibrational structure. If, on the other hand, the resolution is high, the large number of transitions may give rise to an overwhelmingly high density of individual rovibronic lines in the spectrum and make assignment difficult, if not impossible.

Clearly it is sometimes desirable to cool the sample. Cooling the walls of a gas cell by submerging it in a cold bath may be an acceptable solution in some cases. However, an obvious problem with this type of cooling is that, if taken too far, it will lead to condensation of the gas. Most gases will condense at liquid nitrogen temperatures (77 K) and many will condense at far higher temperatures. Thus cell cooling is of limited utility for gas phase studies.

Supersonic jets offer a way of dramatically cooling the internal degrees of freedom of molecules *without excessive condensation*. To see how they work, consider the scenario in Figure 8.1, which shows a gas reservoir located inside a vacuum chamber. A small hole of diameter D links the gas reservoir to the vacuum chamber. Suppose the reservoir is filled with an inert gas such as argon or helium. Furthermore, assume that the vacuum chamber is evacuated by a high speed pump capable of maintaining a low pressure regardless of the amount of gas escaping into the chamber. There are two extreme pressure limits that we will now consider.

If the pressure in the gas container is relatively low then the escaping atoms are unlikely to undergo any collisions with other atoms as they pass through the orifice. Quantitatively, this limit corresponds to $\lambda \gg D$, where λ is the *mean free path* of the gas.[1] If the reservoir contains gas at thermal equilibrium with its surroundings, i.e. there is a Maxwell–Boltzmann distribution of speeds, then the distribution of speeds in the escaping gas will also have the same form. The departing atoms are said to form an *effusive* gas jet.

At the other extreme, if the gas pressure in the reservoir is sufficiently high such that $\lambda \ll D$, then the departing atoms will undergo many collisions as they pass through the

[1] The mean free path of a gas is the average distance a gas particle travels between collisions. It is inversely related to the gas pressure.

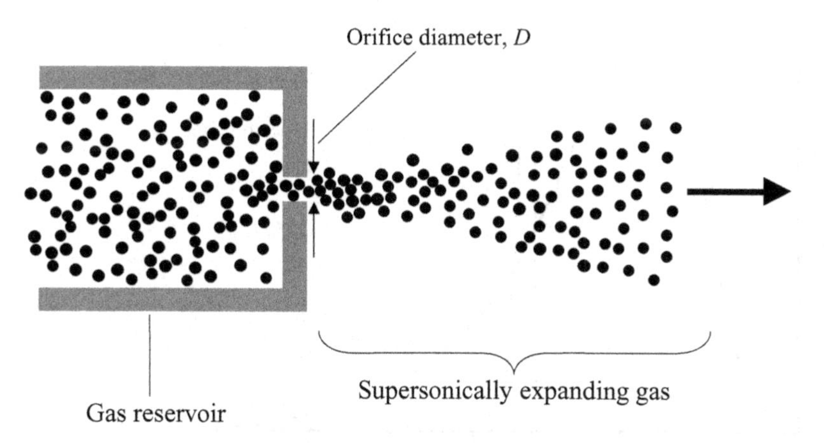

Figure 8.1 Formation of a supersonic jet. Gas in the reservoir is pressurized, usually to a pressure exceeding 1 bar. The supersonic jet expands into a vacuum chamber evacuated by a high speed pump (not shown).

orifice. This is the regime of the supersonic jet. An atom initially moving rapidly towards the orifice will be slowed down by collisions with slower atoms heading in the same direction, while an atom initially moving slowly towards the orifice will be hurried along by collisions with more energetic partners. These collisions will tend to order the departing velocities of the atoms into a narrow range as the atoms 'squeeze' through the orifice and out into vacuum. If one tries to imagine the view from one of the atoms as it moves along in the jet downstream of the orifice, the atoms in the immediate vicinity would appear to be virtually stationary compared with their speeds in the gas reservoir. The translational temperature, as described by the *distribution* of speeds, will therefore be very low and can in fact be lower than 1 K. The thermal energy of the reservoir has been converted into directed gas flow with near uniform gas atom speeds, as illustrated in Figure 8.2.

The average speed of the gas atoms will have increased compared with that in the container. The ratio of the average speed of the gas particles to the *local* speed of sound is called the Mach number. If one took the ratio of the average speed of the atoms in the jet to the speed of sound at room temperature, the Mach number would be modest (on the order of 1.3). However, the local speed of sound in the jet is much lower because the speed of sound decreases as the temperature of the gas falls (it is proportional to $T^{1/2}$). Consequently, since the gas is cooled dramatically by the expansion, the Mach number can be very high, with values >50 not being unusual. This is the origin of the term *supersonic jet*. The orifice separating the gas reservoir from the vacuum is frequently referred to as a *nozzle*.

Cooling of the translational degrees of freedom is not, in itself, particularly interesting for the spectroscopist. However, the cooling of internal degrees of freedom in molecules is also possible. In the region immediately downstream of the orifice each atom or molecule will undergo a moderate number of collisions, typically 10^2–10^3, before the collision rate drops rapidly towards zero because of the low translational temperature and because of the divergence of the jet. Prior to this point energy can be transferred from internal degrees of

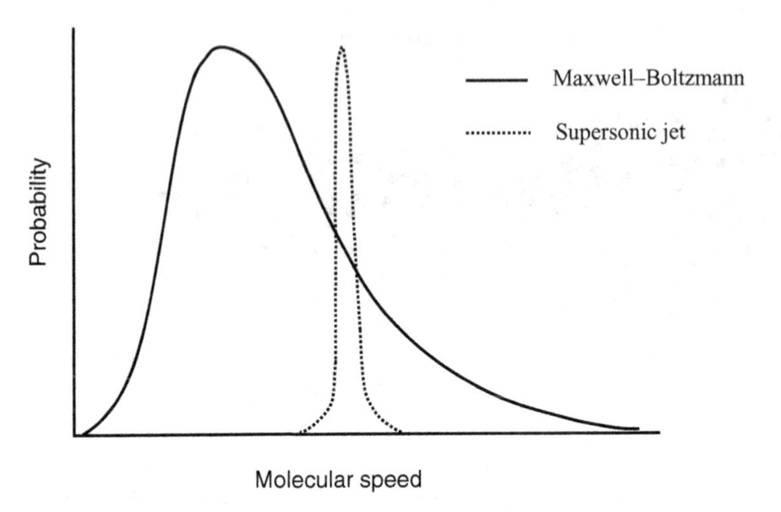

Figure 8.2 Comparison of the speed distributions in the gas reservoir (Maxwell–Boltzmann distribution) and in the supersonic jet far downstream of the orifice.

freedom to the cold (and cooling) translational bath through collisions. Suppose a small proportion of some molecular gas is mixed in with the inert carrier gas.[2] As the gas expands into vacuum, the molecules can undergo collisions with the inert gas atoms in which vibrational and rotational energy is converted into translational energy. The translational motion is then cooled rapidly by the mechanism described above.

The cooling efficiency is different for the rotational and vibrational degrees of freedom, with the former tending to be far more efficient than the latter. The source of this differential cooling is the difference in energies between adjacent quantum states for rotational versus vibrational motion. The more energy that has to be transferred, the lower the chance of success. In fact, to a reasonable approximation, the probability of energy transfer on collision falls off exponentially as the size of the energy mismatch increases between the 'giving' and 'receiving' degrees of freedom. This differential cooling effect can be quantified in terms of the different 'temperatures' of the various degrees of freedom. The rotational temperature, as determined by the relative populations of the rotational levels assuming a Boltzmann distribution, can approach the translational temperature: values as low as 1 K are attainable. The lower efficiency of intermolecular vibrational → translational energy transfer means that the vibrational populations may not alter significantly from their reservoir values. However, since most molecules are normally in the zero point vibrational level before expansion, this is rarely a problem. Thus the dominant cooling of internal degrees of freedom is of the rotational levels, and this has proved to be highly beneficial in spectroscopy, as will be illustrated in several Case Studies later.

2 The molecular species is said to be *seeded* into the inert carrier gas. Typical proportions would lie within the range 0.1–10% by volume of molecular gas, the balance being inert carrier gas. Higher proportions are likely to lead to substantial condensation of the molecular gas in the expansion.

As a final general point about supersonic jets, we return to the issue of condensation at low temperatures. It should be clear from the above that a supersonic jet is a non-equilibrium gas source. As the cooling proceeds, the collision rate drops dramatically until, at a relatively small distance from the nozzle, there are virtually no further collisions at all between gas particles. Condensation is avoided for *kinetic* rather than thermodynamic reasons.

8.2.2 *Pulsed supersonic jets*

An ideal supersonic jet requires a very low pressure in the vacuum chamber. If this is not attained, then collisions of the expanding gas with background gas molecules in the chamber degrade the jet properties.[3] Continuous supersonic jets have a very high gas throughput and therefore a satisfactory vacuum can only be achieved by using large vacuum pumps coupled to large vacuum chambers. This is an expensive option and quite unnecessary for many spectroscopic applications.

If the gas is introduced into the chamber in short bursts, the total gas throughput per unit time can be dramatically reduced. As well as reducing consumption of potentially expensive gases, much smaller (and cheaper) vacuum pumps can be employed without any major loss in the performance of the supersonic jet. This is particularly significant for experiments that use pulsed lasers as light sources (see later). A typical pulsed laser used in electronic spectroscopy may output 20 pulses per second, each pulse having a duration of 10 ns. Since the total on-time of the laser is only 200 ns in every second, it would clearly be very wasteful to use a continuous supersonic jet in this situation. Pulsed jets can be obtained by inserting a pulsed gas valve between the gas reservoir and the vacuum chamber. Some research groups have employed modified pulsed injection valves from cars, but nowadays there are relatively cheap commercial pulsed valves designed specifically for use in spectroscopic and related experiments. These can have opening times as short as a few microseconds, although they are more commonly used with opening times of several hundred microseconds in spectroscopy experiments. The opening time of the valve needs to be synchronized with the firing time of the pulsed laser, and this can be done straightforwardly with electronic timing devices.

8.2.3 *Production of free radicals, clusters, and ions in supersonic jets*

Collisions in a gas can be classified as either two-body or three-body (chemists may be more familiar with the alternative names, bimolecular or termolecular). The collision rate for two-body collisions will necessarily be far higher than that for three-body collisions, since the latter require the simultaneous collision of three distinct entities. Two-body collisions are responsible for cooling in a supersonic jet. On the other hand, it is three-body collisions that lead to the formation of molecular or van der Waals complexes, since the third body

[3] The 'ideal' properties of the supersonic expansion will be maintained for a finite distance before collisions with the background gas cause a shock front. The position of this shock front, which is called the *Mach disk*, is given by $X_m = 0.67D(P_r/P_c)^{1/2}$, where P_r and P_c are the reservoir and chamber pressures, respectively. If the chamber pressure is very low then the hypothetical Mach disk may exceed the vacuum chamber dimensions, the ideal scenario.

can collisionally stabilize the complex before it falls apart (cf. the need for a third body in the recombination reactions of free radicals, such as $CH_3 + CH_3 + M \rightarrow C_2H_6 + M$).[4]

The number of two-body collisions downstream of the nozzle is proportional to P_rD, where P_r is the pressure in the gas reservoir behind the nozzle and D is the diameter of the orifice. The three-body collision rate depends on $P_r^2 D$, and so complex formation is favoured by high reservoir pressures. This idea has been widely exploited by gas phase spectroscopists to study van der Waals complexes. For example, the addition of noble gas atoms to simple species, such as metal atoms or small molecules, or onto larger molecules such as benzene, tetrazine, and azulene, has been achieved. The study of complexes involving inert gas atoms is important because it provides detailed information on van der Waals forces, and the low temperature environment in a supersonic jet is excellent for studying these very weakly bound species. Other types of complexes, such as hydrogen-bonded dimers and trimers, have also been prepared in the gas phase by this means [1].

Many other fascinating species can be formed in supersonic jets. For example, free radicals may be produced by photolysis. The usual method is to cross the jet with an ultraviolet laser beam close to the nozzle so that subsequent cooling in the expanding gas is possible. Many different free radicals have been investigated by this route, ranging from simple diatomic and triatomic species, such as CH, CH_2, HCO, OH, to larger radicals such as cyclopentadienyl (C_5H_5) [2]. The simplification of the spectra of these molecules brought about by supersonic expansion has led to remarkable advances in our knowledge of the structures and properties of these important chemical intermediates.

Molecular ions can also be studied in supersonic jets. One way to make these is by use of an electrical discharge (which can also be used to make free radicals). A possible arrangement for a discharge/supersonic jet experiment is shown in Figure 8.3.

Finally, it is also possible to make highly reactive molecules in the region just upstream of the nozzle, i.e. just prior to expansion, and then entrain these molecules in a supersonic jet. This idea has been widely exploited, most notably in the production of metal-containing molecules. Metal atoms can be ablated from metal surfaces using high intensity pulsed lasers, such as Nd:YAG or excimer lasers (see Chapter 10), and can then be carried to the point of expansion by a suitable carrier gas. If a reagent is seeded into the inert carrier gas, other species can be made by chemical reactions, such as metal hydrides, metal carbides, metal halides, and organometallics.

8.3 Matrix isolation

Inert solid hosts provide an alternative environment for investigating individual molecules. The noble gas solids are the best examples since they are virtually chemically inert and have no absorption bands in the infrared, visible, and near-UV regions. The basic idea is to mix the molecules of interest with an excess of noble gas and this mixture is then condensed on

[4] Examples of cluster formation through two-body collisions are known. In these cases, it is thought that the initial two-body collision leads to the formation of a reasonably long-lived orbiting complex. Providing the lifetime of this complex is sufficiently long, it can be stabilized by another two-body collision leading to the formation of a stable cluster.

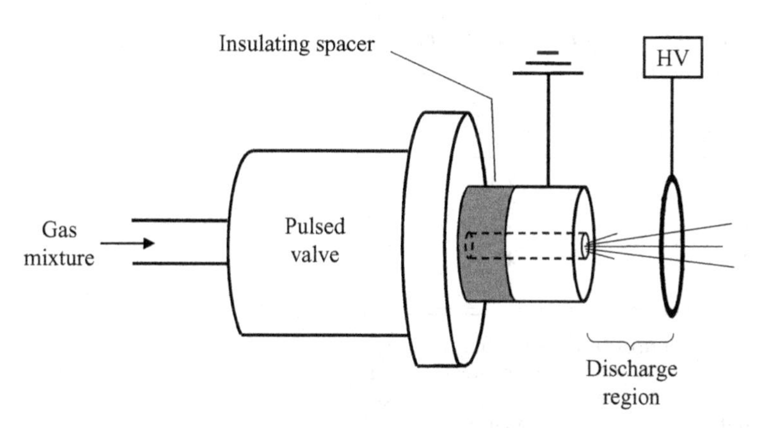

Figure 8.3 A pulsed discharge nozzle for the production of highly reactive molecules in a supersonic jet. A high voltage ring electrode is separated from the main nozzle assembly (which is at earth potential) by a small distance. When the valve opens the presence of gas in the region immediately downstream of the nozzle orifice leads to electrical breakdown and the formation of a discharge.

to a cooled window (see Figure 8.4). Extremely low temperatures are required to solidify the gas, as can be seen from Table 8.1. In the ideal scenario, isolated molecules will be trapped at a specific lattice site within the host matrix and will be distant from any other molecule in the solid. To guarantee this separation a large excess of inert gas is used, typical guest:host ratios being $1{:}10^3$ to $1{:}10^4$. Diffusion must also be minimized to prevent reaction, and this is achieved by using temperatures well below the freezing point of the noble gas host. Any spectra recorded will then be due almost entirely to isolated guest molecules held rigidly within the host matrix.

There are several attractive features of the matrix isolation technique. Providing the matrix is at a sufficiently low temperature it can be maintained almost indefinitely. Consequently, a wide variety of spectroscopic techniques, including some that are relatively insensitive, can be employed. Highly reactive species such as free radicals and molecular ions can be trapped and investigated, as can weakly bound complexes such as hydrogen-bonded or van der Waals bonded species.

However, there are also many disadvantages to the matrix isolation approach. With the exception of some diatomics, the trapping sites are too small to allow molecules to rotate. Consequently, it is impossible to observe rotational structure. Furthermore, the host matrix is never truly inert. Interactions between the noble gas atoms and guest molecules tend to have a very modest impact on the vibrational motion of molecules. However, excited electronic states are often severely perturbed by the noble gas host, especially for the heavier noble gases. This manifests itself in substantial shifts of electronic absorption bands compared with the gas phase. Furthermore, these bands tend to be much broader than in the gas phase. There are two reasons for the broadening. One is that the molecules may occupy several different types of sites within the solid, both substitutional and interstitial. In addition, the guest–host interaction leads to excitation of lattice vibrations (so-called phonon modes) in the solid when the guest molecule is electronically excited. In many instances this makes it impossible even to resolve vibrational structure in the electronic spectra.

Table 8.1 *Maximum operating temperatures (T_{max}) of inert solid matrices*

Substance	T_{max}/K
Ne	7.3
Ar	25
Kr	35
Xe	48
N_2	19

T_{max}, which is one-third of the freezing point, defines the upper limit at which the solid should be relatively rigid and diffusion slow. However, even lower temperatures are required if no diffusion is to be guaranteed.

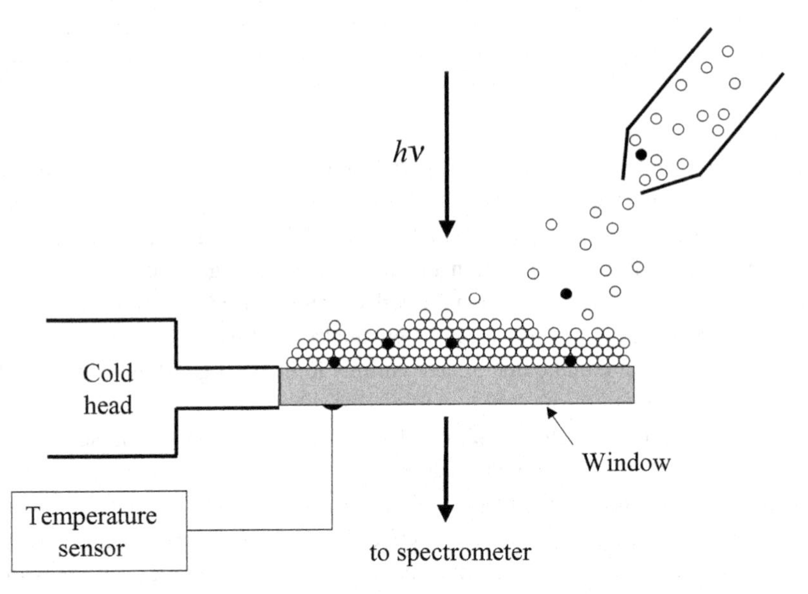

Figure 8.4 Schematic of a matrix isolation experiment. A gas mixture composed of the target molecules (•) diluted in noble gas (○) is sprayed onto the surface of an ultracold window. The cold head is cooled either by a closed cycle helium cryostat or by a static liquid helium cryostat. Various spectroscopic techniques can be applied. In an absorption experiment the transmission of the light beam through the window is measured using standard instrumentation.

Neon is the preferred host for electronic spectroscopy because it produces the smallest perturbations. However, neon is expensive and therefore argon is more commonly used.

References

1. See, for example, the following special issue; *Chem. Rev.* **100** (2000) 3863–4185.
2. S. C. Foster and T. A. Miller, *J. Phys. Chem.* **93** (1989) 5986.

9 Broadening of spectroscopic lines

It is common to refer to each transition as giving rise to a *line* in a spectrum. No line is infinitesimally sharp, and indeed some lines in spectra may be very broad. Before considering the sources of this broadening, it is important to be able to agree on a definition of the width of a transition. The most commonly used is the *full-width at half-maximum* (FWHM), the definition of which is illustrated in Figure 9.1.

The spectrometer itself will always make a contribution to the linewidth, and in many cases this may be the major factor limiting the spectral resolution. Discussion of instrumental resolution will be encountered in appropriate chapters later in this part. However, it is important to realise that the width of a spectral line is not only a function of the quality of the spectrometer. Indeed, with appropriate equipment, the instrumental resolution could be orders of magnitude higher than the observed resolution in an experiment. It is therefore important to be aware of non-instrumental sources of line broadening, and some of the more important ones are briefly considered below.

9.1 Natural broadening

Natural (or lifetime) broadening is a consequence of an uncertainty relationship similar to the well-known Heisenberg uncertainty principle. It arises because of the finite lifetimes (τ) of quantum states. In particular, the following inequality holds,

$$\tau \cdot \Delta E \geq \hbar/2 \tag{9.1}$$

where ΔE is the uncertainty in the energy of the state. Thus a state with a short lifetime will give rise to a large energy uncertainty, while a state with a long lifetime may have a very precisely defined energy. For spectroscopic purposes it is useful to convert from energy to frequency in order to calculate the frequency spread caused by the lifetime:

$$\tau \cdot \Delta\nu \geq 1/4\pi \tag{9.2}$$

In almost all cases the lifetime of the upper state in a spectroscopic transition is much shorter than that for the lower state, and so the former makes the dominant contribution to any natural broadening. All excited states are unstable with respect to spontaneous emission, one source of the finite lifetimes. Non-radiative routes may also be available for depopulating an excited state. In the absence of non-radiative pathways, excited electronic states have typical

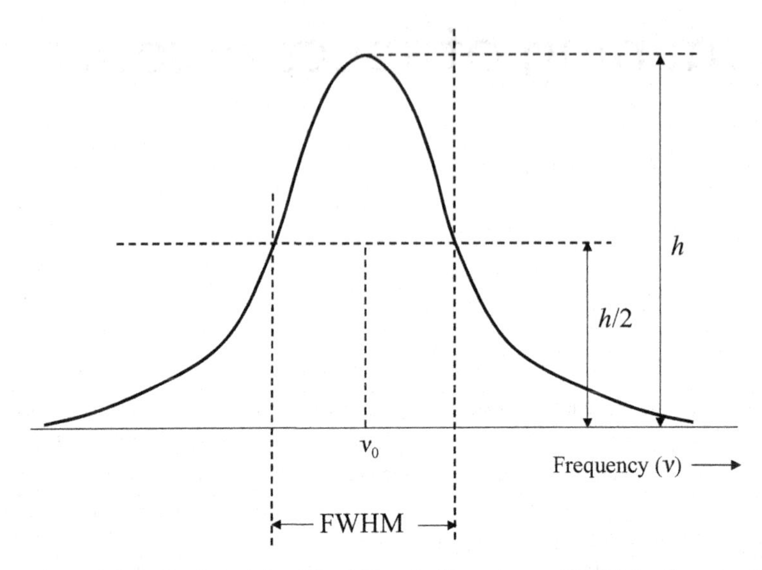

Figure 9.1 Definition of the full-width at half-maximum (FWHM) of a spectral line. The position of the line is normally quoted as ν_0, which is the mid-point of the FWHM region. In this picture ν_0 coincides with the peak maximum but in 'noisy' spectra this need not be the case.

lifetimes in the 10–1000 ns region if the spontaneous emission corresponds to an allowed transition. From (9.2) the corresponding range of natural linewidths is 0.08–8.0 MHz, which is very narrow ($\sim10^{-6}$–10^{-4} cm^{-1}). Consequently, natural broadening can be neglected in most spectroscopic measurements. The exception to this statement is when there are rapid non-radiative decay processes available. It is not unusual for these to produce lifetimes of <1 ps, thus producing natural broadening in excess of 8 GHz (>0.25 cm^{-1}). A commonly encountered example of this is *predissociation* (see Section 11.2).

9.2 Doppler broadening

Doppler broadening is often the most important non-instrumental source of line broadening. Its origin is relatively straightforward to grasp. If a molecule has a velocity component in the direction of a light source, then there will be a shift in the absorbed frequency compared with that of the stationary molecule. Consider first a stationary atom or molecule with absorption frequency ν_0 and imagine a light source which is producing light of this precise frequency. If the molecule now moves towards the light source, it will experience an apparent light frequency *higher* than that when at rest. In order for the radiation to be absorbed, the frequency of the light source must be *lowered* so that the *apparent* frequency seen by the moving molecule is ν_0. The opposite situation will pertain if the atom or molecule is moving away from the light source.

If the gas is at thermal equilibrium, the gas particles will possess a Maxwell–Boltzmann distribution of velocities. The one-dimensional Maxwell–Boltzmann distribution, in contrast to the three-dimensional distribution of speeds (see Figure 8.2), is symmetrical about the rest position. Thus the linewidth of the spectroscopic transition, if dominated by Doppler broadening, will have the same profile as the one-dimensional Maxwell–Boltzmann distribution. It can be shown that the linewidth (FWHM in MHz) is then given by

$$\Delta\nu = 7.15 \times 10^{-7} \nu_0 \sqrt{\frac{T}{M}} \tag{9.3}$$

where M is the molar mass of the molecule (in g mol^{-1}) and T is the temperature. According to equation (9.3), Doppler broadening is smaller for heavier molecules at a given temperature (because they have narrower velocity distributions), is reduced by lowering the temperature, and is directly proportional to the frequency of the incident radiation. The last factor is important in electronic spectroscopy because of the high frequency of visible and ultraviolet radiation. For example, in the near-ultraviolet the Doppler width will be in the region of several gigahertz (where 30 GHz $\approx$ 1 cm^{-1}) for a room temperature sample and could be the major factor limiting the resolution.

9.3 Pressure broadening

Pressure (or collisional) broadening is caused by the depopulation of molecules in excited states brought about through collisions. Since the lifetime of an excited state is reduced by collisional relaxation, this effect is an extension of lifetime broadening. Clearly it will depend strongly on the gas pressure. For pressures $<10^{-3}$ mbar, which are common in many branches of electronic spectroscopy, pressure broadening can be neglected. Wall collisions can also cause a similar effect and can be minimized by increasing the size of the cell. Pressure broadening is relatively unimportant in electronic spectroscopy.

10 Lasers

Crucial to any spectroscopic technique is the source of radiation. It is therefore pertinent to begin the discussion of experimental techniques by reviewing available radiation sources. Although there are many different types of light sources, of which some specific examples will be given later, in many spectroscopic techniques *lasers* are the preferred choice. Indeed some types of spectroscopy are impossible without lasers, and so it is important to be familiar with the properties of these devices. Consequently, before describing some specific spectroscopic methods, a brief account of the underlying principles and capabilities of some of the more important types of lasers is given.

10.1 Properties

Since their discovery in 1960, lasers have become widespread in science and technology. Laser light possesses some or all of the following properties:

(i) high intensity,
(ii) low divergence,
(iii) high monochromaticity,
(iv) spatial and temporal coherence.

Each of these properties is not unique to lasers, but their combination is most easily realized in a laser. For example, a beam of light of low divergence can be obtained from a lamp by collimation via a series of small apertures, but in the process the intensity of light passing through the final aperture will be very low. On the other hand, lasers naturally produce beams of light with a low divergence and so the original intensity is not compromised. Likewise, highly monochromatic radiation can be obtained from a continuum lamp by suitable filtering of unwanted wavelengths, e.g. by a high resolution grating monochromator, but in the process most of the light from the lamp is rejected and the final intensity will be very low. With lasers, very narrow linewidths, in some cases better than $<10^{-4}$ cm^{-1}, can be obtained with all of the light intensity concentrated into this narrow wavenumber range.

Although several different types of lasers have been used as light sources in electronic spectroscopy, by far the most important have been dye lasers. The significance of the dye laser is that it can produce tunable radiation across the whole of the visible region and extending into the near-ultraviolet and near-infrared. This is, of course, precisely the region of interest in much of electronic spectroscopy. Consequently, our discussion of specific types

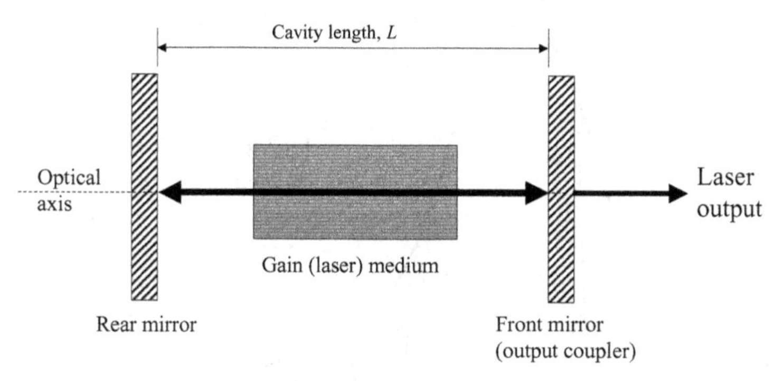

Figure 10.1 A simple laser cavity.

of lasers, which follows a description of the underlying principles in the next section, is deliberately biased towards providing a framework for understanding dye lasers. However, brief mention will also be made of other tunable lasers and several important fixed-frequency lasers.

For a detailed description of the properties of lasers the reader is referred to the books by Svelto [1], Siegman [2] or Silvast [3].

10.2 Basic principles

The name laser is an acronym derived from *l*ight *a*mplification by the *s*timulated *e*mission of *r*adiation. As the acronym implies, laser action is based on stimulated rather than spontaneous emission. The basic idea follows from the discussion given in Section 7.1.1. Consider a material of some sort, which might be solid, liquid, or gas, in which spectroscopic transitions can occur. We will call this material the *laser medium*. If the laser medium is at thermal equilibrium, then for any pair of energy levels in a particular type of atom or molecule, the population of the lower level (1) is greater than that of the upper level (2), i.e. $N_1 > N_2$. Thus if the system is bathed in radiation of the correct wavelength to excite the transition $1 \leftrightarrow 2$, then net absorption will occur. However, if $N_1 < N_2$ could be obtained, a situation known as a *population inversion*, then stimulated emission would dominate over absorption, i.e. the sample could act as a radiation amplifier, at least for a time. A population inversion is essential for laser operation and it will be shown later how this non-equilibrium population distribution can be produced.

However, a population inversion by itself is not enough to make a laser. Uncontrolled stimulated emission would yield light travelling in all directions, as in a light source based solely on spontaneous emission. However, stimulated emission can become strongly directional if the laser medium is placed in a highly reflecting cavity, such as the plane mirror cavity illustrated in Figure 10.1. Any radiation with normal, or very close to normal, incidence on the mirrors will be subjected to many passes along the cavity. For all other angles of incidence the radiation will quickly disappear from the cavity. This geometric constraint ensures that stimulated emission is favoured along the optical axis of the cavity.

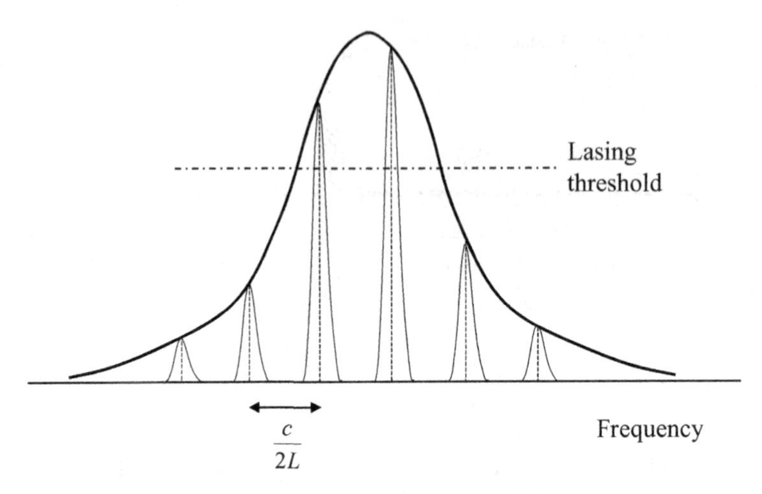

Figure 10.2 Longitudinal cavity modes superimposed onto the line profile of the spectroscopic transition responsible for laser action. Various losses in the cavity create a finite threshold that must be exceeded in order for lasing to occur. In this particular figure two cavity modes exceed the threshold, so lasing is limited to these two modes only.

Laser action works as follows. First a population inversion is produced by some means (see below). Spontaneous emission follows, and one of the photons produced may go on to cause stimulated emission from an atom or molecule to produce two photons, namely the original plus that from the stimulated emission. Owing to the coherent nature of stimulated emission, the two photons will be in phase. This is the beginning of a cascade process in which the number of photons increases exponentially as the stimulation process spreads throughout the cavity. However, high stimulated emission intensities are normally obtained only after many passes of the light backwards and forwards along the cavity, stimulating the same volume, and as mentioned earlier this is only achieved for photons reflecting backwards and forwards along the cavity axis. This is known as *positive feedback* and automatically limits the amplification to light paths along the cavity optical axis and it is this that produces the low beam divergence. In practice of course, most applications of lasers require the laser light to be directed out of the cavity and this is achieved by making one of the end mirors partially transmitting.

The monochromaticity of lasers derives from a combination of two factors. One is the existence of longitudinal cavity modes, which only allow feedback at frequencies satisfying the relationship

$$\nu = \frac{nc}{2L} \tag{10.1}$$

where n is an integer, c is the speed of light, and L is the length of the cavity. Cavity modes are the result of interference along the cavity axis, which requires that standing waves must form. Cavity modes alone do not produce monochromatic radiation since the number of modes is, potentially, infinite. However, in practice the number of modes is severely limited by the width of the spectroscopic transition(s) of the laser medium, as illustrated in Figure 10.2. If there is only a modest population inversion, and if the broadening is small, then only a single mode may be supported. Clearly this will produce highly monochromatic radiation. Even multimode laser operation may yield radiation with fairly narrow linewidths, and certainly $<1\ \text{cm}^{-1}$.

10.3 Ion lasers

Noble gas ion lasers have found widespread use as visible laser sources. The most common is the argon ion laser, which is based on electronic transitions in Ar^+. Details of the operating mechanism can be found elsewhere (for example see Reference [1]). For our purposes, it is only necessary to recognize that both the argon ions, and the population inversion between electronic energy levels in these ions, are produced by an electrical discharge in a sealed argon-containing tube. Mirrors are placed at both ends of the tube, one being partially transmitting to allow a small proportion of the radiation to exit as the output laser beam.

Population inversions can be obtained between several different energy levels, and as a consequence the argon ion laser can produce radiation at a number of wavelengths in the blue and green, the most prominent lines being at 488.0 and 514.5 nm. Although they are sometimes used on their own as spectroscopic light sources, most notably in Raman spectroscopy, the principal use of argon ion lasers in electronic spectroscopy is as pump lasers to drive *continuous* tunable dye lasers (see below). In this application typical output powers of the argon ion laser in the 1–10 W range are employed.

10.4 Nd:YAG laser

Another laser which is used by spectroscopists mainly as a pump laser is the Nd:YAG laser. Both continuous and pulsed Nd:YAG lasers are commercially available, but the principal use of Nd:YAG lasers in spectroscopy is to pump *pulsed* dye lasers. The laser medium is composed of Nd^{3+} ions trapped in a rod of *y*ttrium *a*luminium garnet, or YAG for short. YAG is a glass-like material that has good mechanical and thermal stability, and is transparent to visible and near-infrared light. Population inversion in the Nd^{3+} ions is achieved by optical pumping from a flashlamp, as illustrated in Figure 10.3. The output laser wavelength, 1.06 μm, is in the near-infrared.

To achieve the highest possible output intensity, a pulsed Nd:YAG laser is equipped with a Q-switch. This is an electro-optical device that acts as a very fast shutter in the cavity. When the flashlamp is fired, the Q-switch is initially set to block feedback in the cavity. The pulse of light from the flashlamp lasts for several milliseconds, allowing a build-up of population in the upper laser level. In fact the upper laser level has an average (spontaneous emission) lifetime of about 0.23 ms, and so if the Q-switch is allowed to block feedback for about the first 0.2 ms of the flashlamp firing period, the population inversion reaches a maximum. If the Q-switch is then opened to allow feedback, the maximum possible intensity is obtained and the resulting laser pulse is often referred to as a *giant pulse*. Typical durations for these giant pulses are 5–10 ns, and pulse energies of up to several joules can be extracted

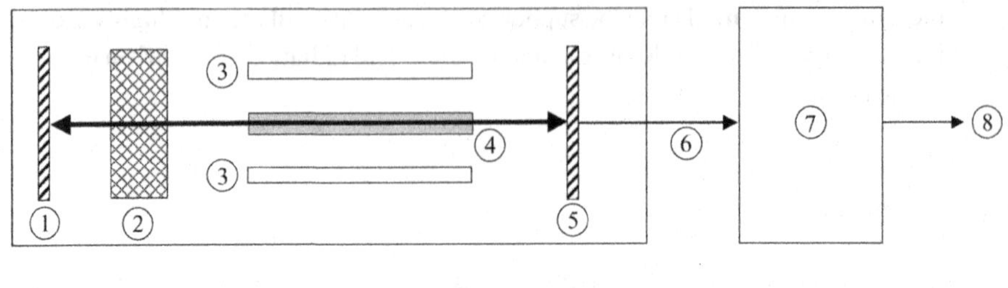

1	End mirror	5	Output mirror (coupler)
2	Q-switch	6	1064 nm fundamental output
3	Flashlamps	7	Harmonic generator (HG)
4	Nd:YAG rod	8	2nd (532 nm), 3rd (355 nm) or 4th (266 nm) harmonic from HG unit

Figure 10.3 Schematic layout for a pulsed Nd:YAG laser. The Q-switch is a Pockels cell, an electro-optical switch that is normally closed but opens a short time into the flashlamp pulse to release a 'giant' pulse of laser light. See text for further details.

at 1.06 μm with quite modest-sized lasers. A pulse of 1 J for 5 ns corresponds to a peak power (the power when the laser is emitting light) of 200 MW!

As will be seen shortly, dye lasers must be pumped by laser light with a shorter wavelength than the dye laser output wavelength. Thus in order to generate visible dye laser radiation the pump laser must have either a visible or ultraviolet output. The 1.06 μm output wavelength of the Nd:YAG laser is clearly inappropriate. It may seem, therefore, that Nd:YAG lasers would be useless for pumping dye lasers. However, this is not the case, since the high intensity at 1.06 μm makes it possible to generate higher harmonics efficiently ($\lambda = (1.06\ \mu\text{m})/n$ where $n = 2, 3, 4, \ldots$) through non-linear optical methods. This entails passing the 1.06 μm radiation, the laser *fundamental*, through crystals with the correct non-linear optical properties for generating higher harmonics. In the case of the Nd:YAG laser, a crystal of potassium dihydrogen phosphate, or KDP for short, is commonly used. It is possible to generate high intensities of the second (532 nm), third (355 nm), and fourth (266 nm) harmonics by this means. The second and third harmonics are employed to pump dye lasers while the fourth harmonic is quite often used as a photolysis light source.

10.5 Excimer laser

Excimer lasers are gas lasers based on transitions in molecules which are bound only in excited electronic states. Important examples are ArF, KrF, and XeCl. In their ground electronic states, the noble gas atoms show no tendency to form chemical bonds with free halogen atoms. However, excited states can be quite strongly bound. This can be understood by considering what would happen if one of the electrons in the outer *p* orbital of the noble gas atom is excited up to a vacant *p* orbital. If this is done, the atom now has unpaired electrons with which it can form a covalent bond to the halogen atom (which of course also has an unpaired *p* electron).[1] Strictly speaking, a heteronuclear diatomic molecule of this type is known as an exciplex, the term excimer being reserved for the homonuclear analogue. However, the name *excimer* has captured the imagination of laser manufacturers and the resulting laser systems are now universally called excimer lasers.

The important point about excimers is that, when they are formed, a population inversion between the upper electronic state and the ground state is automatically obtained since the ground state is unbound (and therefore has zero population). Thus, providing the transition to the ground state is optically allowed, a laser can be constructed based on excimer formation. Actual excimer lasers utilize a high voltage gas discharge through a noble gas/halogen mixture to generate excimers. By changing the gas mixture, the laser wavelength can be altered. The output wavelengths of the most commonly used excimers are 193 nm (ArF), 248 nm (KrF), and 308 nm (XeCl). The output is pulsed, with durations in the 10–15 ns range. XeCl excimer lasers are frequently used alternatives to Nd:YAG lasers for pumping dye lasers, although they are usually more costly to operate due to the requirement for expensive gases.

10.6 Dye lasers

Dye lasers are by far the most important type of laser used in electronic spectroscopy. Their key feature is wavelength tunability, which covers the whole of the visible and parts of the near-infrared and near-ultraviolet, i.e. 330–900 nm. A brief overview is given here.

The laser medium is a solution of an organic dye in a solvent such as methanol. Organic dyes tend to be quite large molecules containing conjugated π systems. The important properties of dyes for laser operation are:

(i) strong absorption and emission bands in the visible or UV;
(ii) broad absorption and emission bands, extending over perhaps 30 or 40 nm.

The importance of these properties can be appreciated by consulting Figure 10.4. The ground electronic state of all organic dyes is a spin singlet, designated S_0. The first excited singlet electronic state is denoted S_1 and it is $S_1 \leftarrow S_0$ transitions that give the dye its colour. The rovibrational levels in each of these states are so close together that, in effect, they form a continuum, as illustrated schematically in Figure 10.4. The continuous nature is caused by two factors. First, organic dye molecules, being relatively large, have a very high density of rovibrational energy levels. Furthermore, each level is collisionally broadened by the very rapid collision rate in solution such that the small gaps between them effectively disappear.

When optically excited into the S_1 state, collisional quenching is rapid and almost complete relaxation to the zero point level in the S_1 state normally occurs before emission gets underway. Optical pumping, using a flashlamp or another laser, is used to produce this excitation of the dye solution. The population inversion is between the zero point level of S_1 and any of the rovibrational levels in S_0 lying above the populated levels. Franck–Condon

[1] An alternative viewpoint is that electronic excitation of the noble gas lowers its ionization energy, thus facilitating formation of an ionic bond to the electronegative halogen atom.

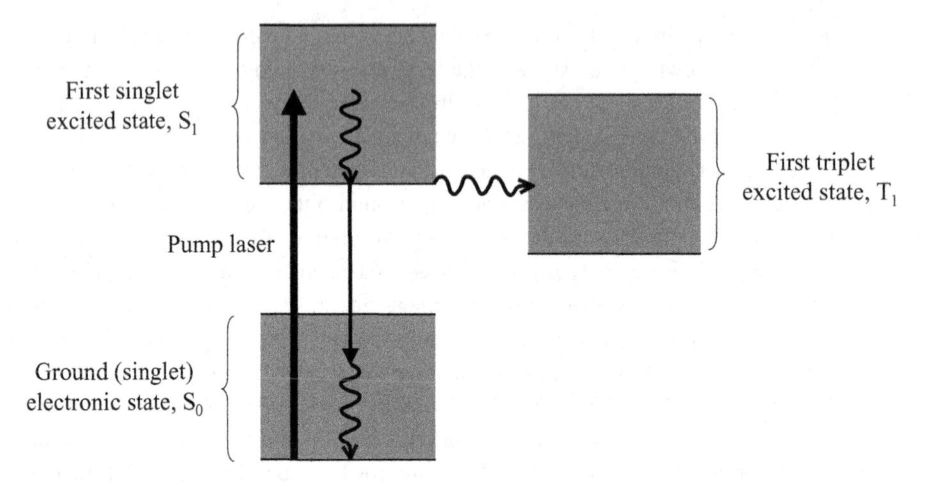

Figure 10.4 Schematic illustration of low-lying singlet and triplet electronic states in a typical dye molecule. The non-radiative processes in the singlet manifolds, shown by the curly arrows, are predominantly collison-induced and are very rapid. The proportion of molecules transferred into the first triplet state (T_1), by intersystem crossing, is small. However, this is detrimental for dye laser operation, especially for continuous dye lasers.

factors favour emission to a wide range of levels in S_0, i.e. the emission band, like the absorption band, will be broad but the former will be shifted to longer wavelengths than the latter. To obtain laser action at a specific wavelength, it is necessary to employ an optical filter or selector so that feedback can be limited to the chosen wavelength rather than be spread over the whole of the broadened emission band.

In pulsed dye lasers, control of the feedback wavelength is achieved by employing a diffraction grating as the rear mirror. A typical arrangement using optical pumping from another laser is shown in Figure 10.5. The wavelength of the reflected light is controlled by rotating the diffraction grating relative to the optical axis of the laser cavity: only light at a specific wavelength is reflected for a given angle (θ). The dye solution is placed in a transparent cell within the cavity and is either stirred (low pump pulse energies) or is flowing (high pump pulse energies). Notice that a beam expander is used to enlarge the laser spot size so that most of the grating surface is exposed: this helps both to narrow the linewidth and to prevent damage to the grating. With this arrangement laser linewidths in the region of 0.2 cm^{-1} can be achieved. An order of magnitude improvement is possible if an additional optical element, an etalon, is inserted into the cavity, as shown in Figure 10.5.

Continuous dye lasers are of a different design to pulsed lasers. One important difference concerns the delivery of the dye solution, which is sprayed as a jet through the pump laser beam. This is necessary to minimize competition from triplet–triplet transitions. The other significant difference is the wavelength selection process, which is not controlled by a diffraction grating. Instead, tuning is obtained by using one or more intracavity filters. Coarse tuning can be achieved with a Lyot (birefringent) filter, while for finer tuning one or more etalons may be inserted.

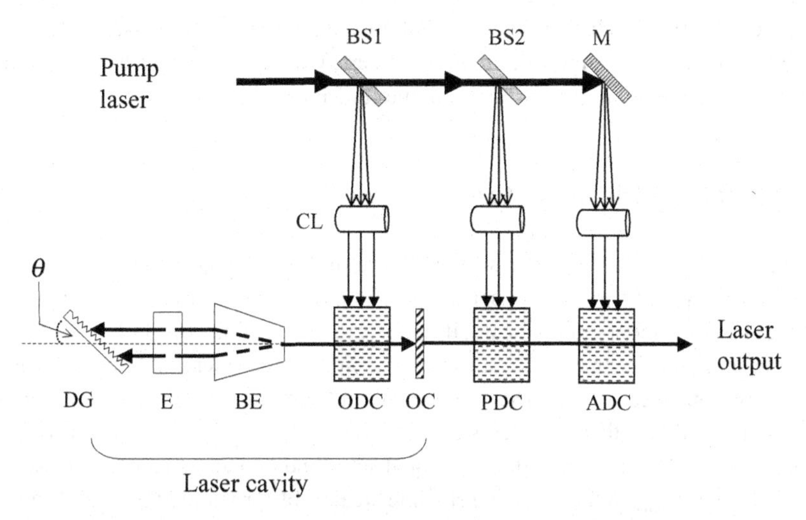

Figure 10.5 Optical arrangement of a tunable pulsed dye laser. The dye laser is pumped by pulsed radiation from another laser. Abbreviations are as follows: BS, beamsplitter; M, mirror; CL, cylindrical lens; OC, output coupler (end mirror); ODC, oscillator dye cell; PDC, preamplifier dye cell; ADC, amplifier dye cell; BE, beam expander; E, intracavity etalon (optional); DG, diffraction grating. The preamplifier and amplifier dye cells are used to increase the intensity of the dye laser beam produced in the laser cavity. This amplification process can increase the intensity by more than two orders of magnitude.

The output wavelength can be extended outside of the traditional dye operating ranges using non-linear optical techniques. The most commonly used is frequency doubling, in which the dye laser fundamental is passed through a suitable crystal to generate the second harmonic ($\nu_{out} = 2\nu_{in}$). This crystal must possess the correct non-linear optical properties, as well as being able to withstand very high laser intensities. β-barium borate is one of the best materials currently available, with KDP as a cheaper alternative for some wavelength ranges. Efficient harmonic generation requires correct *phase matching* of the fundamental and higher harmonic beams. Phase matching is the process by which the refractive indices of the input and output beams are equalized, and this requires a specific orientation of the crystal relative to the incoming laser beam. Frequency doubling allows coverage of the whole of the near-ultraviolet (205–400 nm), and more advanced techniques can extend the wavelength into the vacuum ultraviolet region (<200 nm). At the long wavelength end, tunable radiation beyond 1 μm can be generated using *difference frequency* generation [5].

10.7 Titanium:sapphire laser

The Ti:sapphire laser is a tunable solid state laser based on transitions of Ti^+ ions doped in a sapphire host. The crystalline lattice broadens the electronic energy levels of Ti^+ to such an extent that tunability far exceeding that of a single laser dye is achieved. However, the Ti:sapphire laser is not really a competitor to the dye laser since their tunability ranges only partially overlap. One of the strengths of the Ti:sapphire laser is that much of its tunability range, 660–1180 nm, is in a difficult region for dye lasers. It also possesses better frequency stability and a narrower linewidth than dye lasers. Output in the near-ultraviolet and blue regions is possible by frequency doubling the fundamental output.

10.8 Optical parametric oscillators

These are tunable laser sources that offer the promise of eventually superseding dye lasers. Tunability in optical parametric oscillators (OPOs) is achieved by non-linear optical processing of a single input (pump) beam. It is useful to think of this as the opposite of frequency doubling in a non-linear crystal. In essence, a single high intensity laser beam is passed through the non-linear crystal. The input beam can 'split' into two output beams, one known as the signal and the other the idler, such that $\nu_{in} = \nu_{signal} + \nu_{idler}$. The exact reverse of frequency doubling would correspond to equal idler and signal frequencies. However, any combination of ν_{signal} and ν_{idler} is, in principle, achievable providing the sum equals ν_{in}, and a particular combination can be amplified if the mixing process is carried out in a tunable laser cavity. By combining the tunability of a diffraction grating in the laser cavity, and the orientation of the crystal for optimum phase matching, efficient generation of tunable radiation over a wide spectral range is possible. Commercial OPOs are available which operate over the whole of the visible region and these can be extended into the near-ultraviolet by frequency doubling.

References

1. *Principles of Lasers*, O. Svelto, New York, Plenum Publishing Corporation, 1998.
2. *Lasers*, A. E. Siegman, Mill Valley, California, University Science Books, 1986.
3. *Laser Fundamentals*, W. T. Silvast, Cambridge, Cambridge University Press, 1996.
4. R. H. Lipson, S. S. Dimov, P. Wang, Y. J. Shi, D. M. Maxo, X. K. Hu, and J. Vanstone, *Instrum. Sci. Technol.* **28** (2000) 85.
5. A. S. Pine, *J. Opt. Soc. Am.* **70** (1980) 1568.

11 Optical spectroscopy

Consider a beam of light of intensity I_0 incident on some absorbing sample. Providing only a small fraction of the light is absorbed,[1] and assuming that losses caused by light scattering are negligible, the transmitted light intensity, I, is governed by the familiar Beer–Lambert law,

$$A = \log_{10}\left(\frac{I_0}{I}\right) = \varepsilon(\nu)cl \tag{11.1}$$

where A is known as the *absorbance*. The absorbance is dependent upon the concentration of absorbing species, c, the optical path length, l (distance travelled by the light through the sample), and the *molar absorption coefficient*, ε. The molar absorption coefficient is a measure of the intrinsic absorbing power of the sample and is frequency dependent, which is why it has been written as $\varepsilon(\nu)$. It is customary to give c in units of mol dm^{-3} and l in cm, and so ε is often quoted in the rather strange mixture of units dm^3 mol^{-1} cm^{-1}. As one might expect, ε is related to the Einstein B coefficient introduced in Chapter 7.

The absorbance is an important quantity because it is directly proportional to the concentration. If monochromatic radiation is passed through a material of known thickness and known molar absorption coefficient, the concentration of the absorbing species can be determined from a measurement of the absorbance. This is a widely used feature of absorption spectroscopy.

11.1 Conventional absorption/emission spectroscopy

A schematic of an absorption spectrometer is shown in Figure 11.1. Ideally, the light source is continuous over the wavelength region of interest and shows no major variations in intensity. Resistively heated filaments are good sources of near-continuum light. One example is a white-hot tungsten filament, which will cover the whole of the visible and parts of the near-ultraviolet and near-infrared. A wavelength selector is central to the spectrometer and is usually a monochromator built around a diffraction grating, thus allowing tunability. In order to obtain a spectrum, light intensity transmitted through the monochromator is

[1] If the fraction of light absorbed is large, then the light intensity varies strongly as the sample is traversed and the Beer–Lambert law no longer holds.

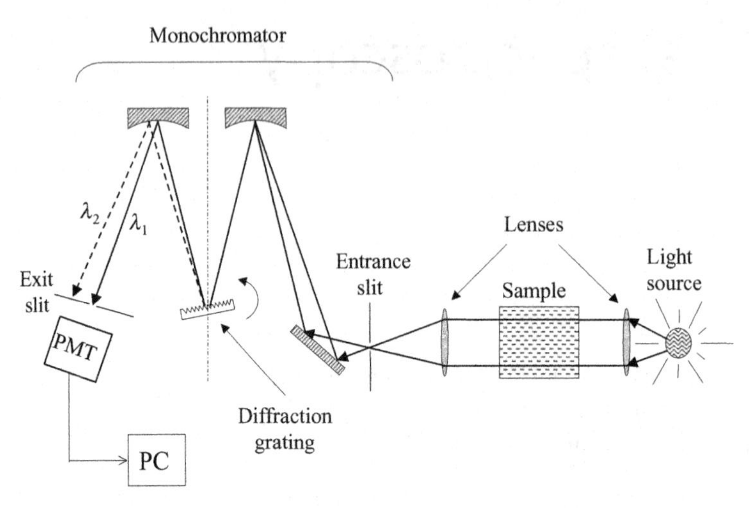

Figure 11.1 Schematic of a conventional grating-based absorption spectrometer. The monochromator is of the Czerny–Turner type in which the entrance and exit slits are placed at the focal points of curved mirrors.

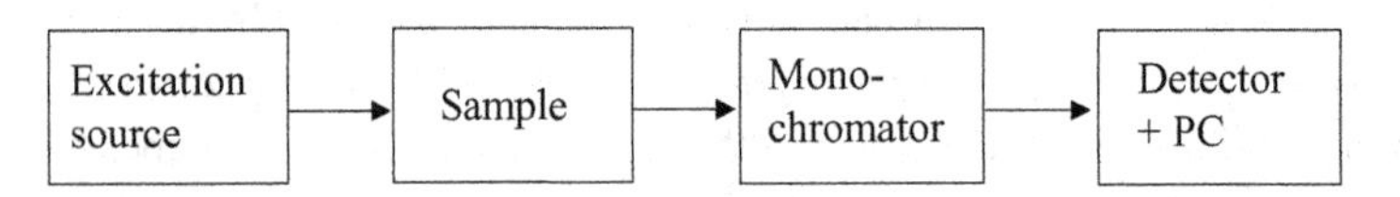

Figure 11.2 Block diagram of a standard emission spectrometer.

measured as a function of wavelength. The light intensity is measured by a photomultiplier tube (PMT), a photodiode, or some other light-detecting device.

In an emission spectrometer, the sample must be driven up to excited quantum states in order for emission to occur. This is normally achieved by an electrical discharge, although broadband optical excitation is also possible. As indicated in Figure 11.2, the monochromator is now used to select a specific emission wavelength from the sample and the intensity at this wavelength is measured by imaging the light onto a detector such as a PMT. An emission spectrum is obtained by recording the PMT signal as a function of emission wavelength.

Monochromators such as that shown in Figure 11.1 have both entrance and exit slits. These are crucial to the wavelength selection process. Narrowing the entrance and exit slits can improve the spectral resolution, but it does so at the expense of sensitivity because of the reduced light throughput. Improvements can be made that make more efficient use of the available light. For example, the exit slits in an emission spectrometer can be dispensed with if a multichannel detector is available. Examples are photodiode arrays and charge-coupled devices (CCDs). These measure the light intensity as a function of position on the detector surface and so are able to record a large portion of the spectrum simultaneously. Another alternative is Fourier transform spectroscopy, which does away with both the entrance and exits slits as well as the diffraction grating. Fourier transform spectroscopy is described later in this chapter.

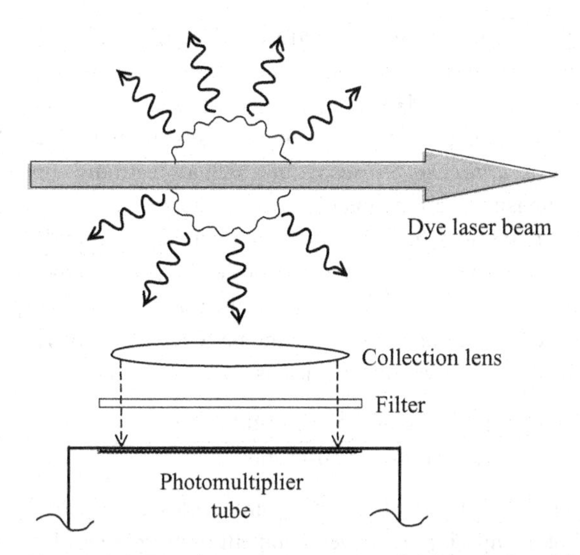

Figure 11.3 Experimental arrangement for laser-induced fluorescence spectroscopy. Fluorescence radiates in all directions, a portion of which is collected by the lens and transmitted to the detector, a photomultiplier tube. The filter, which is used to reduce the amount of scattered laser light reaching the detector, is optional. The arrangement shown is for *laser excitation* spectroscopy. For *dispersed fluorescence* spectroscopy the filter is replaced with a scanning monochromator.

11.2 Laser-induced fluorescence (LIF) spectroscopy

This is one of the principal techniques for studying electronic transitions of both neutral molecules and molecular ions at high sensitivity and at high resolution. In LIF spectroscopy an electronic transition of the molecule is excited using a tunable laser and any fluorescence generated is monitored. There are two complementary methods that parallel, respectively, conventional absorption and emission spectroscopy.

Suppose the wavelength of a tunable laser is scanned through the electronic absorption band of a molecule. Absorption will occur at resonant wavelengths and could be monitored by measuring the intensity of the transmitted laser beam. The high intensity of a laser can greatly increase the probability of absorption compared with low intensity non-laser light sources and thus it might be thought that laser absorption spectroscopy would be very sensitive. Unfortunately, this is not the case because the fractional absorption by a sample will still normally be very low. Thus a small change in intensity is superimposed on a large background signal. When fluctuations in intensity of the laser beam and noise from the light detector are factored in, this approach turns out to have a very limited sensitivity.

However, instead of measuring absorption directly it can be monitored indirectly by detecting fluorescence from the excited electronic state. The experimental arrangement is remarkably simple, and is outlined in Figure 11.3. A tunable laser is passed through the sample and any fluorescence produced is collected off-axis, usually at right angles to the laser beam, by a collection lens. The light is then detected by a photosensitive device, most usually a PMT. PMTs have phenomenal sensitivities and are even capable of detecting single photons in some cases. When the laser is off-resonance, no fluorescence will be produced, and therefore the PMT registers no signal. However, at resonant wavelengths fluorescence is possible and so absorption can be registered by detecting emission from the excited state. This is the basic idea of *laser excitation spectroscopy*, in which a spectrum is obtained by measuring the fluorescence intensity as a function of laser wavelength.

There are several important points to note about laser excitation spectroscopy. First, while there is a clear similarity between laser excitation spectroscopy and absorption spectroscopy, there is also an important difference. The intensity of peaks in a laser excitation spectrum depends on both the absorbance of the sample *and* the fluorescence quantum yield of the excited state. The fluorescence quantum yield is defined as

$$\Phi_f = \frac{\text{rate of photon emission by excited state}}{\text{rate of photon absorption}} \tag{11.2}$$

A fluorescence quantum yield of unity implies that all molecules excited to the upper electronic state relax via photon emission. However, competition from other decay routes (see below) may not only lower Φ_f, but may also cause it to change from one excited state level to another. As a result absorption and fluorescence excitation spectra may look very different.

The high sensitivity of LIF spectroscopy arises from the low background signal received by the PMT at off-resonance laser wavelengths. Even though any fluorescence produced may be very small, it is easily detected by the PMT and therefore if the off-resonance signal is much smaller still then an extremely high signal-to-noise ratio can be achieved. In practice the off-resonance signal is never zero. The principal cause is scattered light from the laser. This can be minimized by keeping potential scattering sites out of the path of the laser. Furthermore, scattered laser light can be virtually eliminated if at least a portion of the fluorescence is at longer wavelengths than the laser. If this condition is satisfied, and it often is for many molecules, then an optical filter, which will only transmit wavelengths longer than that of the laser, can be inserted in front of the PMT.

In laser excitation spectroscopy the fluorescence serves only as a means of detecting the absorption process. However, the fluorescence itself clearly contains spectroscopic information since it arises from emission to lower energy levels. If the emission is dispersed in a monochromator, the spectrum obtained will be the emission spectrum originating from a specific (laser-excited) upper state. This type of spectroscopy goes by several names, including *dispersed fluorescence spectroscopy*, *laser-excited emission spectroscopy*, and *single vibronic level fluorescence spectroscopy*; we will use the first of these throughout this text.

In laser excitation spectroscopy the resolution is often limited by the linewidth of the laser. For pulsed dye lasers, linewidths of $\sim$0.03 cm^{-1} can be obtained relatively straightforwardly. If narrower linewidth lasers are used, such as CW dye lasers or specialized pulsed dye lasers, other factors may begin to limit the resolution, such as Doppler broadening. If steps are taken to minimize Doppler broadening, a resolution of better than 0.001 cm^{-1} can be attained. With such a high resolution, rotationally resolved electronic spectra of quite large molecules can be tackled.

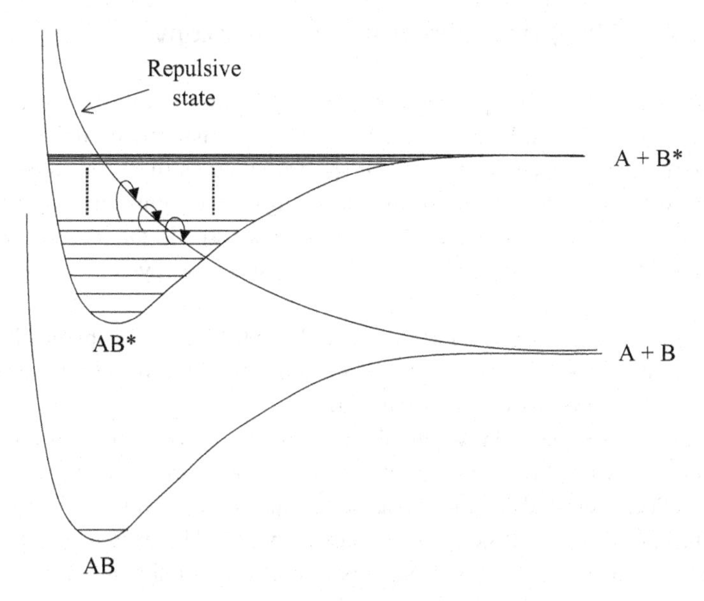

Figure 11.4 Predissociation caused by the crossing of two potential energy curves. Notice that only those energy levels above the crossing point can undergo predissociation.

In dispersed fluorescence spectroscopy, the use of a scanning monochromator is normally the principal factor limiting the resolution. Even a large monochromator may only have a resolution of about 1 cm^{-1}. Thus dispersed fluorescence spectroscopy is normally concerned with vibrationally resolved emission spectra.

The principal disadvantage of LIF is the need for a fluorescent excited state. Fast non-radiative decay routes may reduce the fluorescence quantum yield to zero and in these cases LIF cannot be used. An example of a non-radiative decay process is predissociation, which is illustrated in Figure 11.4. Predissociation results from a crossing of potential energy surfaces of two excited electronic states, one of which is repulsive (dissociative). If the molecule is excited to the bound potential energy curve, it may hop over onto the repulsive curve at the crossing point and will then undergo dissociation. If the probability of predissociation is not too high, there may still be sufficient fluorescence for LIF detection. In such cases, the occurrence of predissociation manifests itself by a broadening of spectral lines, since the effect of predissociation is to decrease the lifetime of the level and hence increase the lifetime broadening.

Depopulation mechanisms such as predissociation are particularly troublesome for large molecules because of their high density of rovibrational energy levels. Usually the coupling mechanism, the process which actually brings about the interaction between the electronic states, will be restricted by symmetry in the same way that symmetry restricts electric dipole transitions. However, the importance of symmetry restrictions decreases as the overall point group symmetry of a molecule is lowered, and large molecules tend to have low symmetry. It is for these reasons that LIF is a particularly powerful technique for investigating small molecules, but is more limited in scope for large molecules.

11.3 Cavity ringdown (CRD) laser absorption spectroscopy

Direct laser absorption electronic spectroscopy is appealing for several reasons. First, the narrow linewidths of lasers can be exploited. Second, it does not rely on the occurrence of a secondary process for detection, as in LIF spectroscopy. Third, the absorbance can be directly related to the concentration of the absorbing species, thus allowing absolute concentration measurements to be made.[2] However, as discussed in the previous section, when done in the conventional manner laser absorption spectroscopy is a low-sensitivity technique.

Cavity ringdown spectroscopy is a form of laser absorption spectroscopy in which the absorbance is determined but in a rather ingenious manner. It is a relatively new technique, first appearing in 1988, but is based on a simple idea.

Suppose a gas is placed between two highly reflecting mirrors which act as an optical cavity. If a pulse of laser light is injected into the cavity, as shown in Figure 11.5, then laser light will reflect backwards and forwards and, if the spacing between the mirrors is relatively small, interference will occur as a consequence of the coherence of the laser beam. However, the coherence of a laser beam is restricted to a finite distance known as the *coherence length*.[3] The finite coherence length is brought about by uncertainty in the frequency of the light, which in turn is a result of the non-zero linewidth. The coherence length, l_{coh}, is given by

$$l_{\mathrm{coh}} = \frac{c}{\Delta\nu} \tag{11.3}$$

where $\Delta\nu$ is the linewidth (FWHM) of the laser. For typical pulsed dye lasers without intracavity etalons, the linewidth is $0.2\,\mathrm{cm}^{-1}$ and so equation (11.3) yields $l_{\mathrm{coh}} = 5$ cm. Consequently, if the mirror separation is significantly larger than 5 cm, interference is not an issue; this is the starting point for cavity ringdown spectroscopy.

If the mirrors are able to transmit a small proportion of the incident light, then each time the laser light pulse impinges on a mirror some is lost from the cavity. Gradually, at a rate determined by the mirror reflectivities, the intensity of the light trapped within the cavity will decay to zero. In fact the decay is exponential, and the time taken for the intensity to decay to $1/\mathrm{e}$ of its initial value is known as the *ringdown time*. It can be measured by placing a sensitive light detector, usually a photomultiplier tube, behind one of the mirrors, as shown in Figure 11.5.

Now suppose that an absorbing sample is placed inside the cavity. Absorption of the laser light by the sample will accelerate the ringdown process, resulting in a faster ringdown time. The larger the absorbance, the shorter the ringdown time. Hence it is possible to record something akin to an absorption spectrum by measuring the *change in ringdown time* as a function of laser wavelength. In fact there is a simple and exact relationship linking

---

[2] It is difficult to deduce *absolute* concentrations of an absorbing species from LIF spectroscopy, although changes in relative concentrations can easily be measured.

[3] The coherence length is a measure of the distance over which the phase relationships between the constituent waves in a light source are maintained. For optical path differences exceeding this difference, the phase relationships are lost and so interference effects become negligible.

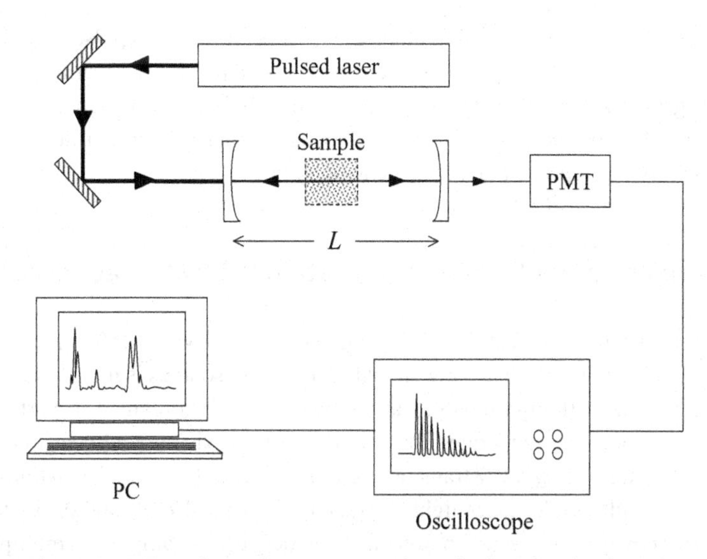

Figure 11.5 Experimental arrangement for pulsed cavity ringdown laser absorption spectroscopy. The cavity is defined by the two plano-concave mirrors. Concave mirrors are preferred over plane mirrors because the former can produce a so-called *stable* optical cavity, making it easier to 'trap' the radiation within the cavity.

ringdown time to the sample absorbance, A, which takes the form

$$A = \frac{L}{c}\left(\frac{1}{\tau} - \frac{1}{\tau_0}\right) \tag{11.4}$$

where L is the cavity length, c is the speed of light, and τ_0 and τ are the ringdown times in the absence and presence of an absorber, respectively. Thus the cavity ringdown spectrum can easily be converted into a conventional absorption spectrum.

To achieve high absorption sensitivity, high mirror reflectivities are required since these lengthen the ringdown times and therefore make it easier to observe small changes. Mirrors with reflectivities better than 99.995% are available in the visible and near-ultraviolet. Of course, one problem with such high mirror reflectivities is that only a tiny proportion of the light from the laser is injected into the cavity in the first place since the laser beam enters through one of the end mirrors! However, with sensitive detectors such as PMTs this does not cause a significant problem.

The discussion so far has focussed on cavity ringdown using pulsed lasers. However, it is also possible to record CRD spectra with continuous lasers (CW-CRD). Typically, a narrow linewidth tunable diode laser is employed as the light source. It is still necessary to inject a pulse of light into the cavity. One way this can be achieved is to scan the cavity length by mounting one of the end mirrors on a piezoelectric transducer. If the cavity length does not match one of the longitudinal modes of the cavity, no significant light can be injected. This restriction is normally unimportant in pulsed laser CRD because the relatively broad laser linewidths mean that there is always some radiation that matches longitudinal modes of the cavity. As the mirror is moved in CW-CRD, at some stage the standing wave condition will be met and light will be injected into the cavity. An electro-optical switch, known as an acousto-optical modulator, is then used to block the laser beam so that a pulse of laser light remains in the cavity. A ringdown profile is then measured in the normal manner.

CW-CRD is growing in importance. One reason for this is that it is capable of much higher spectral resolution than pulsed laser CRD. Much higher pulse repetition rates can also be employed giving improved detection sensitivity [1].

11.4 Resonance-enhanced multiphoton ionization (REMPI) spectroscopy

Highly excited electronic states can be studied by vacuum ultraviolet (VUV) absorption spectroscopy. One of the problems in working with VUV light sources is the low resolution achieved in this region. Although tunable laser radiation can be obtained in parts of the VUV, this is not as routine to generate, nor is it as cheap, as visible and near-ultraviolet laser sources. Fortunately, many VUV transitions can be accessed by *multiphoton* transitions using visible or near-ultraviolet laser light. Resonance-enhanced multiphoton ionization (REMPI) spectroscopy is a particularly powerful and widely used example of a multiphoton spectroscopic technique.

REMPI is a two-stage process. In the first step, molecules are promoted to an excited electronic state by the absorption of one or more photons. It may at first sight seem strange to suggest that more than one photon can be absorbed in a spectroscopic transition, since we normally regard them as single-photon resonant processes. However, there is nothing intrinsically impossible in using two or more photons of lower energy to achieve the same task, providing (i) their combined energy satisfies the resonance condition, e.g. for two photons having the same frequency, $E_2 - E_1 = 2h\nu$, and (ii) all selection rules are satisfied (see later).

The principal reason why multiphoton transitions are not normally considered is that such processes are extremely improbable at normal light intensities. The photons must arrive at the molecule at virtually the same instant in time in order to be simultaneously absorbed. With ordinary light sources, such as lamps or low intensity lasers, this hardly ever happens. However, if extremely high light intensities are employed, as is the case with powerful pulsed lasers, then multiphoton transition probabilities need no longer be negligible. Even so, it is easy to appreciate that the probability will rapidly decrease as the number of photons to be absorbed increases.

Once the molecule has reached the excited electronic state by absorption of one or more photons, it may absorb one or more further photons to climb above the ionization limit. This is a REMPI process. Compare this with direct (non-resonant) multiphoton ionization. Clearly REMPI and direct (non-resonant) multiphoton ionization have the same overall photon order, i.e. the same total number of photons is absorbed. However, in REMPI the ionization is achieved by two steps of lower photon order, each with a much higher probability (many orders of magnitude) than the non-resonant multiphoton ionization process. In other words, the ionization probability is dramatically increased by breaking the ionization process down into two separate, sequential steps.

This suggests a means of detecting electronic transitions. If the laser is tuned to a wavelength that is not resonant with an energy level in the excited electronic state manifold of the neutral molecule, then ionization is only possible by the non-resonant route, and

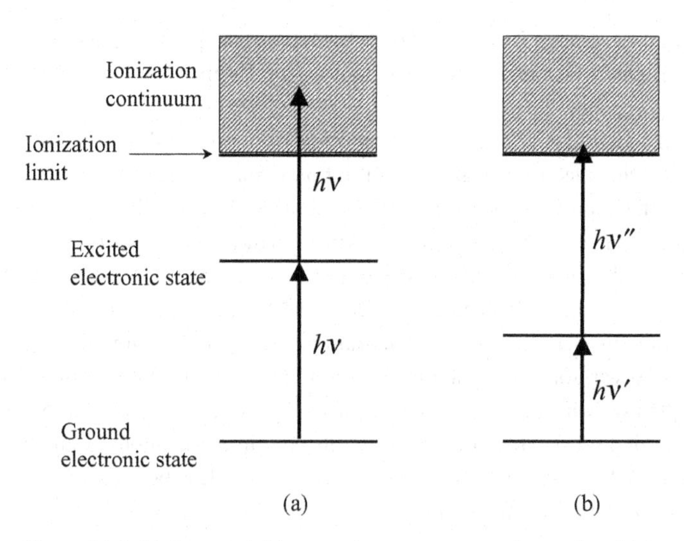

Figure 11.6 (a) One- and (b) two-colour resonance-enhanced multiphoton ionization processes.

therefore has a very low probability. As the wavelength is scanned, when resonance with an intermediate rovibronic level occurs the ionization probability dramatically increases and this can be observed by detecting ions. This is the essence of REMPI spectroscopy, namely the ion current is measured as a function of laser wavelength.

Various experimental arrangements can be used. In the simplest, a single laser is used to excite both the first and second steps, as shown in Figure 11.6(a). This is known as single-colour REMPI. However, two pulsed lasers operating at different wavelengths could be used in a so-called two-colour experiment, one to excite the molecule to the intermediate electronic state, and the second to produce ionization. The two-colour method is important when the wavelength required for exciting the resonant transition is unsuitable for the subsequent ionization step. An example is illustrated in Figure 11.6(b), where the first photon accesses a relatively low-lying electronic state. Absorption of a second photon from this laser will not exceed the ionization limit, but a second laser with a much shorter wavelength can be used to ionize the molecule.

The examples shown in Figure 11.6 use a single photon to access the intermediate state. The single colour process in this case is sometimes said to be a $(1 + 1)$ process, meaning one photon of the same wavelength is used in both the first and second excitation steps. Similarly, the two-colour process is sometimes written as $(1 + 1')$, the prime indicating a different colour is being used for the one-photon ionization step. However, it should be recognized that more than one photon could be used in the initial and ionization steps if sufficiently intense light sources are employed. For example, $(2 + 1)$, $(2 + 1')$, $(2 + 2)$, and $(3 + 1)$ processes are not uncommon in REMPI experiments.

Ion formation can be detected by measuring the ion current between two conducting parallel plates of opposite polarity. Although adequate for many purposes, this approach is less than ideal for the study of mixtures since REMPI signals from more than one type of molecule are possible, thus causing potential confusion. A solution to this problem is to employ a mass spectrometer for detecting the ions, since this allows the mass of the ion, and therefore the carrier of the spectrum, to be identified. Indeed, this is an extremely important advantage REMPI has over LIF spectroscopy. The mass spectrometer may be a time-of-flight device or a quadrupole mass filter. Further details may be found in Reference [2].

We have seen that visible or near-ultraviolet photons from powerful pulsed lasers may be used to access high-lying electronic states by multiphoton transitions. Of course, this is only possible providing the appropriate selection rules are satisfied. A detailed discussion of the selection rules, and in particular their derivation, is beyond the scope of this text. However, in general the selection rules are the result of a sequential application of single-photon selection rules.[4] For example, for linear molecules, for cases where the electronic orbital angular momentum quantum number, Λ, is a good quantum number, the one-photon selection rule is $\Delta\Lambda = 0, \pm 1$. However, for a two-photon transition the selection rule becomes $\Delta\Lambda = 0, \pm 1, \pm 2$. Consequently, whereas transitions from a Σ electronic state to a Δ state are forbidden in single-photon spectroscopy, they are allowed in a two-photon transition. This often means that new electronic states can be observed by REMPI spectroscopy, and this is another interesting aspect of this technique.

11.5 Double-resonance spectroscopy

Double-resonance spectroscopy is the study of any spectroscopic transition using two sequential resonant steps. REMPI could be regarded as an example of double-resonance spectroscopy. However, whereas the second resonant step in REMPI involves excitation into the ionization continuum, one could equally well excite an atom or molecule into a bound state below the ionization continuum. Why would anyone wish to carry out such an experiment, and how would it actually be done? To some extent the answer to the first question has already been stated in the previous section describing REMPI spectroscopy. Any double-resonance absorption transition that uses, for example, two visible photons induces the same energy change in a molecule as a single photon transition in the ultraviolet. Ultraviolet light may be difficult to obtain at the desired wavelengths, or it may be that the linewidth of the ultraviolet source is much higher than that of the visible light sources used in the optical–optical double-resonance experiment. Another facet of a double-resonance experiment is the modified selection rules already discussed in the REMPI case.

How would an optical–optical double-resonance experiment be carried out? In general two lasers are required, both being independently tunable. Care must be taken to ensure that they overlap spatially and, if they are pulsed lasers, that they also overlap temporally. Detection of transitions is usually achieved by observing fluorescence (either from the intermediate state or from the final state), or ions after absorption of a further photon.

The two resonance transitions need not both be 'upwards'. An important example where one of the transitions is 'downwards' is the technique known as *stimulated emission pumping* (SEP). This form of spectroscopy is illustrated in the energy level diagram in Figure 11.7. A photon from one laser, termed the PUMP laser, is used to drive a molecule to a fluorescent

---

[4] We concern ourselves solely with the n-photon resonant step. Selection rules for the ionization step are different (in fact less stringent) because the departing electron may take away angular momentum.

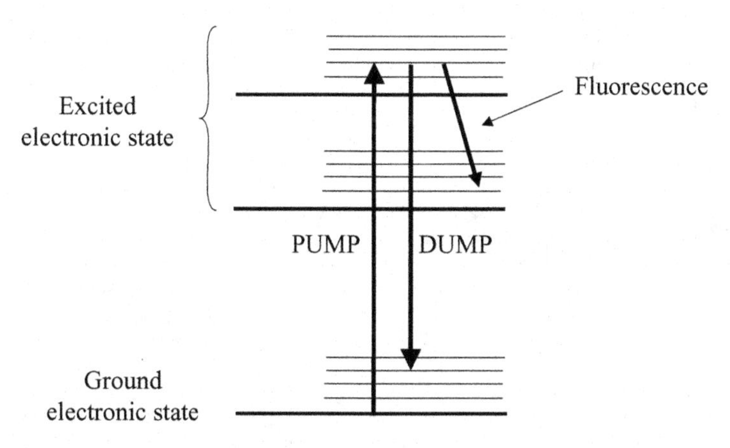

Figure 11.7 Stimulated emission pumping (SEP) spectroscopy. Two lasers are employed. The PUMP laser excites the molecule to a particular rovibrational level in an excited electronic state and fluorescence from that upper level is monitored. The DUMP laser drives molecules back down to a specific rovibrational level in the ground electronic state by stimulated emission. A spectrum is obtained by monitoring the fluorescence intensity as a function of the DUMP laser wavelength; successful stimulated emission is registered as a dip in the fluorescence intensity.

excited electronic state, and the fluorescence is monitored by a photomultiplier tube. If a second laser, known as the DUMP laser, is added with the correct frequency to excite transitions resonantly back down to the lower electronic state, then *stimulated emission* can occur. This will necessarily reduce the fluorescence (which is, of course, spontaneous emission) seen by the PMT since the stimulated emission will follow the path of the DUMP laser. Thus an SEP spectrum can be recorded by fixing the PUMP laser wavelength, scanning the DUMP laser wavelength, and recording the *dip* in fluorescence intensity as a function of the DUMP laser wavelength.

SEP spectroscopy can be compared with dispersed fluorescence spectroscopy (see Section 11.2). In the latter, the resolution is limited primarily by the monochromator, and is often poor. In SEP no monochromator is required and the resolution is limited primarily by the laser linewidth. The much higher resolution of SEP makes it possible to obtain rotationally resolved emission spectra, and also allows the investigation of very dense vibrational manifolds in low-lying electronic states such as those seen near to dissociation limits.

11.6 Fourier transform (FT) spectroscopy

The spectroscopic techniques considered so far all work in the *frequency domain*. In other words, the exciting radiation and/or the emitted radiation is selected according to its frequency. A spectrum is then recorded by controlled variation of this frequency.

Fourier transform (FT) spectroscopy adopts a very different approach. It is based on interference effects produced by radiation of different frequencies. In NMR and microwave spectroscopy the interference phenomena are observed in the *time domain*. However, this is not possible for infrared, visible, and ultraviolet radiation because the frequencies are

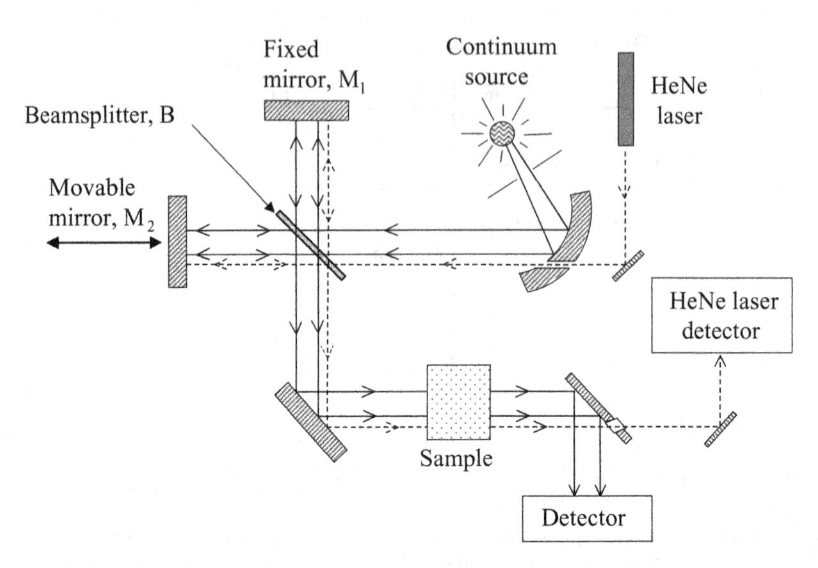

Figure 11.8 Schematic of an FT absorption spectrometer, showing the Michelson interferometer at its heart. The helium–neon laser beam, represented by the dashed line, takes a parallel path to the light from the continuum source and is used to measure the distance moved by mirror M_2.

too high. In these regions of the electromagnetic spectrum the *length domain* is employed and an *interferogram* is generated. The interferogram contains information on the complete spectrum (or at least a large part of it) in an 'encoded' form, which can then be converted into a normal frequency domain spectrum.

The heart of an FT spectrometer is the Michelson interferometer. This is the device that generates an interferogram. To see how it works, consider the apparatus in Figure 11.8. Light from a continuum source is passed into the Michelson interferometer and through a sample cell. However, let us simplify the situation to begin with by imagining that the light source is monochromatic and that the sample cell is absent. The first part of the interferometer that the light encounters is the beamsplitter, B, which sends a portion of the beam towards mirror M_1 and the remainder towards mirror M_2. After reflection by the mirrors the two beams return to the beamsplitter and interference takes place. Whether this interference is constructive or destructive depends on the optical path difference for the beams in the two arms of the interferometer, i.e. $2BM_1 - 2BM_2 = \delta$. The quantity δ is referred to as the *retardation*. If the retardation is an integer multiple of complete wavelengths, i.e. $\delta = n\lambda$, then constructive interference occurs and the light intensity reaching the detector will be relatively high. If, on the other hand, $\delta = n\lambda/2$, then complete destructive interference occurs and no light reaches the detector. At retardations between these two extremes, the detector signal level depends on the degree of constructive versus destructive interference.[5]

[5] Newcomers to FT spectroscopy and the Michelson interferometer are often troubled by two points. (i) How can an incoherent light source, such as a lamp, give rise to the phase coherence necessary for observable interference effects? (ii) Where does the light go when destructive interference occurs if it does not go to the detector? The answer to question (i) is straightforward. An incoherent light source can be thought of as being composed of numerous independent waves, or wavelets. Although there is no phase relationship between the wavelets, each

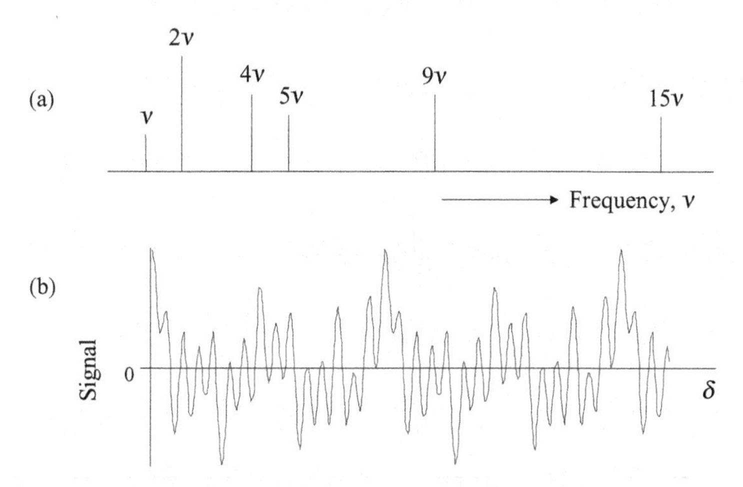

Figure 11.9 (a) Hypothetical stick spectrum and (b) corresponding interferogram showing the beat pattern formed by interference of several cosine waves of different frequencies and relative intensities. The quantity δ is the retardation – see text for further details.

An interferogram is obtained by varying the retardation. This is achieved by moving one of the mirrors and recording the detector signal as a function of mirror position. For a monochromatic light source, the interferogram will consist of a single cosine wave and the wavelength of the light can be measured directly from the interferogram, providing the retardation is known to sufficient precision at all points in the moving mirror motion. In order to be able to distinguish peaks from troughs in the waveform, which is clearly essential for the measurement of the wavelength, the uncertainty in mirror position must be $<\lambda/2$. When dealing with visible or ultraviolet light this is quite a technical challenge but is feasible and has been achieved.

Figure 11.9 simulates a more complicated situation, where five different radiation frequencies of differing intensities interfere to produce an interferogram. A pattern is still discernible in this more complicated case but, in the limit of a continuum light source, fully constructive interference occurs only at $\delta = 0$ and the signal rapidly decays either side of this position. The strong interferogram at and near $\delta = 0$ shows what is known as a *centre burst*.

When an absorbing sample is placed in the spectrometer, a situation somewhat intermediate between the two extreme cases of monochromatic and complete polychromatic (continuum) radiation sources occurs. The intensity of light entering the interferometer at certain wavelengths is reduced when the sample is present due to absorption. The result is an interferogram which is dominated by a centre burst but which also shows interference fringes extending out from $\delta = 0$ (see Figure 11.10). These interference fringes contain, potentially, all of the information about the absorption by the sample. In other words, it is possible to extract the complete frequency domain absorption *spectrum* from the interferogram.

individual wavelet acts as a 'mini' coherent light source. Thus the interference effects seen after splitting and recombining the beam arise from interference of light originating within these individual wavelets. The answer to (ii) is also straightforward: when there is a drop in intensity at the detector due to interference, this is because destructive interference redirects the light back towards the source.

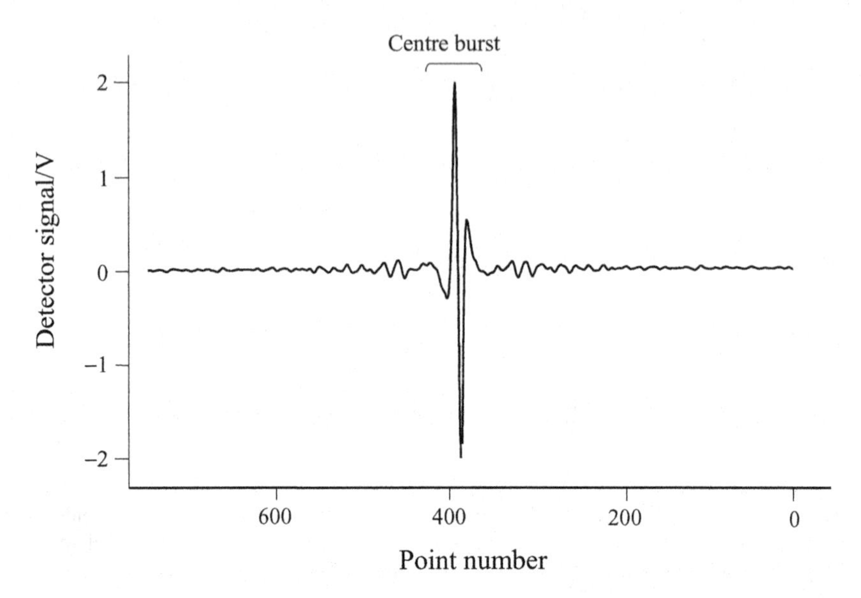

Figure 11.10 Interferogram showing a dominant centre burst and the formation of interference fringes away from the centre burst due to absorption of infrared radiation by the sample. The retardation, δ, is actually monitored at a series of discrete moving mirror positions given by the 'point number' on the horizontal axis. Zero retardation corresponds to a point number just below 400 in this example.

How might this be done? One route would be a trial and error process, in which we guessed at the reduction of light intensity at certain wavelengths caused by absorption, and simulated the interferogram by superimposing the electromagnetic waves at various values of the retardation. This simulated interferogram would then be compared with the actual interferogram and, if agreement was not obtained, a new guess would be made at the absorption spectrum and a new simulation would be attempted. With the help of a computer this approach is just about conceivable to deduce the frequency domain spectrum in Figure 11.9(a) from the interferogram in Figure 11.9(b). However, for more complicated cases, as would be found in real laboratory work, this would be a hopelessly long-winded process even for a computer.

In practice, the length → frequency domain conversion is achieved by the mathematical transformation process known as *Fourier transformation*. Fourier transformation allows information in one domain to be converted to that in an inverse domain. In the case of the Michelson interferometer, the interferogram is measured in the *length domain*, i.e. as a function of the retardation. The inverse of length is the wavenumber ($\overline{\nu}$), and so wavenumber and length are complementary Fourier variables. In other words, if the light intensity at the detector is measured as a function of retardation, Fourier transformation can convert this into intensity versus wavenumber, i.e. into a spectrum.

Fourier transformation is an integral transformation given by

$$I(\overline{\nu}) = 2\int_{0}^{\infty} I(\delta)\cos(2\pi\overline{\nu}\delta)\,\mathrm{d}\delta \qquad (11.5)$$

where $I(\delta)$ is the interferogram signal and $I(\overline{\nu})$ is the spectrum. This compact expression may mislead the reader into thinking it is easy to evaluate. However, $I(\delta)$ is not an analytical function and so the integral must be evaluated numerically. Furthermore, the integral must be calculated at each value of $\overline{\nu}$ in order to construct a spectrum. Thus Fourier transformation is a major computational task and, in the early days of FT spectroscopy, it was a severely limiting factor. Nowadays, with much higher computer speeds coupled with development of the fast Fourier transform algorithm in the mid 1960s [3], Fourier transformation rarely takes more than a few seconds on a high performance PC.

FT spectrometers have a much greater light-gathering power than grating instruments because both entrance and exit slits are eliminated. Consequently, a spectrum of signal-to-noise ratio comparable to that of a grating spectrometer can be obtained in a much shorter measurement time with an FT spectrometer.

Another important advantage of FT spectroscopy is the high accuracy of wavenumber measurements. The accuracy of the wavenumber measurement is determined, in principle, by the accuracy with which the moving mirror position is known at all points in its motion. In real FT spectrometers, the relative position of mirror M_2 relative to M_1 is also measured interferometrically. This is achieved by sending a reference laser beam, usually from a low power helium–neon laser ($\lambda = 632.8$ nm) along the same path as the signal beam (see Figure 11.8). The moving mirror generates interference fringes from the laser beam beyond the beamsplitter and, if the wavenumber of the laser is accurately known, then the relative position of the mirror is easily deduced by fringe counting (this is done electronically). Absolute wavenumber accuracies of better than 0.001 cm^{-1} are possible.

The final point to note about FT spectroscopy is the resolution. It turns out that the wavenumber resolution is the inverse of the maximum retardation, δ_{max}. Thus for a maximum mirror displacement of 10 cm, which corresponds to a maximum retardation of 20 cm, a resolution of 0.05 cm^{-1} is obtained The highest resolution commercial instruments currently on the market have a maximum mirror displacement of about 1 m, giving a best resolution of 0.005 cm^{-1}.

Fourier transform spectroscopy is commonplace in the infrared region. Its extension into the visible and ultraviolet came later but there are now several commercial manufacturers of UV/Vis FT spectrometers.

References

1. G. Berden, R. Peeters, and G. Meijer, *Int. Rev. Phys. Chem.* **19** (2000) 565.
2. *Lasers and Mass Spectrometry*, ed. D. M. Lubman, New York, Oxford University Press, 1990.
3. J. W. Cooley and J. W. Tukey, *Math. Comp.* **19** (1965) 297.

12 Photoelectron spectroscopy

12.1 Conventional ultraviolet photoelectron spectroscopy

The basic principles of conventional photoelectron spectroscopy were described in Section 1.1. To recap, the molecules of interest are illuminated by ultraviolet photons with sufficient energy to ionize them.

$$M + h\nu \rightarrow M^+ + e^-$$

The photon energy must equal or exceed the ionization energy of molecule M in order for the above process to take place. Ignoring the kinetic energy of the recoiling ion, which is negligible owing to the large mass disparity between the ion and the electron, the excess energy from photoionization can appear either as electron kinetic energy, ion internal energy (vibrational and rotational), or a combination of the two.

From conservation of energy, as summarized in equation (1.2), measurement of the electron kinetic energy spectrum for a fixed ultraviolet wavelength provides spectroscopic information on the ion. The ionization energy depends on which electron is being removed, and thus the most weakly bound will give rise to electrons with the highest kinetic energy while those more tightly bound will yield lower energy electrons. This gives rise to coarse band structure, with each band representing a different ionization process. However, each band contains structure arising from the population of different vibrational and rotational levels within the particular electronic state of the ion, and this additional structure provides a great deal of important information. This structure can only be observed if the resolution of the electron spectrometer is sufficiently high and, as will be seen shortly, the resolution in conventional ultraviolet photoelectron spectroscopy is relatively poor.

Most readers will know that highly electropositive elements, such as the alkali and alkaline earth atoms, have relatively small first ionization energies. Their first ionization energies mostly fall in the range 4–7 eV because the *s* electrons in the outer shell are quite weakly bound to the nucleus. The first ionization energies of the majority of molecules, and indeed other elements, tend to be higher, usually exceeding 9 eV. Consequently, just to reach the first ionization limit requires ultraviolet light of wavelengths $\leq$140 nm, and to access higher ionic states much shorter wavelengths may be required. These wavelengths fall in the vacuum ultraviolet, and this is a difficult region in which to generate monochromatic light with usable intensities. Indeed, this difficulty was not resolved until the early 1960s through the introduction of noble gas resonance lamp sources.

In VUV noble gas resonance lamps, a high voltage DC discharge along a capillary tube[1] is employed to drive noble gas atoms up to excited electronic states. The electronic transition back to the ground state is then responsible for the radiation. For helium, the principal emission line is at 21.218 eV ($\lambda = 58.4$ nm) and arises from the transition $^1P(1s^12p^1) \rightarrow$ $^1S(1s^2)$. This line is referred to as the HeIα line, the I signifying emission from neutral helium and the α designating that this is the first of a series of possible $np \rightarrow 1s$ transitions. Other transitions do occur, not only from neutral helium but also He^+ (these are labelled HeII transitions), but they are normally much weaker than the HeIα line. Other gases can be used. For example, neon gives two NeI lines, one at 16.671 and the other at 16.848 eV. However, helium is the most commonly used both because it is cheaper than neon and because the higher photon energy means that the valence orbitals of most molecules can be photoionized with the HeIα line.

The electron kinetic energy spectrum is obtained by passing the ejected electrons through an energy analyser. This analyser is based on an electric or magnetic field, usually the former, to distinguish the electrons according to their kinetic energies. There are two main analyser types, retarding field and deflection analysers.

Retarding field devices transmit only those electrons that have energies higher than the retarding potential, and to obtain a spectrum the retarding potential is scanned. This type of analyser is rarely used nowadays and we shall discuss it no further.

Deflection analysers, as the name implies, separate electrons by forcing them to follow different paths according to their velocities. There are a number of different types, including the parallel plate analyser (this uses an electric field applied between two parallel plates) and the cylindrical mirror analyser (containing two charged coaxial cylinders). However, the only one that we will discuss in any detail is the hemispherical analyser, since it is simple to understand and is widely used.

The basic geometry of the hemispherical analyser is illustrated in the overall schematic of a photoelectron spectrometer in Figure 12.1. The name derives from the use of two concentric hemispherical electrodes, both charged to a potential with the same magnitude but opposite signs; the inner one is positive and the outer negative. The entrance and exit to the analyser are restricted by slits that define the range of acceptable entrance and exit trajectories of the electrons. Electrons that pass through the entrance slits after photoionization may traverse the analyser and out through the exit slits only by following a specific curved path, but they will do so only if they have the correct energy (determined by the selected voltages on the hemispheres). The fate of electrons with higher or lower kinetic energies is clear from the figure; the electric field is either too weak or too strong, respectively, to allow them to follow the correct trajectory and they are lost in collisions with the walls. An electron kinetic energy spectrum is obtained by measuring the electron current at the detector as a function of the voltage applied to the hemispheres. The voltage can be used to calculate the electron kinetic energy.[2]

1 The capillary serves two purposes. First it helps to collimate the radiation. Second, it helps to minimize the amount of sample gas passing into the discharge region, since there are no suitable window materials for wavelengths shorter than 100 nm.

2 In practice one cannot extract a particularly accurate electron kinetic energy by calculations based solely on the applied voltage. This is because the electron energy also depends on the local charges on any surfaces it passes,

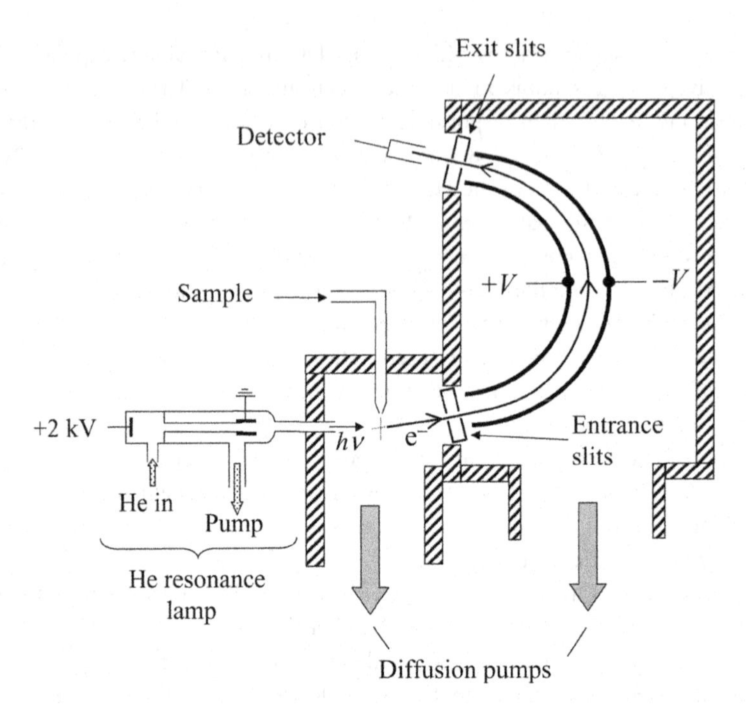

Figure 12.1 Schematic of a photoelectron spectrometer with a hemispherical electrostatic electron energy analyser. Electrons of the correct energy traverse the path shown between the charged hemispherical plates. Electrons at higher or lower energies will either strike the walls of the plates or the exit slits and are not detected. A spectrum is recorded by varying the potential difference between the plates.

The electrons from the analyser are usually detected by electron multipliers. These are devices coated with a material which, when hit by an electron, produce secondary electron emission (typically two or three electrons per incident electron). They thus serve as electron amplifiers and, when placed in series so that the secondary electrons from one are accelerated into the next, can produce amplifications in excess of 10^7. The actual electron current produced after amplification may still be small but it can be measured with picoammeters or other sensitive current-measuring devices.

A photoelectron spectrometer must be kept under vacuum and indeed the quality of the vacuum is crucial. A typical spectrometer will have at least three separate pumping regions, the resonance lamp, the sample chamber, and the analyser chamber (see Figure 12.1). The pressure of the sample must be sufficiently high for it to be detectable, but at the same time it must be low enough to allow the great majority of electrons to escape unimpeded into the analyser. The usual compromise is a pressure of 10^{-4}–10^{-5} mbar. The analyser chamber must be kept at a considerably lower pressure since the electrons must travel much further in this chamber than in the ionization chamber. Thus pressures of $<10^{-5}$ mbar are typically required there.

and any contamination on the inner walls of the spectrometer always has some effect of this type. Consequently, the energy scale is established by mixing the desired sample with one or more calibrants of known ionization energy.

The two most important properties of a photoelectron spectrometer are its resolution and sensitivity. Although intrinsic factors do play a role, especially Doppler broadening, the major factor affecting resolution is instrumental in origin. The main limitations on instrumental resolution are the dimensions of the analyser, the widths of both the entrance and exit slits, as well as other factors, such as the presence of outside electric or magnetic fields and local charges inside the spectrometer (e.g. from surface contamination). The resolution can be improved by decreasing the entrance and exit slit widths, but this necessarily impairs the sensitivity. Thus a trade-off between good sensitivity and acceptable resolution is necessary, and the compromise that is normally taken yields a resolution in the 10–30 meV range ($\sim$80–240 cm^{-1}). This is clearly much worse than that routinely obtained in optical spectroscopy, and is such that it may even be a struggle to achieve full vibrational resolution for small molecules. Rotationally resolved spectra are not practical with conventional photoelectron spectroscopy.

12.2 Synchrotron radiation in photoelectron spectroscopy

There have been a number of important experimental developments in photoelectron spectroscopy over the years. One of the most significant has been the widespread use of synchrotron radiation. In fact synchrotron radiation has many other applications in science and technology. Synchrotron radiation is produced from electron storage rings. In outline, a burst of electrons is injected into a storage ring and confined to a near-circular path by a series of magnets. The electrons, travelling at speeds close to that of light, generate intense radiation as they accelerate around the ring and this radiation can be extracted for various experiments. The construction of synchrotrons requires major financial investment. They are essentially large particle accelerators and it is therefore only feasible to operate them as central facilities. Experimental stations, known as *beamlines*, are located at various points around the storage ring, as illustrated in Figure 12.2. The investigator travels to the synchrotron to carry out experiments and will use the radiation output, together with any other imported or permanent equipment, at one of the beamline stations.

The key properties of synchrotron radiation for photoelectron spectroscopy are: (i) it is continuous over a wide wavelength range (10^{-10}–10^{-5} m); (ii) it is highly intense; (iii) the radiation is plane-polarized. A specific wavelength is necessary for photoelectron spectroscopy and so a suitable monochromator is placed in front of the spectrometer. The plane-polarized nature of synchrotron radiation is important in angle-resolved work, i.e. in studies where the intensity of electrons is measured at various angles relative to the plane of polarization. Photoelectron angular distributions can provide important information on photoionization dynamics.

12.3 Negative ion photoelectron spectroscopy

The most weakly bound electron in a singly charged anion has a binding energy equal to the negative of the electron affinity of the atom or molecule. The electron affinity of an anion is analogous to the ionization energy of a neutral species, but the former is normally much

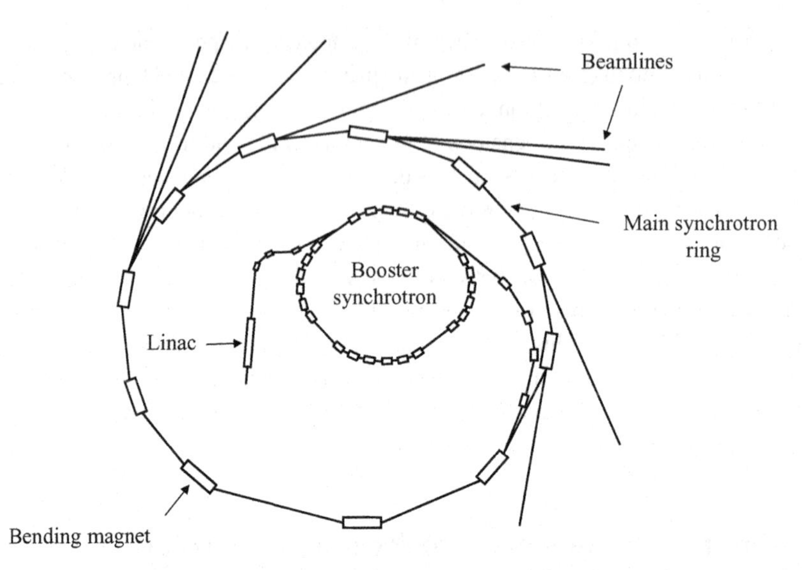

Figure 12.2 Schematic illustration of a synchrotron radiation source. Electrons, produced by thermionic emission from a heated cathode, are injected into the storage ring after acceleration up to very high speeds by a *lin*ear *ac*celerator (linac). A series of bending magnets situated at various positions along the ring force the electrons to adopt a roughly circular path. As the electrons traverse the bending magnets their acceleration produces light emission that can be exploited in a beamline. Experimental stations are located at the end of each beamline.

smaller than the latter. For example, the highly electronegative Cl atom has a first electron affinity of only 3.7 eV, and this is large by the standards of most atoms and molecules. One can immediately see that if photoelectron spectroscopy is attempted on anions, relatively low photon energies can be used to remove an electron. Indeed, visible lasers, such as a selected single line from an argon ion laser, are often used as the light source.

Apart from the light source, the essential components in negative ion photoelectron spectroscopy are similar to those in conventional photoelectron spectroscopy. The only other experimental difference concerns the production of anions, which can be achieved in a number of ways. The most common source is a gas electrical discharge, which will make a variety of species including neutrals, cations, and anions. The low photon energy will only remove electrons from anions, and so the presence of neutrals and cations will not cause any complications in the spectroscopy. Other sources have used dissociative electron attachment (an electron attaching to a neutral molecule followed by rapid dissociation to form a neutral fragment and an anionic fragment) or sputtering of anions off solid surfaces by energetic particle bombardment.

Negative ion photoelectron spectroscopy provides similar information to conventional photoelectron spectroscopy with one important difference. While traditional photoelectron spectroscopy is performed on neutral atoms or molecules and yields their ionization energies together with spectroscopic constants of the corresponding cations, negative ion photoelectron spectroscopy provides electron affinities and spectroscopic constants of the *neutral species*. Case Study 15 considers a specific example in some detail. There are many examples of neutral molecules that were first studied by negative ion photoelectron spectroscopy. In some cases the spectroscopic constants obtained were subsequently used to guide the search for higher resolution optical spectra of these molecules.

12.4 Penning ionization electron spectroscopy

Penning ionization electron spectroscopy is somewhat similar to photoelectron spectroscopy. The principal difference is that instead of using ultraviolet photons to ionize a sample, ionization is brought about by collisions with metastable excited atoms. Metastable species are atoms or molecules in excited states that have long lifetimes, sometimes as long as several seconds, because the transition back to the ground state is forbidden by one or more optical selection rules. For spectroscopic purposes, metastable noble gas atoms are used, especially Ne ($^3P_2$), which has an energy 16.62 eV above ground state Ne. Metastable noble gas atoms can be produced in a carefully controlled electrical discharge or by using electron impact with a high energy (80–100 eV) electron beam. In both cases, care must be exercised to remove charged particles from the metastable atom beam and to prevent light from the discharge reaching the ionization region. On collision with the sample molecules their excess energy is used for ionization:

$$\mathrm{M} + \mathrm{Ne}^* \rightarrow \mathrm{M}^+ + \mathrm{Ne} + \mathrm{e}^-$$

The electrons are then analysed according to their kinetic energies as in photoelectron spectroscopy. The similarity in experimental conditions is such that it is possible to perform Penning and photoelectron spectroscopy in the same apparatus under identical conditions.

12.5 Zero electron kinetic energy (ZEKE) spectroscopy

ZEKE spectroscopy was introduced in 1984 and has developed into an important spectroscopic technique. It is an example of threshold photoelectron spectroscopy, so-called because the aim is to photoexcite molecules to a specific energy level of the ion (an ionization threshold), which will produce electrons with very low (or, in principle, zero) kinetic energy. A tunable light source is necessary for threshold photoelectron spectroscopy.

The basic idea is shown in Figure 12.3. When the photon energy exactly matches the energy difference between a specific level of the neutral and a specific level of the ion, excitation to that energy level of the ion must produce electrons with zero kinetic energy by conservation of energy. However, the ion may also end up in lower energy levels of the ion (if any are available), and the emitted electron will therefore take up the excess energy.[3] Consequently, some electrons will be produced with zero kinetic energy and others with non-zero kinetic energies. We will call the former ZEKE (pronounced 'zee-kee') electrons. At other wavelengths, where the photon energy does not precisely match a neutral–ion energy level

[3] The probability of populating the various vibrational levels in the ion is, to a good approximation, governed by the Franck–Condon principle.

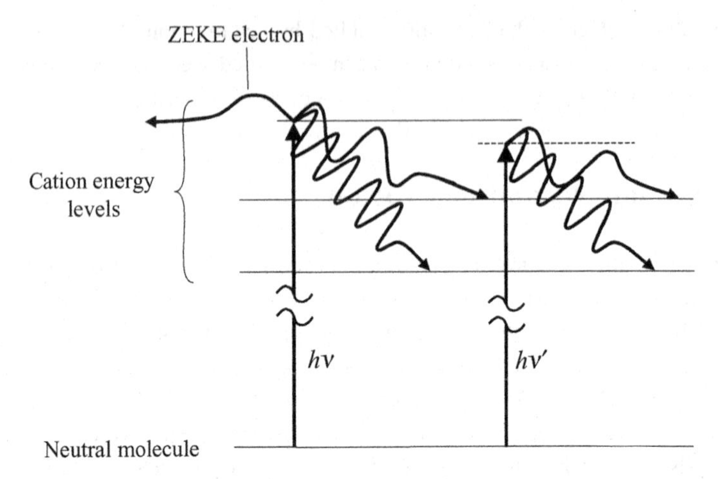

Figure 12.3 Basic principles of threshold ionization. In the process shown on the left, electrons can be produced with both zero and non-zero kinetic energies. However, on the right the energy mismatch between the photon energy and the separation between ion and neutral molecule energy levels means that only electrons with non-zero kinetic energies are produced.

separation, photoionization will still take place but electrons with zero kinetic energy *cannot* be produced. Consequently, if it were possible to preferentially detect ZEKE electrons as a function of the wavelength of the tunable light source, then an ion←neutral excitation spectrum could be obtained. This is the basic idea of threshold photoelectron spectroscopy.

Threshold photoelectron spectroscopy was around for some years before what is now known as ZEKE spectroscopy was introduced. Many clever schemes for discriminating between zero and non-zero kinetic energy electrons were developed, and some of these ideas were subsequently employed in ZEKE spectroscopy. However, while much important work was carried out with the pre-ZEKE forms of threshold photoelectron spectroscopy, there were a number of problems, a notable one being complications caused by autoionization.[4]

The introduction of ZEKE spectroscopy in 1984 combined the use of pulsed lasers for ionization together with a delayed pulsed electrical field method for detecting ZEKE electrons. The basic idea is simple and is illustrated in Figure 12.4. Suppose a tunable pulsed laser is capable of ionizing a molecule. As its wavelength is scanned it will move in and out of resonance with ion←neutral transitions. Any ZEKE electrons produced will be stationary whereas non-ZEKE electrons (i.e. moving) will drift rapidly out of the original ionization volume. It is possible to discriminate between ZEKE and non-ZEKE electrons by applying a pulsed electric field across the ionization volume. This is achieved by sandwiching the ionization volume between two conducting plates. If the electric field is initially zero, and is then pulsed on shortly *after* the laser pulse, then electrons will be attracted towards the more positively charged plate. Furthermore, if there is a small hole in this positive plate then the accelerated electrons can pass out of this region and go on to reach the detector. However,

[4] Autoionization is a spontaneous ionization process that can occur for neutral molecules in excited electronic states lying above the lowest ionization limit.

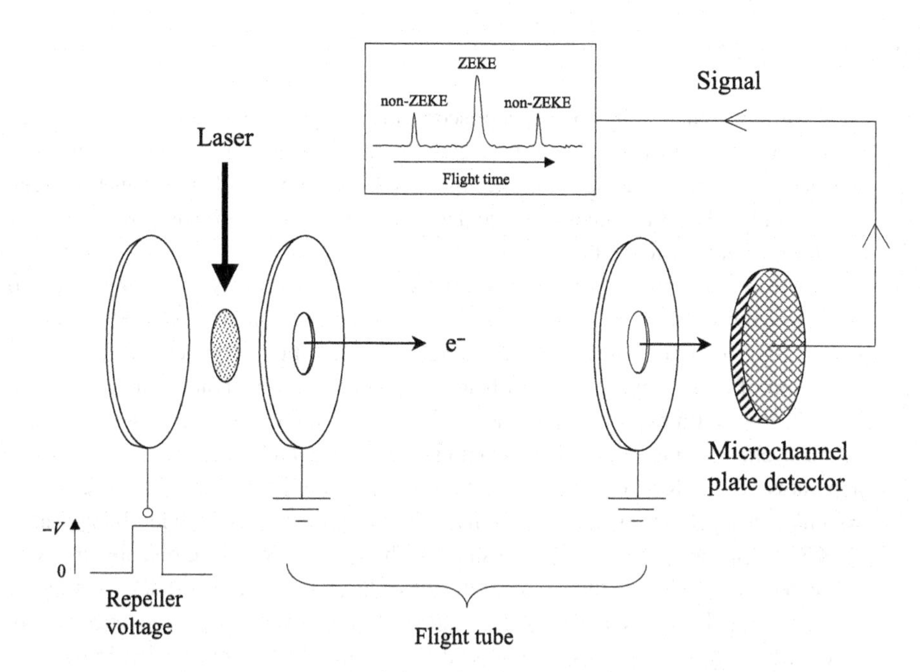

Figure 12.4 Basic arrangement for a ZEKE experiment. This assembly is mounted inside a high vacuum chamber. Non-ZEKE electrons are emitted in all possible directions and drift away from the initial ionization volume. The voltage pulse, applied after a predetermined delay of *c.* 1 μs, sends both ZEKE and non-ZEKE electrons into the flight tube. The non-ZEKE electrons that are detected are those moving towards or away from the flight tube along the central axis prior to application of the voltage pulse: all non-ZEKE electrons with an off-axis velocity component drift too far away from the ionization volume and when the voltage pulse is switched on are unable to pass through the aperture in the middle plate. The non-ZEKE electrons that do reach the detector are easily distinguished from the ZEKE electrons by their different flight times, as shown in the inset.

the movement of the non-ZEKE electrons means that these will start from different positions in the inter-plate region when the electric field is pulsed on compared with ZEKE electrons. Consequently, their arrival times will differ, as illustrated in the inset of Figure 12.4.

Most early ZEKE experiments used resonance-enhanced multiphoton ionization with pulsed tunable dye lasers rather than single-photon ionization. However, there have now been many single-photon ionization ZEKE experiments with VUV laser radiation generated from specialized harmonic generation processes or from synchrotron radiation. In both cases the time delay between pulsed laser ionization and application of the pulsed electric field is normally in the region of 1 μs. Although both ZEKE and non-ZEKE electrons are registered by the electron detector, all non-ZEKE signals are subsequently discarded.

The great advantage of ZEKE spectroscopy is that electron energy analysis is not required. Consequently, the primary cause of the low resolution in conventional photoelectron spectroscopy disappears. In some of the most favourable cases a resolution as good as 0.2 cm^{-1} has been achieved, allowing rotationally resolved structure to be obtained.

This technique may be used for neutrals or anions, but for the former a much more robust variant is possible and is described in the following section.

12.6 ZEKE–PFI spectroscopy

ZEKE–PFI is a variant of ZEKE spectroscopy and nowadays the distinction between the two is often ignored. The role of the pulsed electric field in ZEKE spectroscopy is to accelerate the electrons towards the detector once the ZEKE and non-ZEKE electrons have separated in space. In ZEKE–PFI, the electric field actually causes ionization, hence the abbreviation PFI for pulsed field ionization.

Molecules contain a large number of energy levels, known as Rydberg levels, close to their ionization limit. A molecule in a Rydberg state has an electron in an orbital that is so diffuse that it 'sees' the ionic core as virtually a point charge. Thus a Rydberg orbital is similar to an orbital of atomic hydrogen. Just below the ionization threshold, within approximately 5 cm^{-1}, the lifetimes of molecular Rydberg states become quite long, sometimes tens of microseconds, owing to interactions with small electric and magnetic fields present in the apparatus. These Rydberg states can be ionized by application of a small electric field, which pulls the Rydberg electron free from the ion core. This is the underlying principle of ZEKE–PFI spectroscopy. Instead of using threshold ionization, the molecules are excited up to Rydberg levels close to threshold by the pulsed laser(s). After a short delay, a microsecond or so, a small electric field is pulsed on and the Rydberg states are *field-ionized* to produce zero kinetic energy electrons. Of course direct ZEKE electron production is also possible, but the ZEKE and ZEKE–PFI electrons can be distinguished by their arrival times at the detector by application of a very small dc electric field across the ionization region prior to the field ionizing pulse.

ZEKE and ZEKE–PFI spectra contain essentially the same information. There is also another variant of the ZEKE method that has recently been introduced, mass analysed threshold ionization (MATI), in which ions rather than electrons are detected. As the name implies, this allows one to record a mass-selected ZEKE spectrum.

A detailed account of ZEKE and related spectroscopic techniques has been given by Müller-Dethlefs and Schlag [1].

Reference

1. K. Müller-Dethlefs and E. W. Schlag, *Angew. Chemie Int. Ed.* **37** (1998) 1346.

Further reading

Further information on spectroscopic techniques can be found in the following books.

Laser Spectroscopy, 3rd edn., W. Demtröder, Berlin, Springer-Verlag, 2002.
Photoelectron Spectroscopy, J. H. D. Eland, London, Butterworths, 1984.
Principles of Ultraviolet Photoelectron Spectroscopy, J. W. Rabalais, New York, Wiley, 1977.

Part III

Case Studies

13 Ultraviolet photoelectron spectrum of CO

Concepts illustrated: *vibrational structure and Franck–Condon principle; adiabatic and vertical ionization energies; Koopmans's theorem; link between photoelectron spectra and molecular orbital diagrams; Morse potentials.*

Carbon monoxide was one of the first molecules studied by ultraviolet photoelectron spectroscopy [1]. A typical HeI spectrum is shown in Figure 13.1.[1] The spectrum appears to be clustered into three band systems. The starting point for interpreting this spectrum is to consider the molecular orbitals of CO and the possible electronic states of the cation formed when an electron is removed.

13.1 Electronic structures of CO and CO^+

Any student familiar with chemical bonding will almost certainly be able to construct a qualitative molecular orbital diagram for a diatomic molecule composed of first row atoms. Such a diagram is shown for CO in Figure 13.2. The orbital occupancy corresponds to the ground electronic configuration $1\sigma^2 2\sigma^2 3\sigma^2 4\sigma^2 1\pi^4 5\sigma^2$. The σ MOs actually have σ^+ symmetry but it is not uncommon to see the superscript omitted. Since all occupied orbitals are *fully* occupied, the ground state is therefore a ${}^1\Sigma^+$ state and, since it is the lowest electronic state of CO, it is given the prefix *X*, i.e. $X^1\Sigma^+$, to distinguish it from higher energy ${}^1\Sigma^+$ states of CO.

Consider the electronic states of the cation formed by removing an electron. If the electron is removed from the highest occupied molecular orbital (HOMO), the 5σ orbital, then the cation will be in a ${}^2\Sigma^+$ state. Since this is expected to be the lowest energy state of the cation, it is therefore labelled $X^2\Sigma^+$. Removing an electron from the 1π or 4σ MOs gives ${}^2\Pi$ and ${}^2\Sigma^+$ states, respectively. From the orbital ordering in the MO diagram, our expectation is that these two states are the lowest energy *excited* electronic states of CO^+ and so will be labelled as the $A^2\Pi$ and $B^2\Sigma^+$ states.

[1] HeI radiation has a wavelength of 58.4 nm ($\equiv$ 21.2 eV) – further details can be found in Section 12.1.

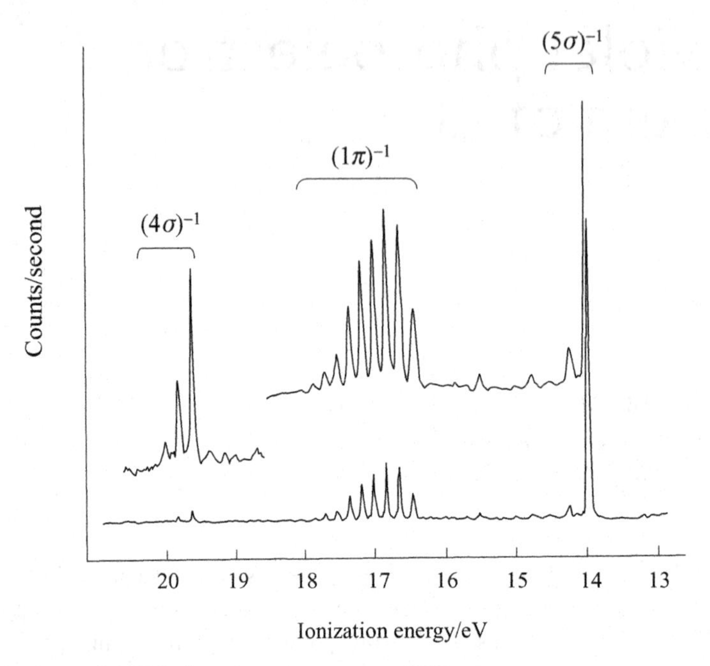

Figure 13.1 HeI photoelectron spectrum of CO.

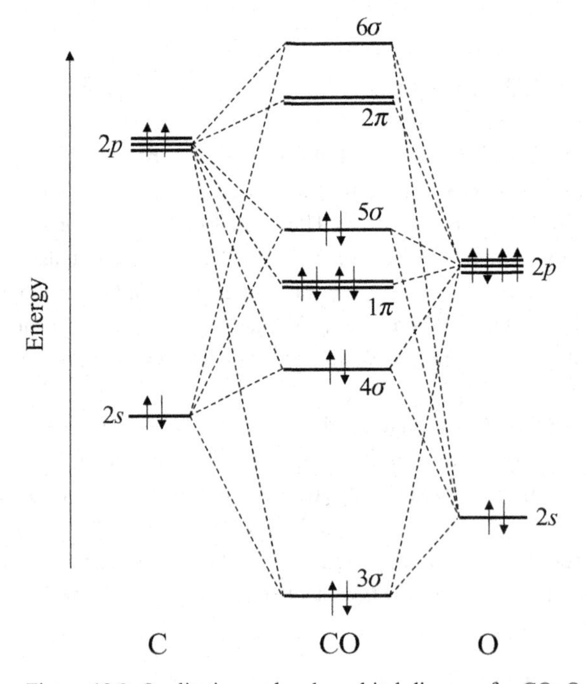

Figure 13.2 Qualitative molecular orbital diagram for CO. Only the valence orbitals are shown, i.e. the 1σ and 2σ orbitals formed by overlap of the $1s$ orbitals of C and O have been omitted since they are core orbitals and cannot be photoionized by HeI radiation.

13.2 First photoelectron band system

From the discussion above, the first photoelectron band envelope is expected to arise from the ionization process

$$CO^+(X\,^2\Sigma^+) + e^- \leftarrow CO(X\,^1\Sigma^+)$$

A strong peak is observed at 14.01 eV and is followed by much weaker peaks at higher ionization energies. The resolution is far too low (*c.* 50 meV $\equiv$ 400 cm^{-1}) to resolve rotational structure, so any structure within this band must be vibrational in origin. The lack of prominent vibrational structure is indicative of little change in the C—O bond length on ionization: this follows from the Franck–Condon principle (see Section 7.2.2) and suggests that ionization is from a non-bonding orbital. In such a case the potential energy curves for the neutral molecule and the cation will look very similar. Consequently, the strongest feature must arise from the 0 $\leftarrow$ 0 transition, where the two numbers refer to the vibrational quantum number in the ion and neutral molecule, respectively.

The weak peak at 14.28 eV is due to the 1 $\leftarrow$ 0 transition. The difference in energy between the first and second peaks corresponds to the energy difference between the $v = 0$ and $v = 1$ vibrational levels in the cation in its $X^2\Sigma^+$ state. Converting to wavenumbers (1 eV $\equiv$ 8066 cm^{-1}) gives a separation of 2180 cm^{-1}. By feeding the quantum numbers into equation (5.14), this separation is found to be equivalent to $\omega_e - 2\omega_e x_e$ for the ion. Without observing further members of the progression, it is impossible to deduce both ω_e and $\omega_e x_e$. However, $\omega_e x_e$ is normally much smaller than ω_e and so it is a reasonable approximation to associate ω_e with the observed vibrational interval.

For the ground state of CO, infrared spectroscopy has yielded $\omega_e = 2170$ cm^{-1}. This is similar to the interval in the first photoelectron band and implies very similar force constants. This in turn is consistent with the suggestion made earlier that the potential energy curves in the ground electronic states of the neutral molecule and the cation are very similar. The conclusion is therefore that the 5σ orbital is mainly non-bonding.

13.3 Second photoelectron band system

The second band system has a very different intensity profile from that of the first. A regular vibrational progression is formed in which the first member, at 16.53 eV, is not the most intense. This immediately indicates a substantial change in the C—O bond length on ionization, and consequently the neutral and cationic potential energy curves are displaced with respect to each other.

The separation between adjacent members of the progression is measured to be ~1530 cm^{-1} which, employing the argument made above, approximates to the harmonic vibrational frequency. There is clearly a large decrease in vibrational frequency upon ionization to the first excited state of the ion, demonstrating that the electron removed is strongly bonding. Assuming the validity of our earlier MO model, the ionization is from the 1π MO and the resulting state of the ion is the $A^2\Pi$ state. The conclusion is that the 1π orbital is strongly bonding.

In principle, further information can be extracted from the vibrational progression. Due to anharmonicity, the vibrational interval should decrease as the ionization energy increases. If the peak positions are measured to sufficient precision, it should be possible to determine both ω_e and $\omega_e x_e$. Using the term value expression given in equation (5.14), a vibrational term *interval* can be derived as

$$\Delta G_{v+1/2} = G(v+1) - G(v) = \omega_e - 2\omega_e x_e(v+1) \qquad (13.1)$$

where v is the vibrational quantum number in the cation for the lower of the two adjacent peaks. Consequently, if $\Delta G_{v+1/2}$ is plotted against $(v+1)$, then ω_e and $\omega_e x_e$ can be obtained from the intercept and slope, respectively. Unfortunately, the resolution is so poor that there is insufficient precision to obtain any more than a rough value of $\omega_e x_e$. Consequently, we will not pursue this any further.

13.4 Third photoelectron band system

The third band in Figure 13.2 bears qualitative resemblance to the first band. There is clearly no major change in C—O bonding on ionization to the second excited state of CO^+, although the Franck–Condon activity is greater than in the first photoelectron band system. The vibrational frequency of the $B^2\Sigma^+$ state of the ion is found to be 1690 cm^{-1} from the short progression in the third band. This is not as low as the $A^2\Pi$ state, but it is substantially below that of the neutral molecule. Assuming that $(4\sigma)^{-1}$ ionization is responsible, the conclusion reached is that the 4σ orbital possesses some bonding character but not as much as the 1π orbital.

13.5 Adiabatic and vertical ionization energies

In the first member of a vibrational progression, the ion is formed in the zero point vibrational level, $v = 0$. The corresponding ionization process is said to be the *adiabatic* ionization transition, so-called because the ion has no excess vibrational energy. The most intense vibrational component is said to be due to a *vertical* ionization, because it most closely corresponds to the vertical transition in a classical picture of the Franck–Condon principle.

In the first and third band systems in the photoelectron spectrum of CO, the adiabatic and vertical ionization energies are one and the same. However, the vertical and adiabatic ionization energies do not coincide for the second band system because of the substantial change in C—O bond length on ionization.

According to Koopmans's theorem, the negative of the ith *vertical* ionization energy (IE_i) can be equated with the energy (ε_i) of the ionizing orbital. This result, which can be derived from Hartree–Fock theory (see Appendix B), is exceedingly useful since it provides a means of quantifying the energy scale on an MO diagram.

However, it is important to recognize the limitations of Koopmans's theorem. First, it applies only to closed-shell molecules. One of the complications with open-shell molecules is that more than one ionic state may result from removal of an electron from a specific orbital. In such circumstances more than one vertical ionization energy is associated with the orbital, making Koopmans's theorem meaningless. Even for closed-shell molecules there are problems with Koopmans's theorem. It assumes that orbital energies are the same in the ion and the neutral molecule. However, this is not the case in practice, and nor would one expect it to be since the loss of an electron will usually reduce the e–e repulsion and lead to more tightly bound orbitals. In a more realistic model the link between ionization energy and orbital energy must be modified to

$$IE_i = -\varepsilon_i + \Delta_i \tag{13.2}$$

where Δ_i is an *orbital relaxation energy* to account for the change in orbital energy from neutral molecule to the ion. Since the relaxation energy may differ from one orbital to another, the HOMO in the neutral molecule may no longer be the HOMO in the ion. In other words the ordering of orbitals in terms of energy may switch on ionization, especially if there are two or more orbitals that have quite similar energies. This does not occur for CO but it is known to occur for N_2, which is isoelectronic with CO. In fact for N_2 accurate Hartree–Fock calculations show that the π orbital is the HOMO, but in the cation this switches and the π orbital lies below the highest occupied σ orbital.

The comments made in this section are intended to provide a sense of perspective. It is convenient to invoke a simple MO model to explain photoelectron (and electronic) spectra, as was done above. However, one must also be prepared to recognize its limitations.

13.6 Intensities of photoelectron band systems

If the relative intensities of the band systems depended solely on the populations of the orbitals from which photoionization occurs, then the second system of CO would be twice as strong as the first and third systems because of the twofold degeneracy of π orbitals. To compare intensities, it is necessary to sum over all vibrational components. In general, areas under each vibrational band should be summed but, if the all the bands have approximately the same widths,[2] then it is sufficient to sum peak heights.

It is found that the first band system is marginally more intense than the second, and both are far more intense than the third. Clearly there are factors influencing the intensities other than just orbital populations. One factor is the transmission of the electron energy analyser, which may be a strong function of electron kinetic energy. For electrostatic dispersion analysers, as used to record the spectrum in Figure 13.1, the ability to transmit electrons to the detector falls markedly as the electron kinetic energy approaches low values.

In addition, there are quantum mechanical effects that influence photoionization probabilities. The transition moment expression (7.13) applies but the upper state wavefunction is more complicated than in electronic spectroscopy because it involves both the molecular ion and the free electron. Factors such as the energy and angular momentum of the free electron can have a major effect on the photoionization probability and it is often found that this is a strong function of the photon energy. For a detailed discussion of photoelectron band intensities the interested reader should consult the book by Rabalais [2].

[2] By widths we mean full-widths at half-maximum. See Figure 9.1 for more details.

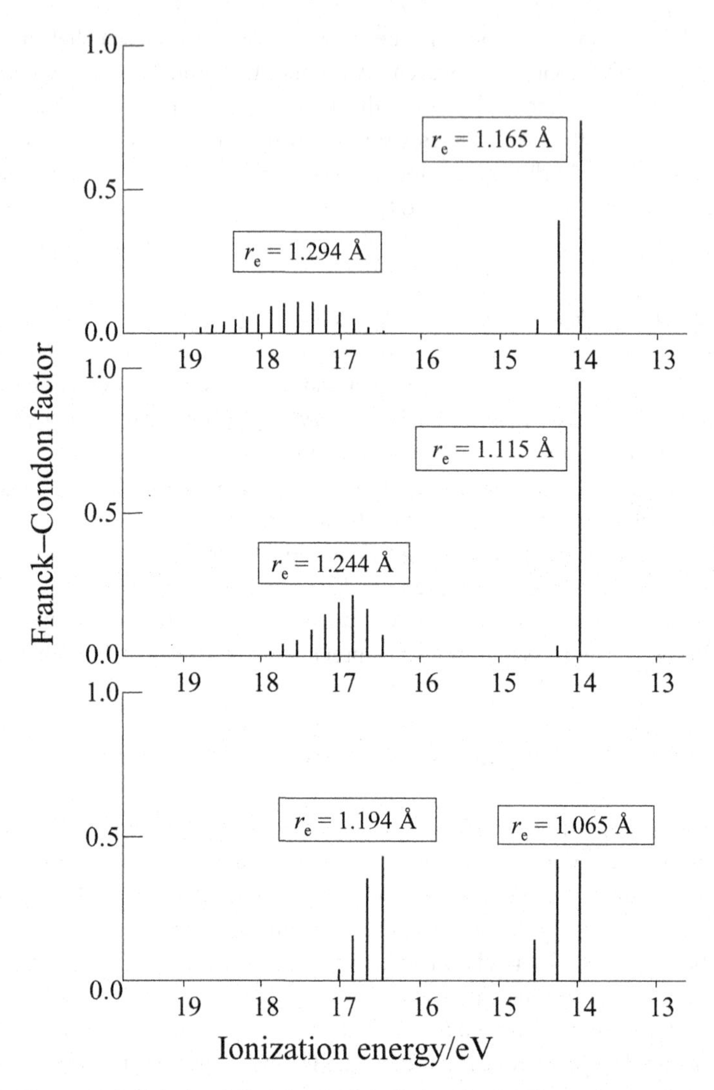

Figure 13.3 Calculated Franck–Condon factors for the first and second photoelectron bands of CO using different values of the equilibrium bond length in CO^+. Literature values for r_e and ω_e in the neutral molecule were assumed. Best agreement with experiment (Figure 13.1) is obtained for the middle spectrum, for which $r_e = 1.115$ Å in the $\tilde{X}$ state of the ion and 1.244 Å for the $\tilde{A}$ state.

13.7 Determining bond lengths from Franck–Condon factor calculations

Although rotational structure cannot be resolved in ordinary photoelectron spectroscopy, it is still possible to deduce the bond length of the ion, albeit with modest precision. This can be achieved by comparing calculated Franck–Condon factors with those determined from experiments.

Suppose that the equilibrium bond length (r_e), the harmonic vibrational frequency (ω_e), and the anharmonicity constant (x_e) are known for electronic states of both the neutral molecule and the ion. This is sufficient information to be able to calculate vibrational wavefunctions for these states, providing the potential energy in each state can be adequately represented by Morse potentials.[3] A Morse potential (equation (5.12)) is completely defined by three parameters, D_e, r_e, and a. It can be shown that D_e and a are linked to ω_e and x_e by the expressions

$$D_e = \frac{\omega_e}{4x_e} \tag{13.3}$$

$$a = 2\pi\sqrt{\frac{2c\mu\omega_e x_e}{h}} \tag{13.4}$$

The vibrational Schrödinger equation can be solved using a Morse potential to determine the vibrational energies and wavefunctions. Although this can be done analytically, it is also trivial to do using a numerical procedure on a computer. The advantage of the numerical approach is that the overlap integral in the Franck–Condon factor (7.14) is also easy to evaluate numerically, e.g. using Simpson's rule.

The quantities r_e, ω_e, and x_e are normally known to high precision for the ground state of a neutral molecule from techniques such as microwave spectroscopy or rotationally resolved infrared spectroscopy. For the ion, the photoelectron spectrum will yield a reasonable estimate of ω_e, as seen for CO^+. The anharmonicity constant may be more difficult to determine, but a precise value for this is not particularly important *unless* transitions to relatively high vibrational levels in the ion have significant probability (since the anharmonicity determines the slope of the curve on the approach to dissociation). Consequently, since ω_e is known and x_e can be estimated, the only unknown is r_e. This can therefore be used as a trial parameter from which Franck–Condon factors are calculated and compared with the actual relative intensities of the vibrational components in a given photoelectron band. When the best possible agreement is found, a good estimate of the bond length of the ion can be obtained. This approach to estimating ion bond lengths is illustrated in Figure 13.3, where calculated Franck–Condon factors are shown for selected values of the bond lengths of the ground and first excited electronic states of CO^+.

References

1. *Molecular Photoelectron Spectroscopy*, D. W. Turner, C. Baker, A. D. Baker and C. R. Brundle, London, Wiley, 1970.
2. *Principles of Ultraviolet Photoelectron Spectroscopy*, J. W. Rabalais, New York, Wiley, 1977.

[3] The potential energy curves of many electronic states are quite good approximations to Morse potentials, except in the region very close to dissociation. However, it is also worth bearing in mind that there are some states where a Morse potential is known to be a poor approximation even in the region near the potential minimum.

14 Photoelectron spectra of CO_2, OCS, and CS_2 in a molecular beam

Concepts illustrated: *supersonic expansion cooling; adiabatic and vertical ionization energies; vibrational structure in the spectra of triatomic molecules; Franck–Condon principle; link between photoelectron spectra and molecular orbital diagrams.*

A severe restriction of conventional photoelectron spectroscopy is its low resolution. The main limitation is instrumental resolution, particularly that caused by the electron energy analyser, as was discussed in Chapter 12. Resolving rotational structure is not a realistic prospect for conventional photoelectron spectroscopy but even vibrational structure may be difficult to resolve. In addition to the instrumental resolution must be added other factors such as rotational and Doppler broadening which, if they could be dramatically reduced, might make a sufficient difference to improve many photoelectron spectra. A potential solution is to combine conventional photoelectron spectroscopy with supersonic molecular beams. Supersonic expansions can produce dramatic cooling of rotational degrees of freedom and, if part of the expansion is skimmed into a second vacuum chamber, can be converted to a beam with a very narrow range of velocities. This is precisely the approach adopted by Wang *et al.* [1], the molecular beam being crossed at right angles by HeI VUV radiation (58.4 nm) to produce a near Doppler-free photoelectron spectrum. The resolution achieved is in the region of 12 meV (100 cm^{-1}).

The ultraviolet photoelectron spectra of CO_2, OCS, and CS_2 in molecular beams are discussed here. These illustrate some of the important concepts involved in the interpretation of the photoelectron spectra of polyatomic molecules. They are clearly related molecules and therefore some similarities in their photoelectron spectra are to be expected. Figure 14.1 shows the overall HeI spectrum for each molecule. It would be inappropriate to discuss every aspect of the spectrum of each molecule. Instead the focus is on the main bands and we shall try to discover what each reveals about both the neutral molecule and the corresponding molecular ion.

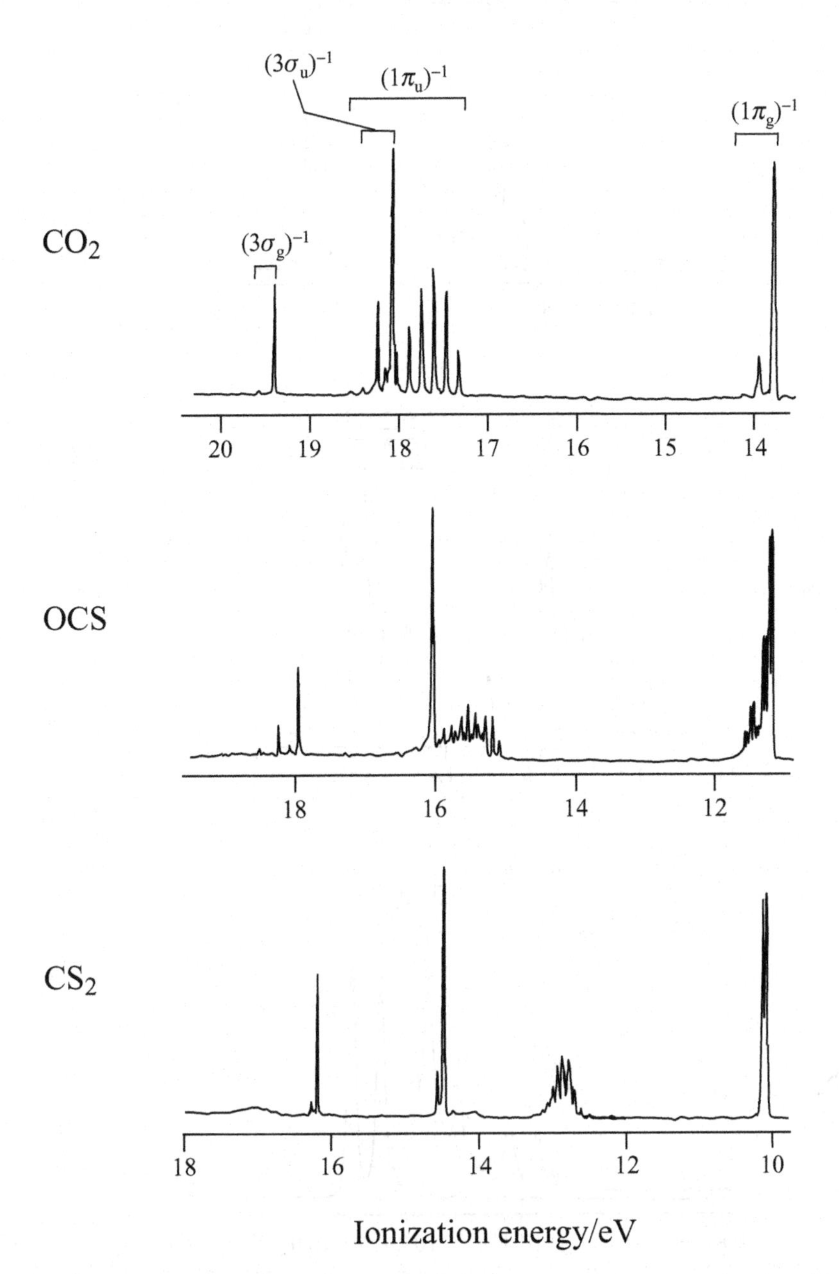

Figure 14.1 Overall view of the HeI photoelectron spectra of CO_2, OCS, and CS_2. Justification for the orbital ionization assignments above the CO_2 spectrum is given in the text. Similar assignments apply to OCS and CS_2, although for the former molecule the g and u subscripts on the orbital symmetries are no longer applicable because OCS lacks a centre of symmetry. (Reproduced from L.-S. Wang, J. E. Reutt, Y. T. Lee, and D. A. Shirley, *J. Elec. Spec. Rel. Phen.* **47** (1988) 167, with permission from Elsevier.)

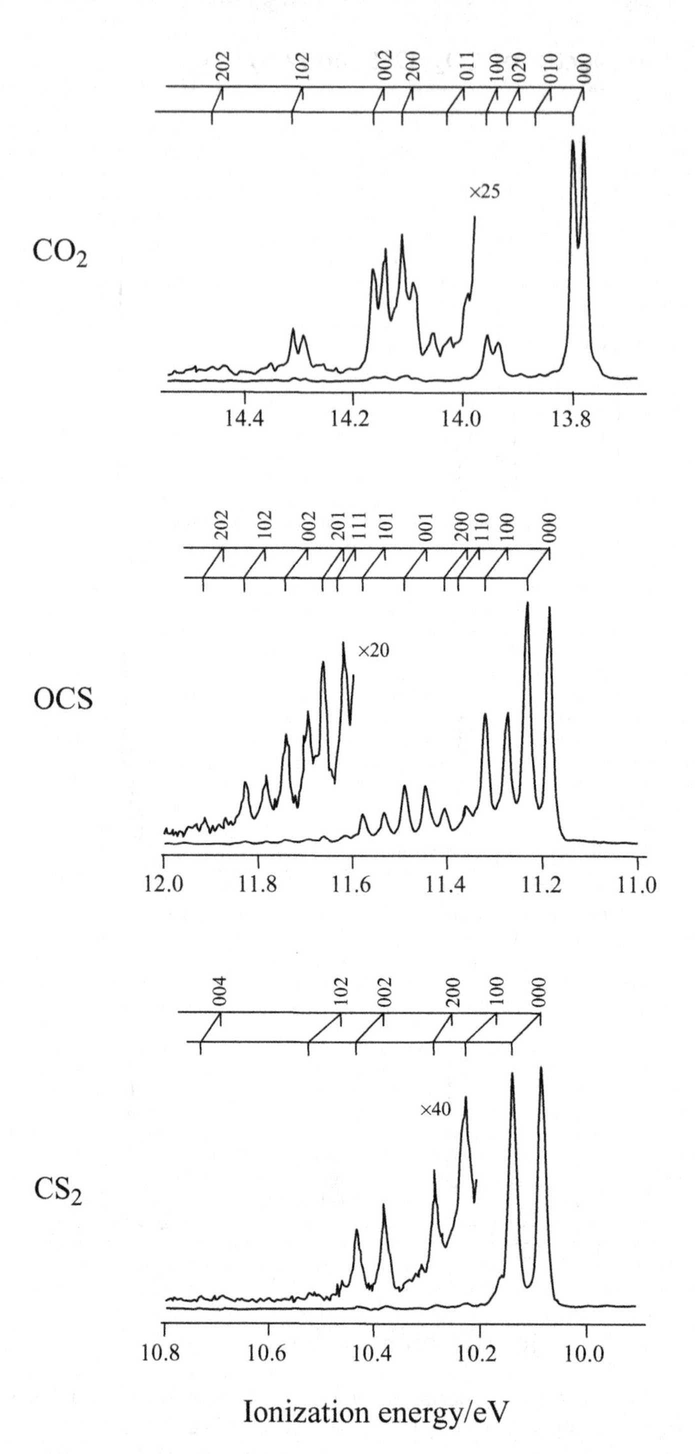

Figure 14.2 Expanded views of the first photoelectron bands of CO_2, OCS, and CS_2. The labels *lmn* above the peaks refer to the vibrational quantum numbers in the ion, where *l* is the vibrational quantum number for mode v_1, *m* for v_2, and *n* for v_3. All peaks originate from the zero-point vibrational levels in the respective neutral molecules, i.e. hot band contributions are negligible in these spectra. (Reproduced from L.-S. Wang, J. E. Reutt, Y. T. Lee, and D. A. Shirley, *J. Elec. Spec. Rel. Phen.* **47** (1988) 167, with permission from Elsevier.)

14.1 First photoelectron band system

Figure 14.2 shows expanded views of the first photoelectron band system of each molecule. Consider, initially, the spectrum of CO_2. There is little vibrational structure associated with the first band system, indicating that the electron removed on photoionization possessed mainly non-bonding character in the neutral molecule. The adiabatic peak, which is also the vertical peak,[1] shows a clear doublet splitting, as do all of the weak vibrational components to higher ionization energy, indicating spin–orbit coupling in the ion (0.02 eV $\approx$ 160 cm^{-1}). The ground electronic state of the neutral CO_2 molecule, which is discussed in more detail later, is a spin singlet and in fact has $^1\Sigma_g^+$ symmetry. In order for spin–orbit coupling to occur the molecule must have electronic orbital angular momentum and so it is reasonable to conclude that a $^2\Pi$ cationic state has been formed upon ionization, i.e. an electron has been removed from a π orbital (but note we cannot deduce whether it has g or u symmetry on the basis of this information alone).

The most prominent vibrational feature in the first photoelectron band system is 0.157 eV ($\sim$1270 cm^{-1}) above the adiabatic ionization energy, as measured from the mid-points of the corresponding spin–orbit doublets. Following the arguments presented in Section 7.2.3, the dominant vibrational features in the electronic and photoelectron spectra of polyatomic molecules are usually from excitation of totally symmetric vibrational modes. Linear CO_2 has only one totally symmetric vibrational mode, the symmetric stretch (see Section 5.2.1), which is normally designated by the shorthand notation ν_1. Other spectroscopic studies have shown that this mode has a harmonic frequency of 1388 cm^{-1} for the ground state of the neutral molecule. This is similar to the main observed vibrational interval in the first photoelectron band, and it is therefore logical to assign that progression to ν_1. The fact that the frequency change is modest is consistent with the lack of extensive vibrational structure, and leads to the conclusion that there is no significant change in bonding, and therefore molecular structure, on photoionization.

There are other very weak peaks in the first photoelectron band system of CO_2. The next member in the progression in ν_1, labelled 200 in Figure 14.2,[2] is observed. Near to the 200 doublet is a weak doublet assigned as double quantum excitation in ν_3, a transition which is Franck–Condon allowed but which we would predict to be very weak, as indeed it is. The combination feature 102, which also has double quantum excitation of ν_3, can also be seen. In addition, notice that there is some evidence of single quantum excitation of ν_2 and ν_3, namely the 010 and 011 transitions, which are formally forbidden. If these assignments are correct, and there is copious evidence from several studies that they are, then they must gain their intensities through vibronic coupling, which represents a breakdown of the Born–Oppenheimer approximation (and therefore the Franck–Condon principle). Vibronic coupling is discussed in more detail later in several Case Studies, e.g. Chapter 25.

1 For definitions of adiabatic and vertical ionization energies, see the previous Case Study.

2 An alternative way of labelling this peak would be as 1_0^2, which indicates that mode ν_1 has zero quanta in the lower state and two quanta in the upper state. The absence of any reference to other modes is taken as implying that there are zero quanta in all other modes in both upper and lower electronic states. The combination feature 102 would be labelled $1_0^1 3_0^2$ in this scheme.

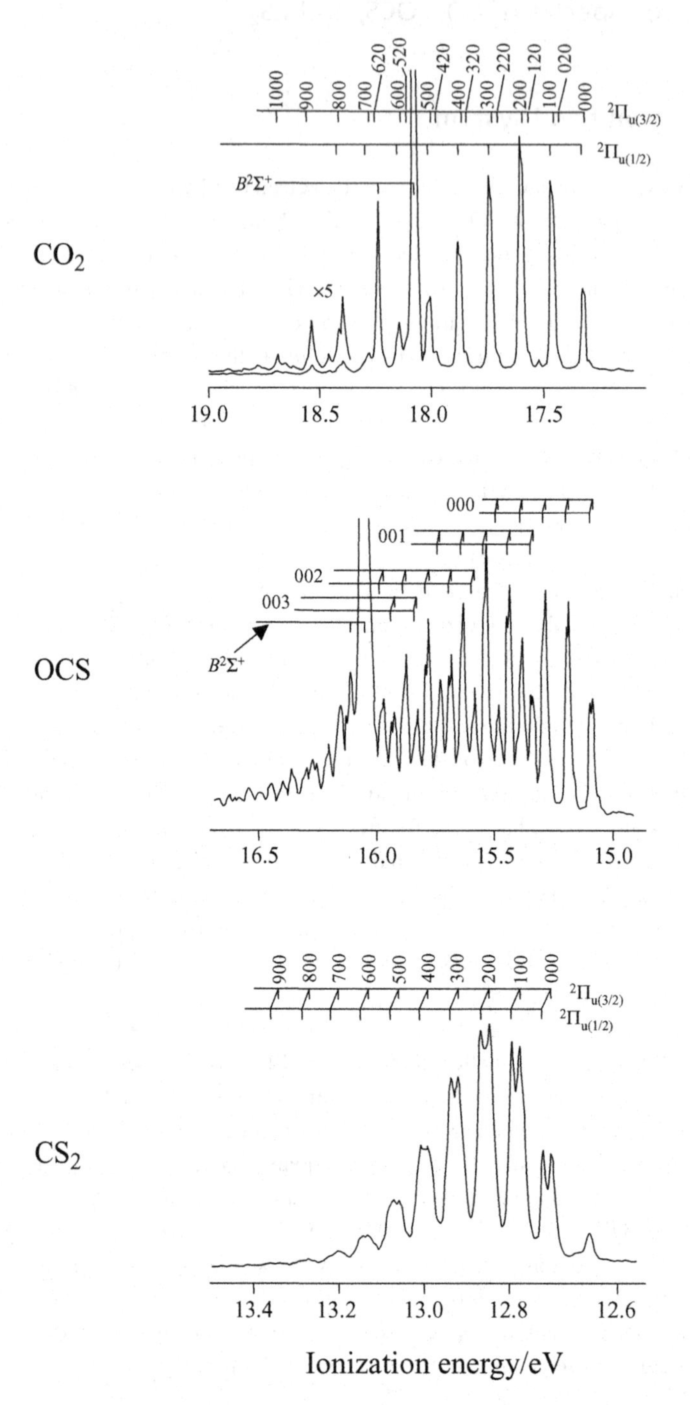

Figure 14.3 Expanded views of the second photoelectron bands of CO_2, OCS, and CS_2. As in Figure 14.2, the labels *lmn* above the peaks refer to the vibrational quantum numbers in the ion, where l is the vibrational quantum number for mode ν_1, m for ν_2, and n for ν_3. For CO_2 and OCS the third photoelectron band system (forming the $\tilde{B}\,^2\Sigma^+$ state of the cation) overlaps with the second photoelectron band system. (Reproduced from L.-S. Wang, J. E. Reutt, Y. T. Lee, and D. A. Shirley, *J. Elec. Spec. Rel. Phen.*, **47** (1988) 167, with permission from Elsevier.)

The first photoelectron band system of CS_2 is simpler than that of CO_2. There is a doubling of peaks attributable to spin–orbit coupling in the ion, but the splitting (440 cm^{-1}) is considerably larger than for CO_2. This is not surprising given the substitution of sulfur for oxygen: atomic spin–orbit coupling increases rapidly as the atomic number increases, and therefore if the unpaired electron density on the sulfur atoms in CS_2^+ is quite high then the molecular spin–orbit splitting will be larger than in CO_2^+. Put in reverse, the increase in spin–orbit splitting from CO_2 to CS_2 reveals that the unpaired electron in the ground state of the ion spends much of its time on the sulfur atoms. The vibrational structure in the first band of CS_2 can be interpreted in much the same way as for CO_2, and this is left as an exercise for the reader.

Turning to OCS, linearity is maintained, and so spin–orbit coupling still occurs in the excited state, with the splitting, 370 cm^{-1}, being somewhat intermediate between CO_2 and CS_2. However, an important difference between OCS and the other two molecules is the effect its lower symmetry ($C_{\infty v}$) has on the vibrational structure. In particular *both* stretching modes, ν_1 and ν_3, are now totally symmetric (see Figure 5.6). Consequently, single quantum excitation in these modes is possible and substantial Franck–Condon activity might occur in both. In fact progressions in both stretching modes are seen in Figure 14.2. The main vibrational features are formed from the strong spin–orbit doublet at 11.273 and 11.319 eV, and the weaker but still prominent doublet at 11.443 and 11.489 eV. If these are due to single quantum excitation of different modes, as indicated in the label above the figure, then one must represent the C=O stretch and the other the C=S stretch given the vibrational selection rules.[3] Assuming the force constants for the two bonds are similar, then the C=S stretch will have the lower frequency on account of its larger reduced mass. Thus fundamental frequencies of 710 and 2080 cm^{-1} are deduced for the C=S and C=O stretches in the ground electronic state of OCS^+. There are several other very weak vibrational peaks of OCS in Figure 14.2, and these are relatively straightforward to assign.

14.2 Second photoelectron band system

The second photoelectron band systems are shown in Figure 14.3. For all three molecules far more extensive vibrational structure is seen than in the first photoelectron band systems, and this time the adiabatic and vertical ionization energies no longer coincide. An immediate conclusion is that a substantial change in equilibrium structure occurs on ionization to the first excited electronic states of the cations. All three bands also show evidence of spin–orbit coupling, although the splitting is not fully resolved for any of the molecules and is only clear for CS_2. Nevertheless, this shows that the cation is, as in the first photoelectron band system, formed in a $^2\Pi$ state. Furthermore, the occurrence of spin–orbit structure is only possible if the cation, like the neutral molecule, is linear at equilibrium. Any significant

[3] The description of the two stretching modes of OCS as being C=S and C=O is only approximate (see Section 5.2.1). Notice also that the labelling of the C=S stretch as ν_1 rather than the C=O stretch is illogical: the normal labelling procedure is to assign the ν_1 label to the mode of highest frequency *and* highest symmetry. However, this notation for OCS has persisted in the literature and is employed in Figure 14.2.

deviation from linearity would quench the orbital angular momentum due to the loss of π orbital degeneracy.

The vibrational structure for CO_2 and CS_2 is particularly simple, being dominated by a fairly long progression in a single mode. The only totally symmetric mode is the symmetric stretch, ν_1, and so the progression is assigned to this mode. The separations between adjacent peaks in CO_2 and CS_2 spectra are *c.* 1130 and 590 cm^{-1}, respectively, considerably smaller than for the ground state of the ion, which indicates a substantial weakening of the C=O and C=S bonds.

As in the first photoelectron band, both C=O and C=S stretching vibrations are Franck–Condon allowed in OCS and we might expect, and actually see (Figure 14.3), substantial progressions in *both* modes.

14.3 Third and fourth photoelectron band systems

These systems are characterized by a lack of extensive vibrational structure (see Figure 14.1) and therefore must, like the first band system, be the result of removing an electron from a molecular orbital with little bonding or antibonding character. There is no evidence of spin–orbit splitting in the bands, and therefore we can tentatively conclude that they arise from removal of electrons from σ orbitals, leading to $^2\Sigma$ states in the cation. Detailed discussions of the structure can be found in the original research papers [1, 2].

14.4 Electronic structures: constructing an MO diagram from photoelectron spectra

The photoelectron data can be used to construct a quantitative molecular orbital diagram for each molecule. The basis for this is Koopmans's theorem, which states that the orbital energy is equal to the negative of the vertical ionization energy for a closed-shell molecule. The formation of double bonds in the ground state of each neutral molecule means that all occupied orbitals are full. These molecules are therefore closed-shell and so Koopmans's theorem will apply. The ground electronic states are $^1\Sigma_g^+$ for CO_2 and CS_2 and $^1\Sigma^+$ for OCS.

The photoelectron spectra show that HeI radiation is capable of photoionizing four MOs in each molecule. According to Koopmans's theorem, there are therefore four MOs with orbital energies > -21.22 eV. The first ionization energy corresponds to removal of an electron from a largely non-bonding orbital, which we deduced earlier to be of π symmetry because of the observation of spin–orbit splitting in the corresponding photoelectron band system. For similar reasons, the next ionization process also involves removal of a π electron, although the extensive vibrational structure, and in particular the substantial decrease in stretching vibrational frequencies upon ionization, suggests that this orbital is strongly bonding. The third and fourth bands correspond to removal of electrons from largely non-bonding σ MOs, as mentioned earlier.

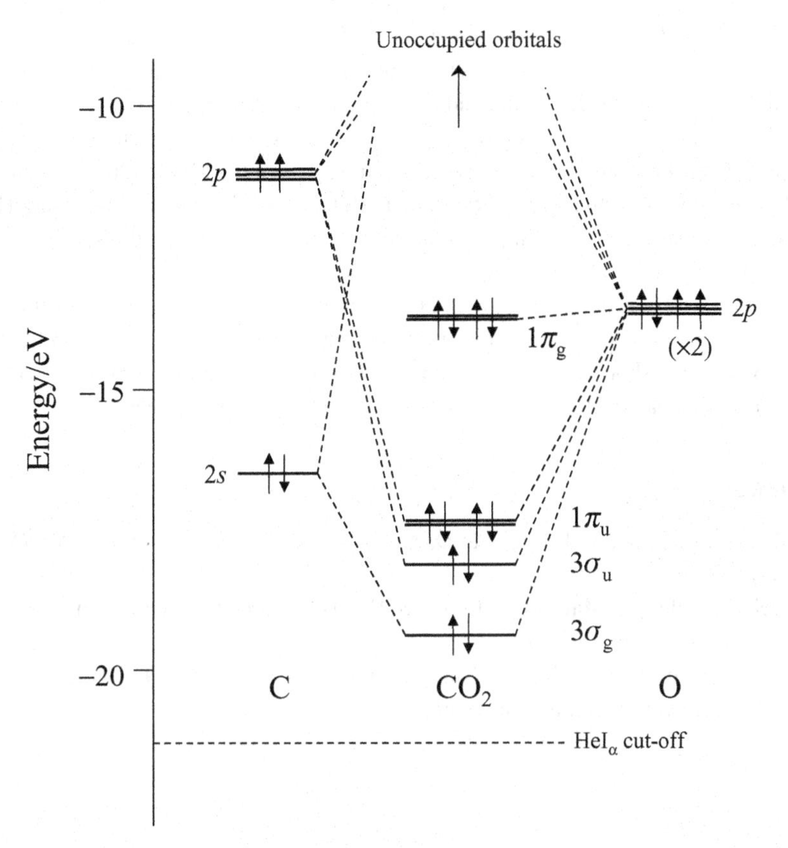

Figure 14.4 Partial MO diagram for CO_2 based on the ultraviolet photoelectron spectrum. The energy scale is the negative of the vertical ionization energies (Koopmans's theorem). The ionization energies for the atoms have been taken from the tables compiled by Moore [3]. Notice that the 2*s* orbital has an ionization energy far beyond the HeIα limit (21.22 eV) and is therefore not shown.

These findings, taken together, provide important clues in the construction of an MO diagram. Such a diagram for CO_2 is shown in Figure 14.4, concentrating on those *occupied* orbitals that are photoionized by HeI radiation. To include the atomic orbitals on the same energy scale, use has been made of atomic energy level data for carbon and oxygen [3]. The HOMO is a largely non-bonding π orbital formed by combining $2p\pi$ orbitals on the two oxygen atoms with opposite phases. This gives a HOMO of π_g symmetry, the $1\pi_g$ orbital, which can be thought of as the lone pairs on each oxygen atom. If the two oxygen atoms have $2p\pi$ orbitals with the same phases then a bonding interaction with C $2p\pi$ orbitals is possible giving rise to the $1\pi_u$ MO.

The next two MOs are both σ orbitals. According to the diagram in Figure 14.4, the highest occupied σ orbital ($3\sigma_u$) looks to be bonding in character. However, the absence of significant vibrational structure in the photoelectron spectrum indicates mainly non-bonding character. The same arguments hold for the $3\sigma_g$ MO. The explanation for this apparent failing in the MO picture is the neglect of the O 2*s* atomic orbitals. Although far more tightly bound than the C 2*s* orbital, and therefore not shown in Figure 14.4, the σ_g and σ_u combinations formed from the two O 2*s* orbitals do make a significant contribution to the $3\sigma_u$ and $3\sigma_g$ MOs. In particular they add antibonding character, approximately cancelling out the bonding character that would result in the absence of O 2*s* contributions.

Of course we have only obtained information on part of the MO diagram, and it would be interesting to probe the more tightly bound orbitals, which could be done using HeII or X-ray radiation. However, the important orbitals in chemical bonding, the valence orbitals, will nearly always fall in the HeI region.

Analogous MO diagrams for OCS and CS_2 can be constructed, although for the former care must be taken to distinguish the different contributions from oxygen and sulfur to specific MOs. One should also be aware that OCS has no centre of symmetry so the g/u notation is inapplicable when labelling orbitals and states of this molecule.

References

1. L.-S. Wang, J. E. Reutt, Y. T. Lee, and D. A. Shirley, *J. Electron. Spectrosc. Rel. Phenom.* **47** (1988) 167.
2. I. Reineck, C. Nohre, R. Maripuu, P. Lodin, S. H. Al-Shamma, H. Veenhuizen, L. Karlsson, and K. Siegbahn, *Chem. Phys.* **78** (1983) 311.
3. *Atomic Energy Levels*, C. E. Moore, National Bureau of Standards, Circ. 467, Washington DC, US Department of Commerce, 1949.

15 Photoelectron spectrum of NO_2^-

Concepts illustrated: *anion photoelectron spectroscopy; electron affinity; vibrational structure and the Franck–Condon principle; link to thermodynamic parameters; molecular orbital information and Walsh diagrams.*

The photoelectron spectroscopy of anions is, in many respects, directly analogous to the photoelectron spectroscopy of neutral molecules. However, an important difference is that an electron in the valence shell of an anion is much more weakly bound than in a neutral molecule. In fact there are some molecules, such as N_2, that are unable to bind an additional electron at all. The binding energy of an electron in an anion, which is known as the *electron affinity* (EA), is the energy difference between the neutral molecule and the anion. The electron affinity is defined as a *positive* quantity if the anion possesses a lower energy than the neutral molecule, i.e. the electron is bound to the molecule and energy must be *added* to remove it.

The photoelectron spectrum of an anion, also known as the photodetachment spectrum, can provide information on both the anion and the neutral molecule. A good example of this is the photoelectron spectrum of NO_2^-, which was first recorded by Ervin, Ho, and Lineberger [1].

15.1 The experiment

The most common method for generating anions in the gas phase is an electrical discharge. Ervin *et al.* produced NO_2^- by a microwave (ac) discharge through a helium/air mixture. A variety of neutral and charged species would be expected under such conditions, including several possible anions and cations. However, unlike neutral molecules, specific ions can be readily separated from a mixture using a mass spectrometer. Ervin *et al.* used this idea to obtain the photoelectron spectrum of NO_2^-.

As will be seen later, NO_2^- has a relatively large electron affinity. Consequently, while it is usually possible to employ visible light to remove an electron from an anion, shorter wavelength light proved necessary for NO_2^-. The actual wavelength used was 351.1 nm, which is in the near-ultraviolet, from a frequency doubled continuous dye laser.

As in all types of photoelectron spectroscopy where the electron kinetic energy is scanned, the resolution is limited primarily by the electron kinetic energy measurements. In the

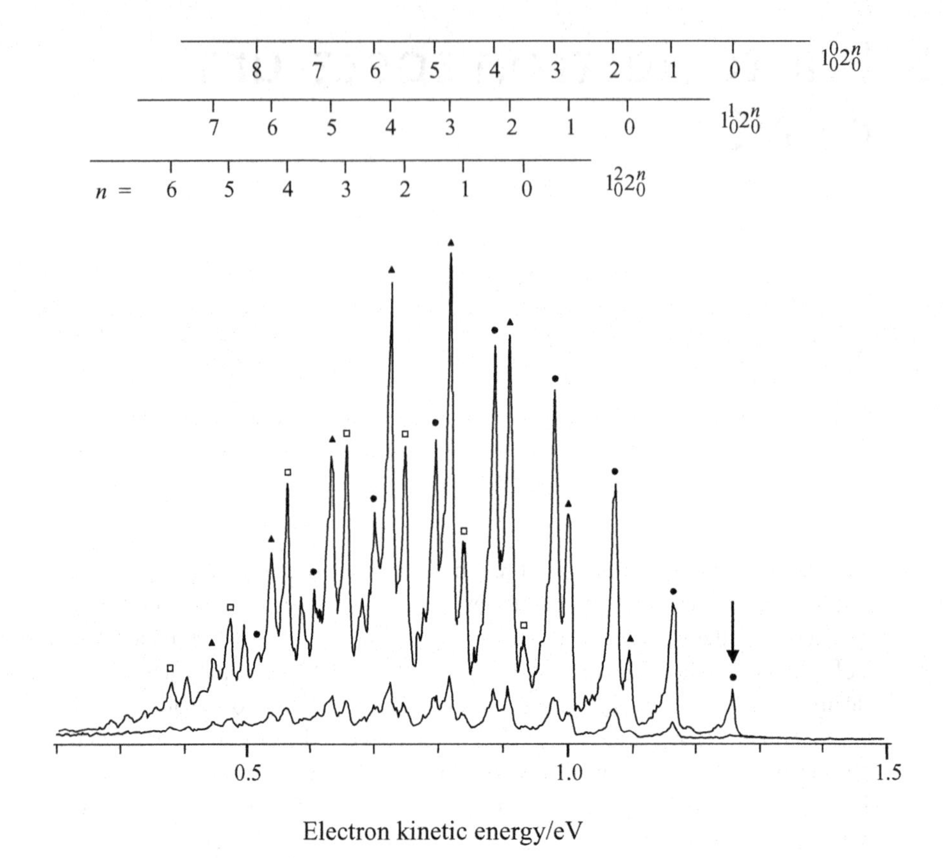

Figure 15.1 The photoelectron spectrum of NO_2^- obtained with 351.1 nm laser photodetachment. Two different scans are shown, the upper one with the laser polarized parallel to the path of electrons entering the analyser, and the lower one oriented perpendicular to this direction. The arrow marks the adiabatic electron detachment process. For an explanation of the vibrational structure labelling shown above the spectrum see text. (Reproduced from K. M. Ervin, J. Ho, and W. C. Lineberger, *J. Phys. Chem.* **92** (1988) 5405, with permission from the American Chemical Society.)

instrument used by Ervin *et al.*, the electron energy analyser was of the hemispherical type (see Section 12.1) with a resolution of approximately 9 meV (~70 cm^{-1}). The observed resolution in the spectrum (FWHM) was 16 meV (130 cm^{-1}), a convolution of instrumental and substantial broadening due to (unresolved) rotational structure. The difference between the photon energy (3.532 eV) and the electron kinetic energy gives the binding energy of the electron to the anion.

15.2 Vibrational structure

Photoelectron spectra of NO_2^- obtained using polarized laser light are shown in Figure 15.1. The more prominent spectrum was obtained with the laser polarization parallel to the path of electrons entering the energy analyser, while the weaker spectrum was obtained with perpendicular polarization. Although the absolute intensities are very different in the two cases, the relative intensities of all observed features are roughly the same. Different responses of parts of the spectrum to a change in laser polarization are likely if more than one photodetachment process contributes to the spectrum. It is therefore reasonable to conclude that a single photodetachment process is responsible for the structure in Figure 15.1, presumably leading to the formation of NO_2 in its ground electronic state (see later).

Extensive vibrational structure is evident in Figure 15.1. If NO_2^- is present in only its zero-point vibrational level, then all the structure will be due to excitation of vibrations in neutral NO_2. Several regular progressions are easily identified and can be explained in terms of two active vibrational modes with intervals of $\sim$1320 and $\sim$750 cm^{-1}, respectively. A lengthy progression in the lower frequency mode is built upon successive quanta in the higher frequency mode, giving rise to three prominent vibrational progressions. Symbols have been added to Figure 15.1 to distinguish these three progressions.

NO_2 and NO_2^- possess three normal vibrational modes, two stretches and one bend. In determining selection rules for these modes, by application of the Franck–Condon principle (see Section 7.2.3), it is necessary to establish whether a particular vibration is totally or non-totally symmetric with respect to the full set of symmetry operations of the molecular point group. Microwave spectra of NO_2 show that it is bent at equilibrium in its electronic ground state with an O—N—O bond angle of 134° and an N—O bond length of 1.194 Å [2]; NO_2 therefore possesses C_{2v} equilibrium symmetry. There will be two totally symmetric (a_1) normal modes, the totally symmetric N—O stretch, ν_1, and the O—N—O bend, ν_2. The remaining mode, the antisymmetric N—O stretch, is designated as mode ν_3 and has b_2 symmetry.

Assuming that NO_2^- does not possess a lower equilibrium symmetry than NO_2, a reasonable assumption, then we can concentrate on the two totally symmetric modes of NO_2 to explain the vibrational structure. The harmonic wavenumbers of ν_1 and ν_2 have been measured previously with very high precision from IR spectra and are known to be 1325.33 $\pm$ 0.06 and 750.14 $\pm$ 0.02 cm^{-1}, respectively [3]. These values are, within experimental error, identical to those determined from the photoelectron spectrum of NO_2^- and confirm the assignment. The vibrational structure in the photoelectron spectrum can therefore be interpreted in terms of various combinations of quanta in modes ν_1 and ν_2. The standard notation for labelling the individual vibrational peaks is $1^m_n 2^p_q$, where 1 and 2 refer to modes ν_1 and ν_2 and the superscripts and subscripts reveal the number of quanta in these modes in the upper and lower electronic states (neutral molecule and anion), respectively.

Three bending progressions have been assigned in Figure 15.1 built upon different degrees of excitation of ν_1, the $1^0_0 2^n_0$, $1^1_0 2^n_0$, and $1^2_0 2^n_0$ progressions. The limited resolution and signal-to-noise ratio prevents other, less prominent, progressions being identified. The $1^0_0 2^0_0$ band, more commonly written as 0^0_0, corresponds to the *adiabatic* photodetachment process, i.e. NO_2 is formed in its zero-point level from NO_2^- in its zero-point level. This transition is marked with an arrow in Figure 15.1. The adiabatic electron affinity is obtained as the difference between the photon energy and the electron kinetic energy. The assignment of the adiabatic transition in any band in which there is extensive vibrational structure should always be viewed with some suspicion since it is possible that this transition will not be observed if its Franck–Condon factor (FCF) is small. However, the assignment made in Figure 15.1 is supported by other data, notably a photodetachment threshold experiment in which an intense tunable laser was used to accurately determine the onset wavelength for electron photodetachment from NO_2^-. The spectrum in Figure 15.1 yields an adiabatic electron affinity of 2.273 ± 0.005 eV [1].

15.3 Vibrational constants

With all the vibrational structure assigned, the next step is to determine the vibrational constants of NO_2^- and NO_2. For NO_2, there is ample vibrational information and adequate resolution in the anion photoelectron spectrum to allow the determination of anharmonicity constants as well as harmonic vibrational frequencies. The transition wavenumbers can be fitted to the vibrational term value expression

$$G(v_1, v_2) = \omega_1\left(v_1 + \tfrac{1}{2}\right) + \omega_2\left(v_2 + \tfrac{1}{2}\right) + x_{11}\left(v_1 + \tfrac{1}{2}\right)^2 + x_{22}\left(v_2 + \tfrac{1}{2}\right)^2 \\ + x_{12}\left(v_1 + \tfrac{1}{2}\right)\left(v_2 + \tfrac{1}{2}\right)$$

where ω_1 and ω_2 are the harmonic frequencies of modes ν_1 and ν_2 and x_{11}, x_{22}, and x_{12} are anharmonicity constants (see also Section 5.2.3). Linear regression yields the constants presented in Table 15.1. These values can be checked against the results from high resolution infrared spectroscopy and show good agreement.

Vibrational constants for NO_2^- are rather more difficult to obtain because, as mentioned earlier, all the main vibrational components in Figure 15.1 arise from transitions out of the zero-point level of the anion. However, a magnified view (see Figure 15.2) of the region beyond the origin transition, i.e. at lower electron binding energies, shows additional peaks arising from hot band transitions. They are transitions out of excited vibrational levels in NO_2^- (hence the name 'hot band', since these grow in significance as the temperature increases) and they therefore provide vibrational information on NO_2^-. The number of peaks is insufficient to determine meaningful anharmonicity constants but approximate harmonic frequencies for the two totally symmetric modes can be extracted. These are listed in Table 15.1.

15.4 Structure determination

The observation of substantial Franck–Condon activity in both ν_1 and ν_2 shows that the equilibrium N—O bond length and the O—N—O bond angle of NO_2^- must both differ significantly from their values in NO_2. It is possible to quantify these changes by calculating Franck–Condon factors (FCFs) for each possible vibrational component and comparing with experiment. In order to calculate FCFs, vibrational wavefunctions are required. These in turn require knowledge of the structures of the anion and the neutral molecule. As mentioned earlier, the structure of the neutral molecule is known to high precision and so the

Table 15.1 *Vibrational and structural constants for* NO_2 *and* NO_2^-

Quantity	NO_2	NO_2^-
ω_1/cm^{-1}	1316.4 ± 9.2	1284 ± 30
ω_2/cm^{-1}	748.0 ± 4.2	776 ± 30
x_{11}/cm^{-1}	3.1 ± 2.6	
x_{22}/cm^{-1}	−0.59 ± 0.44	
x_{12}/cm^{-1}	−2.1 ± 1.3	
r(N—O)/Å	1.194	1.25 ± 0.02
Bond angle/°	133.9	117.5 ± 2

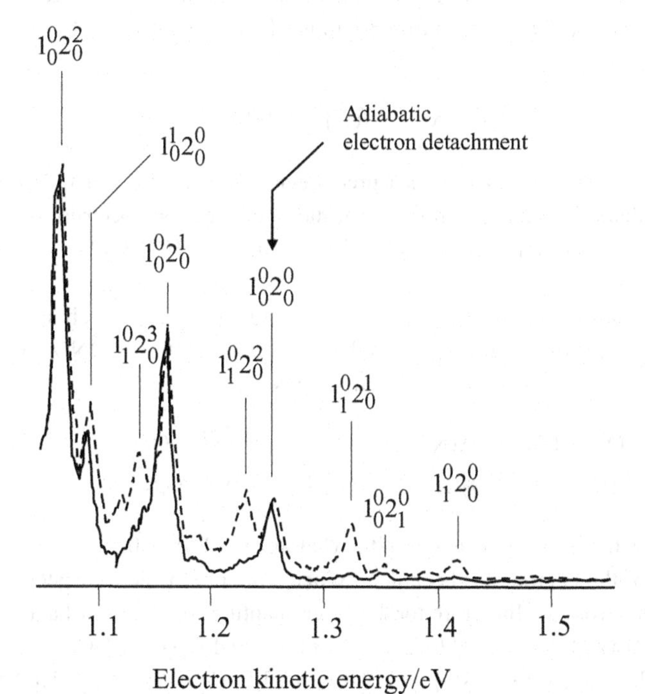

Figure 15.2 Expanded view of photoelectron spectra of NO_2^- near the adiabatic threshold at two different temperatures. The dashed line spectrum corresponds to a warmer NO_2^- sample than that shown by the solid line. The population of excited vibrational levels in NO_2^- is enhanced in the warmer spectrum, giving more prominent *hot bands*. (Reproduced from K. M. Ervin, J. Ho, and W. C. Lineberger, *J. Phys. Chem.* **92** (1988) 5405, with permission from the American Chemical Society.)

N—O bond length and the ONO bond angle of the anion can be used as trial parameters to bring theory and experiment into agreement.

Full details of the FCF calculations are quite involved; the interested reader should consult Reference [1] for further information. It is important to recognize that FCF simulations on their own yield only the magnitude of changes in internal coordinates, not their signs. However, as will be seen later, it is usually possible to draw on other information, perhaps from *ab initio* quantum chemical calculations or even just qualitative bonding arguments, which allow the signs to be deduced as well. Ervin *et al.* found that $r(\mathrm{N{-}O}) = 1.25 \pm 0.02$ Å and $\theta(\mathrm{O{-}N{-}O}) = 117.5 \pm 2.0°$ in NO_2^-. The precision is nowhere near as good as would typically be achieved from high resolution (rotationally resolved) spectra, but so far these have proved elusive.

15.5 Electron affinity and thermodynamic parameters

The photoelectron spectrum allows thermochemical parameters to be determined for NO_2^- that would be difficult to obtain by other means. Among the most important is the enthalpy of formation of NO_2^-, $\Delta_f H°(NO_2^-)$, which from a simple Hess's law cycle can be expressed as

$$\Delta_f H°(\mathrm{NO_2^-}) = \Delta_f H°(\mathrm{NO_2}) - \mathrm{EA}(\mathrm{NO_2})$$

The enthalpy of formation of NO_2 has been measured previously and is 35.93 ± 0.8 kJ mol^{-1} [4]. Combining the adiabatic electron affinity from the photoelectron spectrum (2.273 ± 0.005 eV $\equiv 219.3 \pm 0.05$ kJ mol^{-1}) with $\Delta_f H°(NO_2)$ leads to $\Delta_f H°(NO_2^-) = -183.4 \pm 0.9$ kJ mol^{-1}.

Similarly, the dissociation energy (D_0) of NO_2^- to give O^- and NO, as well as the gas-phase acidity, $\Delta_a H°$, of nitrous acid (i.e. the enthalpy for the reaction HONO $\rightarrow H^+ + NO_2^-$), can also be determined using the Hess's law cycles

$$D_0(\mathrm{O^-{-}NO}) = \mathrm{EA}(\mathrm{NO_2}) - \mathrm{EA}(\mathrm{O}) + D_0(\mathrm{NO{-}O})$$
$$\Delta_a H°(\mathrm{HONO}) = D_0(\mathrm{H{-}ONO}) + \mathrm{IE}(\mathrm{H}) - \mathrm{EA}(\mathrm{NO_2^-})$$

where IE(H) represents the ionization energy of the H atom (1312.05 ± 0.04 kJ mol^{-1} [5]). The electron affinity of the O atom has been determined from photoelectron spectroscopy [6]. Values are available from the literature for the other quantities on the right-hand side of the above equations: $D_0(\mathrm{NO{-}O}) = 300.64 \pm 0.8$ kJ mol^{-1} and $D_0(\mathrm{H{-}ONO}) = 324.6 \pm 1.6$ kJ mol^{-1}. These lead to $D_0(\mathrm{O^-{-}NO}) = 379.4 \pm 0.9$ kJ mol^{-1} and $\Delta_a H°(\mathrm{HONO}) = 1417.4 \pm 1.7$ kJ mol^{-1}.

15.6 Electronic structure

Qualitative molecular orbital arguments can be employed to explain the change in structure between NO_2^- and NO_2. The key is to understand how the energies of the occupied molecular orbitals change as the structure is altered, in particular as the bond angle changes.

The 1*s* orbitals on both N and O can be ignored since they make no significant contribution to the bonding. The valence molecular orbitals, derived from the 2*s* and 2*p* orbitals on each atom, will give rise to a total of twelve MOs. For NO_2^- and NO_2 there are 18 and 17 electrons, respectively, to be distributed amongst the valence MOs.

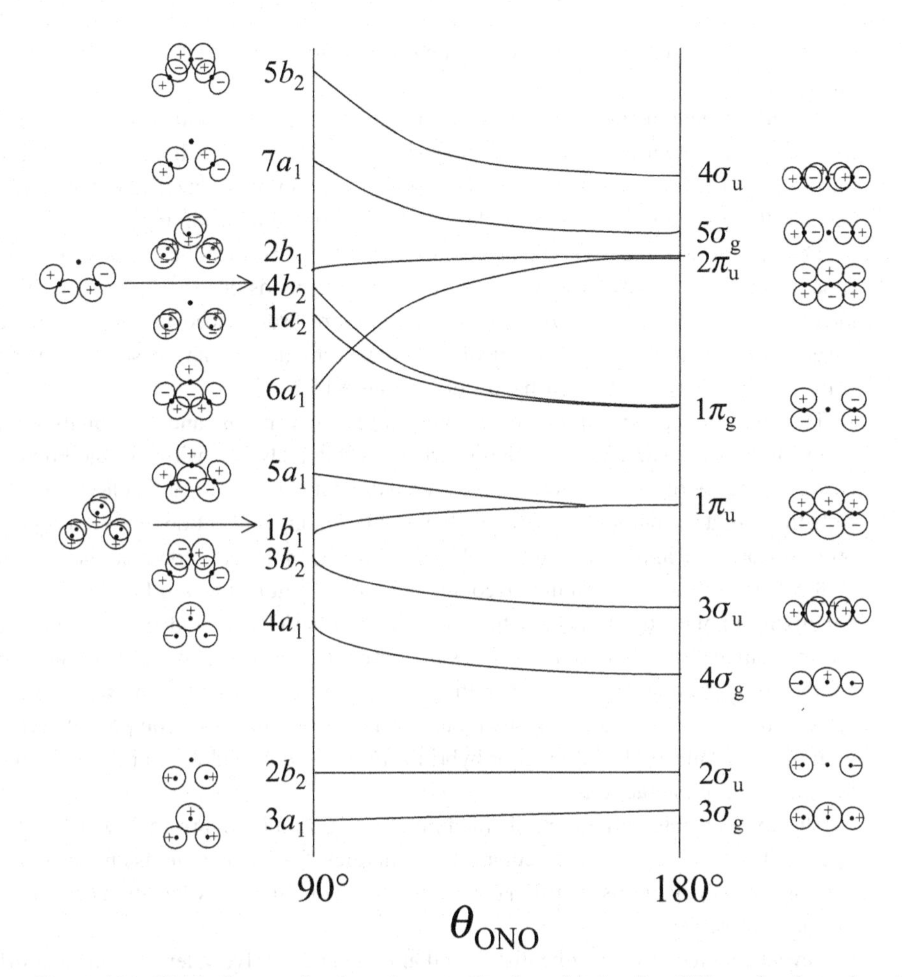

Figure 15.3 Walsh diagram for the valence molecular orbitals of an XY_2 molecule such as NO_2. Approximate atomic orbital contributions to the MOs are shown. Notice that for the π MOs on the right-hand side only the in-plane component is shown. Also, all $\sigma_{g/u}$ orbitals should actually be labelled $\sigma^+_{g/u}$ but the + superscript has been omitted for clarity.

Although we know NO_2 is bent, it is simpler to start by considering a linear geometry. The valence MOs can be divided into two groups, σ and π orbitals. It is easy to envisage that the energies of the σ MOs, in which the electron density points primarily *along* bonds, will not be strongly altered if the molecule bends. Consequently, while the σ MOs will play a major role in the N—O bonding, they do not have a strong influence on the equilibrium bond angle.

The π orbitals in the linear molecule are doubly degenerate MOs. There are three separate π orbitals resulting from $2p$ orbital overlap, one bonding, one non-bonding, and one antibonding. The right-hand side of Figure 15.3 shows the phases of the atomic orbitals for the three π orbitals. As expected, in the bonding MO ($1\pi_u$) all three atoms have $2p$ orbitals with the same phase. In the antibonding MO ($2\pi_u$) the $2p$ orbitals on the O atoms have the same phase but opposite to that on the central N atom. In the non-bonding MO ($1\pi_g$) the

O atoms have opposite phases and are therefore unable to interact with a $2p\pi$ orbital on nitrogen.[1]

As the molecule bends, each π orbital loses its degeneracy and two distinct MOs are formed with different symmetries. As shown in Figure 15.3, the $1\pi_u$ bonding MO is resolved into a_1 and b_1 MOs.[2] The energies of these orbitals are not very sensitive to the bond angle because the bonding interactions are largely unaltered by the bending process.

The π_g non-bonding MO is resolved into a_2 and b_2 MOs when the molecule is bent. The energies of both of these orbitals rise as the molecule bends due to increased antibonding interactions between the O $2p$ orbitals. This is more pronounced for the b_2 component because of the in-plane orientation of the O $2p$ orbitals, as can be seen from the AO contributions shown on the left-hand side of Figure 15.3.

The antibonding $2\pi_u$ MO of linear NO_2 is resolved into a_1 and b_1 orbitals when the molecule is bent. The energy of the b_1 orbital is relatively insensitive to the bond angle. However, the energy of the a_1 orbital is lowered dramatically as the molecule is bent. At first sight this might seem implausible; after all, the antibonding interactions between adjacent $2p$ orbitals would not appear to be removed by bending the molecule. However, there is another factor that needs to be taken into account, and which is not shown in Figure 15.3, namely the mixing-in of nitrogen $2s$ character. Such mixing is strictly forbidden by symmetry at the linear geometry (the $2s$ orbital on N has σ_g symmetry and so cannot interact with the π_u combination of $2p$ orbitals) but when the molecule bends mixing becomes possible (since the $2s$ orbital on N now has a_1 symmetry and can interact with the a_1 component correlating with the π_u orbital). This mixing, or hybridization of the $2s$ and $2p$ orbitals on N, reduces the antibonding interactions.

The above considerations are distilled into the diagram in Figure 15.3, which shows the effect of the bond angle on molecular orbital energies. Such a diagram is known as a Walsh diagram. Walsh diagrams are often used to provide a qualitative explanation of bond angles in small molecules [7].

Seventeen electrons must be distributed amongst the twelve valence molecular orbitals of NO_2. If the molecule is linear in its ground electronic state then these electrons will fill all orbitals up to and including $1\pi_g$. The remaining electron will be in the $2\pi_u$ orbital, the HOMO, which is the π antibonding orbital discussed earlier. Evidently, there will be some energetic gain by bending the molecule so that the unpaired electron is now in the $6a_1$ MO (which correlates with $2\pi_u$ in the linear molecule limit). However, the energies of the $4b_2$ and $1a_2$ orbitals, both of which are doubly occupied, rise as the molecule is bent. Clearly there will be a compromise and the equilibrium bond angle will adopt an intermediate value between the linear and fully bent (90°) limits. The known bond angle is 134°, consistent with this proposition.

---

[1] This follows also from formal symmetry arguments. The combination of $2p\pi$ orbitals on the O atoms with opposite phases gives a *symmetry orbital* having π_g symmetry. This cannot interact with a $2p\pi$ orbital on the N atom since all $2p$ orbitals on the central atom have u symmetry (because the phases on their two lobes have opposite signs). The N $2p$ orbitals therefore make no contribution to the π_g antibonding MO.

[2] When NO_2 is bent the symmetry is lowered from $D_{\infty h}$ to C_{2v}. The resulting symmetry of each orbital can be deduced by consulting the C_{2v} character table and noting how the orbital is transformed under each symmetry operation of the point group.

The additional electron in the ground electronic state of NO_2^- will reside in the $6a_1$ MO. Since this orbital will now be doubly occupied, the Walsh diagram leads us to expect a smaller bond angle for NO_2^- than NO_2. These arguments confirm the sign of the bond angle change deduced by Erwin *et al.*, namely that the bond angle *increases* by *c.* 16° on photodetaching an electron from NO_2^-.[3]

References

1. K. M. Ervin , J. Ho, and W. C. Lineberger, *J. Phys. Chem.* **92** (1988) 5405.
2. Y. Morino, M. Tanimoto, S. Saito, E. Hirota, R. Awata, and T. Tanaka, *J. Mol. Spectrosc.* **98** (1983) 331.
3. W. J. Lafferty and R. L. Sams, *J. Mol. Spectrosc.* **66** (1977) 478.
4. M. W. Chase, C. A. Davies, J. R. Downey, D. J. Frurip, R. A. McDonald, and A. N. Syverud, *JANAF Thermochemical Tables*, 3rd edn., *J. Phys. Chem. Ref. Data, Suppl.* **14** (1987).
5. *Atomic Energy Levels*, C. E. Moore, National Bureau of Standards, Circ. 467, Washington DC, US Department of Commerce, 1949.
6. H. Hotop and W. C. Lineberger, *J. Phys. Chem. Ref. Data* **14** (1985) 731.
7. A. D. Walsh, *J. Chem. Soc.* (1953) 2260; *ibid.* (1953) 2266.

[3] These arguments can also be extended to explain the change in bond length on photodetachment, although a more sophisticated analysis is needed. See Reference [1] for further details.

16 Laser-induced fluorescence spectroscopy of C_3: rotational structure in the 300 nm system

Concepts illustrated: *laser-induced fluorescence spectroscopy; symmetries of electronic states; assignment of rotational structure in spectra of linear molecules; combination differences; band heads; nuclear spin statistics.*

As described in Chapter 11, laser-induced fluorescence (LIF) spectroscopy is one of the simplest and yet most powerful tools for obtaining high resolution spectra. Its high sensitivity is particularly convenient for the investigation of extremely reactive molecules, such as free radicals and ions. In this Case Study we illustrate how LIF spectroscopy can be used to obtain important information on a small carbon cluster, the C_3 molecule. The spectra presented were originally obtained by Rohlfing [1], who produced C_3 by pulsed laser ablation of graphite. This is a violent method for vaporizing a solid and the plasma formed above the graphite surface will undoubtedly contain carbon atoms, clusters such as C_2, C_3, and various cations and anions. To reduce spectral congestion, the laser ablation source was combined with a supersonic nozzle to produce a cooled sample for spectroscopic probing.

The LIF spectrum was obtained by crossing the supersonic jet with a tunable pulsed laser beam and measuring the intensity of fluorescence as a function of laser wavelength. As discussed in Section 11.2, an LIF excitation spectrum is similar to an absorption spectrum but the signal intensity depends not only on the absorption probability, but also the fluorescence quantum yield of the upper state. C_3 has LIF spectra in several regions of the ultraviolet, and one such system, in the 298–311 nm region, is shown in Figure 16.1. Given that this spectrum spans several hundred cm^{-1}, most if not all of the coarse structure must be vibrational in origin. Rotationally resolved scans of individual vibrational components would greatly facilitate the spectral assignment, as well as providing structural information on the molecule. Figure 16.2 shows a higher resolution scan of the strongest band in Figure 16.1, and this will now be considered in some detail.

16.1 Electronic structure and selection rules

Spectra of C_3 in the region shown in Figure 16.1 are very strong. Consequently, the transitions presumably originate from the ground electronic state. *Ab initio* electronic structure

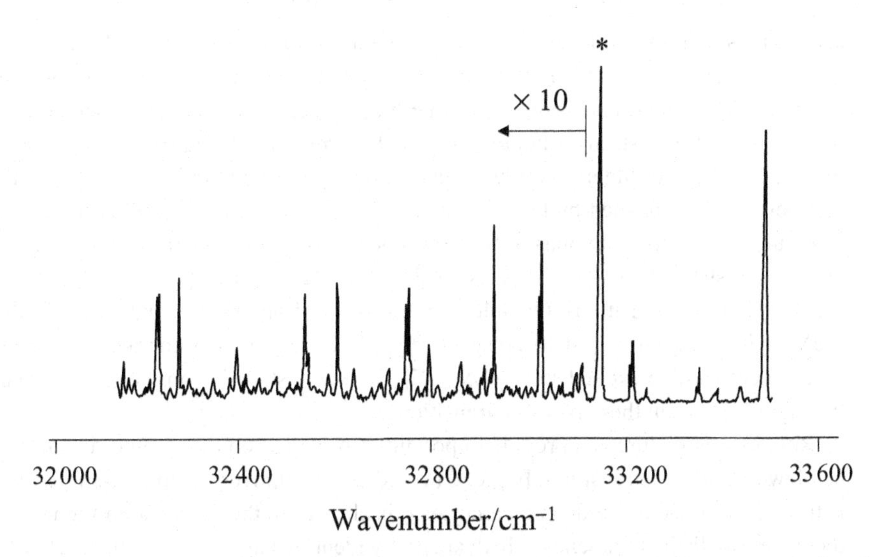

Figure 16.1 Survey LIF excitation spectrum of jet cooled C_3 in the 32 145–33 500 cm^{-1} region. The band marked with an asterisk is rotationally resolved in Figure 16.2. (Reproduced with permission from E. A. Rohlfing, *J. Chem. Phys.* **91** (1989) 4531, American Institute of Physics.)

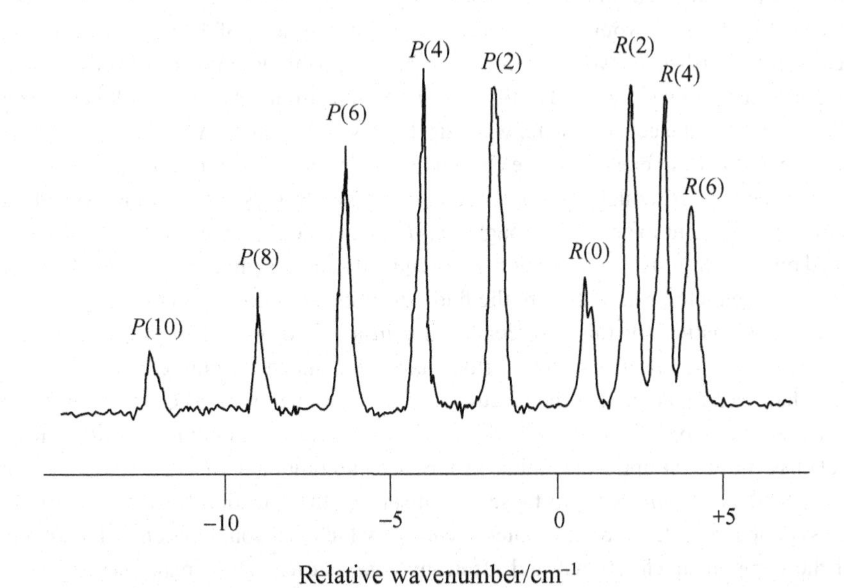

Figure 16.2 Rotationally resolved LIF excitation spectrum of the vibronic band of jet-cooled C_3 at 33 147 cm^{-1}. (Reproduced with permission from E. A. Rohlfing, *J. Chem. Phys.* **91** (1989) 4531, American Institute of Physics.)

calculations on C_3 provide a useful starting point for understanding the electronic spectra. C_3 is a linear molecule with an outer electronic configuration $\ldots 4\sigma_g^2 3\sigma_u^2 1\pi_u^4$, which gives rise to a $^1\Sigma_g^+$ ground electronic state. The highest occupied molecular orbital (HOMO), the $1\pi_u$ orbital, is a strongly bonding molecular orbital produced by the in-phase overlap of C $2p\pi$ atomic orbitals. The lowest unoccupied molecular orbital (LUMO) is the $1\pi_g$ non-bonding MO formed by the $2p\pi$ orbitals on the two carbon atoms at the ends of the molecule having opposite phases. Since this orbital is vacant, electron promotion into this orbital is possible. Only singlet states need be considered given the $\Delta S = 0$ spin selection rule in electronic transitions. One-electron transitions from the $4\sigma_g$ and $3\sigma_u$ orbitals to the LUMO yield $^1\Pi_g$ and $^1\Pi_u$ states, respectively, while the one-electron transition $1\pi_g \leftarrow 1\pi_u$ gives three possible excited states, $^1\Delta_u$, $^1\Sigma_u^+$, and $^1\Sigma_g^-$ states.[1] Can the spectra be used to distinguish between these possible transitions?

Rotational structure can provide important information on the symmetries of the upper and lower states in a transition. In fact C_3 represents a relatively simple example where this is true. Even without any detailed analysis it is clear from the spectrum in Figure 16.2 that there are two distinct branches, which are easily identified as $P(\Delta J = -1)$ and $R(\Delta J = +1)$ branches. There is no trace of any Q branch ($\Delta J = 0$), which shows that the transition must be $\Sigma - \Sigma$ in character, since any other possibility would give rise to a Q branch. Since *ab initio* calculations show that the ground state is a $^1\Sigma_g^+$ state, then the excited state must be a $^1\Sigma_u^+$ state. This follows from the application of symmetry arguments, which show that electric dipole allowed transitions must satisfy $g \leftrightarrow u$, $+ \leftrightarrow +$, and $- \leftrightarrow -$ selection rules (see Section 7.2.1). It would seem therefore that the symmetry of the excited electronic state has been established. However, there is a problem with this assignment because high quality *ab initio* calculations predict that the lowest $^1\Sigma_u^+$ electronic state is ~8 eV (64 500 cm^{-1}) above the ground electronic state, whereas the lowest $^1\Pi_g$ and $^1\Delta_u$ states are calculated to be at about the right energy, 4.13 eV (33 313 cm^{-1}) and 4.17 eV (33 635 cm^{-1}), above the ground state [2]. Although accurate prediction of the energies of electronic excited states is sometimes difficult to achieve through *ab initio* calculations, an error of several eV can be ruled out for a high-level calculation. Consequently, the assignment based on the rotational structure seems to be at odds with the findings of the *ab initio* calculations.

The explanation for this discrepancy lies in a breakdown of the Born–Oppenheimer approximation. So far the selection rules have been stated for pure electronic transitions, i.e. it has been assumed that the electronic and vibrational motions can be fully separated. However, this separation is never exact, and in some cases the mixing is sufficiently large that it is more appropriate to think in terms of a combined *vibronic* state, i.e. a mixed *vib*rational–elect*ronic* state. In these circumstances the selection rules are determined by the symmetries of the vibronic state(s), each of which is a combination of the symmetries of the component vibrational and electronic symmetries. This breakdown of the Born–Oppenheimer approximation is possible in polyatomic molecules by a mechanism known

1 These excited states can be established from direct products of the symmetries of the MOs, as described in Section 4.2.3. For example, the vacancy introduced into the $1\pi_u$ MO as a result of the transition $1\pi_g \leftarrow 1\pi_u$ yields electronic states with spatial symmetries derived from $\pi_g \otimes \pi_u = \sigma_u^+ + \sigma_u^- + \delta_u$. In contrast to MOs, the convention with electronic states is to employ upper case symmetry labels.

as Herzberg–Teller coupling, sometimes also known as vibronic coupling (for more details about Herzberg–Teller coupling see Case Study 25). In C_3, there are three vibrational normal modes, the symmetric stretch, the antisymmetric stretch, and the bending mode, with σ_g^+, σ_u^+, and π_u symmetries, respectively.[2] The symmetries of the resulting vibronic states can be determined from the direct product of the symmetries of the pure electronic and pure vibrational states. If only the stretches are excited, then mixing of either the $^1\Pi_g$ or the $^1\Delta_u$ electronic state with any stretching vibrational state can never yield the required $^1\Sigma_u^+$ vibronic state symmetry.

On the other hand, if there is one quantum in the bending mode when the molecule is in the $^1\Pi_g$ electronic state, then three vibronic states, $^1\Sigma_u^-$, $^1\Sigma_u^+$, and $^1\Delta_u$ states, are possible (as seen by taking the direct product $^1\Pi_g \times \pi_u$). If the symmetries of vibronic states for higher excitation of the bend are calculated it is readily shown that a $^1\Sigma_u^+$ state is produced on exciting odd quanta of the bending mode. Similarly even quantum excitation of the bending mode within the $^1\Delta_u$ electronic state can also produce a $^1\Sigma_u^+$ vibronic state. Unfortunately, the data are insufficient to be able to determine whether it is a transition to the $^1\Delta_u$ or $^1\Pi_g$ electronic state that is being vibronically induced. However, we can at least show that there is no inconsistency between the *ab initio* predictions and the spectroscopic findings.

16.2 Assignment and analysis of the rotational structure

The rotational structure in Figure 16.2 appears to be very simple and indeed it is. The P and R branches have already been pointed out, and clearly lines in the P branch diverge away from the band centre whereas lines in the R branch converge. Our first concern is to extract the rotational constants for the upper and lower states from the spectrum. However, before this is done it is beneficial to derive an estimate of the rotational constant. Assuming C_3 is linear and symmetrical ($D_{\infty h}$ point group), the moment of inertia is $I = 2m_C r^2$, where r is the C—C separation. Since we are not after an exact value at this stage, a typical C=C bond length (1.3 Å) can be used to estimate the rotational constant, yielding $B \approx 0.4$ cm^{-1}.

The interesting thing about the spectrum in Figure 16.2 is that all the lines have been assigned to transitions out of levels with even J. The estimate of the rotational constant allows us to show that this is correct. Accepting the assignment for the moment, the rotational constants in the upper and lower electronic states can be assigned by the method of combination differences, in which lines are identified that originate from, or terminate in, a common rotational level. For example, both the $R(J-1)$ and $P(J+1)$ lines transitions terminate at the Jth rotational level in the upper electronic state and hence the difference $R(J-1) - P(J+1)$ gives the energy difference between the $(J+1)$ and $(J-1)$ levels in the lower electronic state. In a similar way, $R(J) - P(J)$ gives the energy difference between

[2] It is customary to employ lower case symbols to represent the normal coordinates of individual vibrations, as it is to represent individual molecular orbitals. Upper case symbols are used for symmetries of overall electronic and vibrational states.

Table 16.1 *Approximate line positions and combination differences in the LIF spectrum of the C_3 molecule*

J	$P(J)$	$R(J)$	$\frac{R(J-1)-P(J+1)}{J+1/2}$	$\frac{R(J)-P(J)}{J+1/2}$
0	33 147.8			
			1.7333	
2	33 145.2	33 149.1		1.5600
			1.7429	
4	33 143.0	33 150.1		1.5778
			1.7636	
6	33 140.4	33 150.9		1.6154
			1.7733	
8	33 137.6			
10	33 134.6			

the $(J+1)$ and $(J-1)$ levels in the upper state, because both transitions originate from the same lower state level. Consequently, these two combination differences are given by

$$\begin{aligned}\Delta_2 F''(J) &= R(J-1) - P(J+1) = B''(J+1)(J+2) - B''(J-1)J \\ &= 4B''\left(J+\tfrac{1}{2}\right) \\ \Delta_2 F'(J) &= R(J) - P(J) = B'(J+1)(J+2) - B'(J-1)J \\ &= 4B'\left(J+\tfrac{1}{2}\right)\end{aligned}$$

in the rigid rotor limit, where the $'$ and $''$ labels refer to quantities in the upper and lower states, respectively. The observed lines and the combination differences are given in Table 16.1. As can be seen, the combination differences, when divided by $J+\frac{1}{2}$, give a nearly constant value, as expected from the relationships above; the slow increase is due to the neglect of centrifugal distortion, which is normally very small at low J.

The rotational constants in the lower and upper states can be derived from the intercept of a line fitted through these points by linear regression. As the intercepts equal $4B''$ and $4B'$, the rotational constants are $B'' = 0.438 \pm 0.005$ and $B' = 0.396 \pm 0.007$ cm^{-1}. The corresponding C—C bond lengths are therefore 1.263 ± 0.007 Å and 1.328 ± 0.012 Å, respectively. The increase in bond length on electronic excitation is consistent with the idea of an electron moving from a bonding MO to a non-bonding MO.

The rotational constants extracted from the spectral analysis are reasonably close to our earlier estimate. If the assignments in Figure 16.2 were wrong, and instead the P and R branch lines with odd rotational quantum numbers were not missing, then rotational constants roughly twice as large as the above values would be obtained. This would clearly be physically unreasonable, and we would have to conclude that the assignment was incorrect.

The absence of lines from odd J levels is the result of nuclear spin statistics, which is important in molecules where two or more atoms are in equivalent locations. C_3 is a good example since the terminal carbon atoms are equivalent; other examples would include C_2, O_2, H_2O, and NH_3. Interested readers can find a brief discussion on the origin of nuclear spin statistics and its impact on the rotational levels of molecules in Appendix F. The effect for C_3 is that, because the nuclear spin of the most abundant isotope, ^{12}C, is $I = 0$, only even J levels exist in the $^1\Sigma_g^+$ ground state and odd J levels in the excited vibronic $^1\Sigma_u^+$ state. As a result every second line is missing from the spectrum.

16.3 Band head formation

The convergence of the R branch and divergence of the P branch is a consequence of the difference in rotational constants between the upper and lower states. To see how this arises, consider the expression for the position of R branch transitions:

$$\begin{aligned}\nu(R(J)) &= F'(J+1) - F''(J) = \nu_0 + B'(J+1)(J+2) - B''J(J+1) \\ &= \nu_0 + 2B' + (3B' - B'')J + (B' - B'')J^2\end{aligned}$$

In the above expression ν_0 represents the transition frequency in the absence of rotational structure. If B' and B'' differ then the quadratic term in J may be significant, particularly at high J. If $B'' > B'$ then $\nu(R(J))$ will reach a maximum at some value of J and begin to decrease as J continues to increase, i.e. the branch forms a so-called *band head*. Beyond the band head the branch reverses direction and diverges as J continues to increase. In contrast, the P branch will simply diverge as J increases. This is clearly the behaviour observed for the spectrum of C_3 in Figure 16.2. If $B' > B''$ it is the P branch that has a band head and the R branch that diverges.

Band heads are not always observed if $|B' - B''|$ is very small, since the turning point occurs for transitions out of a high rotational level and this may have an insignificant population at the given temperature. Clearly band head formation is an indicator of the sign and magnitude of the difference $B' - B''$.[3] The band head in the C_3 spectrum occurs at $R(6)$; higher R branch transitions are hidden under the stronger (low J) R branch transitions.

References

1. E. A. Rohlfing, *J. Chem. Phys.* **91** (1989) 4531.
2. J. Römelt, S. D. Peyerimhoff, and R. J. Bunker, *Chem. Phys. Lett.* **58** (1978) 1.

[3] Even if the rotational structure is not fully or even partially resolved, the shape, or *contour*, of the band still provides information on the rotational constants. Even if no individual peaks were resolved in a low resolution version of Figure 17.2, the overall band would clearly be asymmetric with a tail on the long wavelength side. Such a band is referred to as being *red-shaded*, and this red-shading immediately reveals that $B' < B''$.

17 Photoionization spectrum of diphenylamine: an unusual illustration of the Franck–Condon principle

Concepts illustrated: *MATI spectroscopy; vibrational wavefunctions; Franck–Condon principle and Franck–Condon factors.*

The photoionization spectrum of diphenylamine provides an unusual and interesting illustration of the Franck–Condon principle. Diphenylamine (DPA), illustrated in Figure 17.1, is a relatively large molecule to study by gas phase spectroscopy and it might be thought that the vibrational structure in its electronic spectra would be highly congested and difficult to interpret. After all, this is a molecule with 66 vibrational modes! However, it was shown in Section 7.2.3 that only totally symmetric modes generally need to be considered in interpreting electronic spectra. Also, there is the further simplification that not all of the totally symmetric modes need be Franck–Condon active, i.e. will give a significant progression. DPA is an excellent example of this, with the main structure arising from a single vibrational mode.

Before spectra are considered, the experimental procedure, carried out by Boogaarts and co-workers [1], will be outlined. Mass-analysed threshold ionization (MATI) spectroscopy was employed. This technique, which was briefly described in Section 12.6, is essentially the same as ZEKE spectroscopy but employs ion rather than electron detection. It has the advantage over ZEKE spectroscopy in that ions can be separated according to their mass, which in most cases enables the spectral carrier to be determined with confidence. By analogy with ZEKE spectroscopy, a cation $\leftarrow$ neutral molecule electronic absorption spectrum is effectively obtained.

In the experiments on DPA this molecule was promoted from its ground electronic state, which is a spin singlet (S_0), to its first excited singlet state (S_1), using the output from a pulsed dye laser. Different vibrational levels of the S_1 state can be accessed by appropriate choice of the dye laser wavelength, λ_1. A pulse from a second dye laser was then employed to ionize DPA from its S_1 state, with the ion signal being detected as a function of the wavelength, λ_2, of this dye laser. Actually the experiment is a little more complicated in that only threshold ions are detected, i.e. those ions for which the corresponding electron

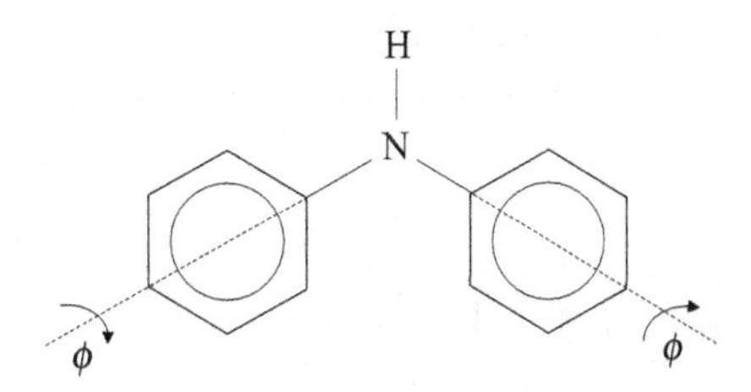

Figure 17.1 The structure of diphenylamine (DPA). The angle ϕ is the torsional coordinate and corresponds to twisting (in opposite senses) of the two phenyl rings about the C—N bonds. $\phi = 0°$ corresponds to a planar arrangement of the two phenyl rings. When planar, DPA has C_{2v} point group symmetry, as in the ground electronic state, but when $\phi \neq 0$ the symmetry is lowered to C_2. It is the torsional vibration that is responsible for the bulk of the vibrational structure in Figure 17.3.

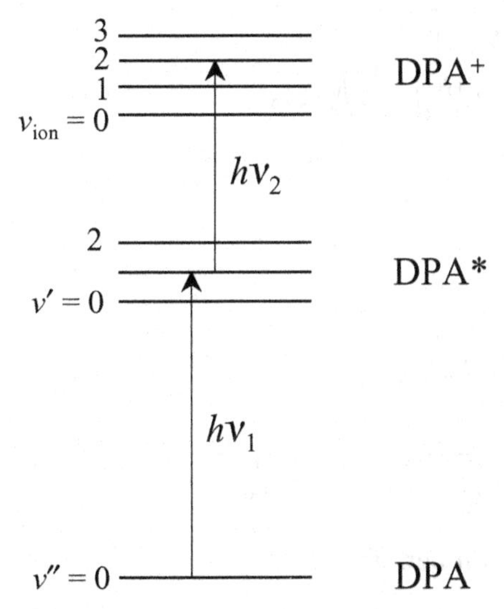

Figure 17.2 Laser excitation process employed in the MATI spectroscopy of DPA. The first laser pulse, of wavelength λ_1, excites the DPA from the zero-point level in S_0 to a specific vibrational level in S_1. A very short time later (~1 ns) a second dye pulse, of wavelength λ_2, produces resonant excitation to a specific vibrational level of the DPA cation in its ground electronic state. A single active vibrational mode is assumed in this simple diagram.

kinetic energy is zero (see Sections 12.5 and 12.6 for more details). The excitation process is summarized in Figure 17.2. It is important to recognize that the delay between the light pulses from the two dye lasers must be carefully controlled and kept very short, on the order of nanoseconds, otherwise the S_1 state will depopulate by mechanisms other than photoionization, e.g. by fluorescence back down to the ground electronic state.

Since the ionization process is channelled through a resonant intermediate state, S_1, the photoionization spectrum can be treated as if originating from that state. As mentioned above, it is possible to vary the specific vibrational level of the S_1 state from which ionization takes place by appropriate choice of λ_1. Figure 17.3 shows spectra originating out of $v' = 0, 1$, and 2 of S_1, where v' refers to the vibrational quantum number of the torsional mode of DPA

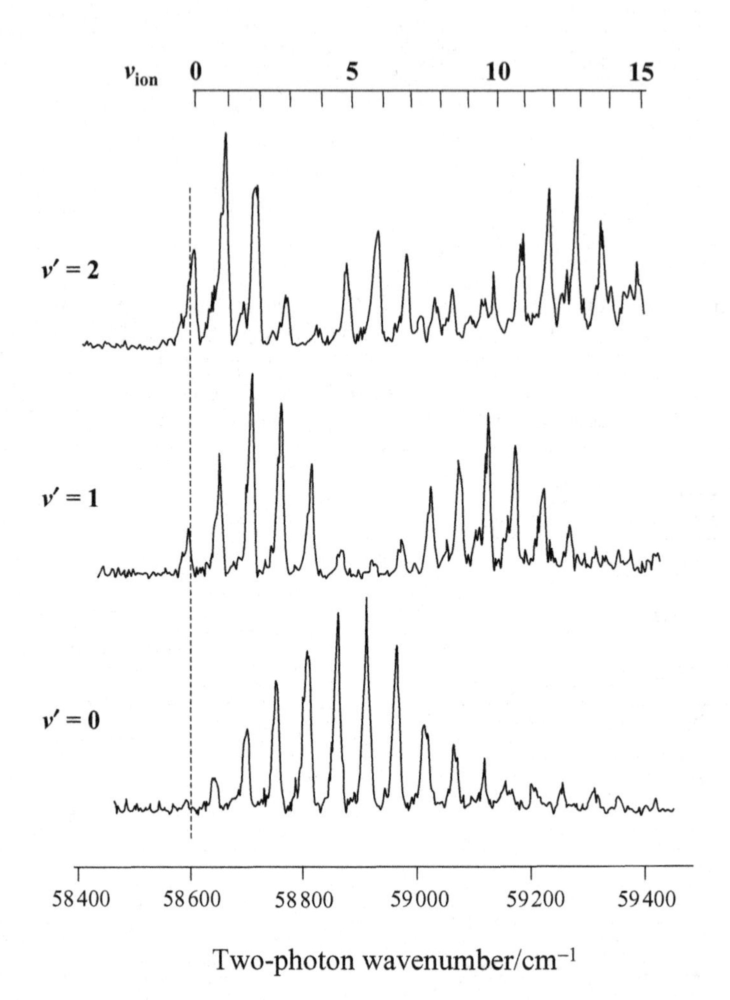

Figure 17.3 Photoionization spectra of DPA recorded by resonance-enhanced photoionization via the (a) $v' = 0$, (b) $v' = 1$, and (c) $v' = 2$ torsional levels of the S_1 state. The total energy needed for adiabatic ionization of DPA is indicated by the dashed vertical line and is 58 600 $\pm$ 5 cm^{-1}. (Reproduced from M. G. H. Boogaarts, P. C. Hinnen, and G. Meijer, *Chem. Phys. Lett.* **223** (1994) 537, with permission from Elsevier.)

in S_1. The torsional mode involves twisting of the two phenyl rings relative to each other, and corresponds to the motion in the angle ϕ in Figure 17.1. If DPA were planar in both the S_0 and S_1 states then Franck–Condon arguments would preclude significant torsion vibrational structure in the $S_1 \leftarrow S_0$ spectrum. The fact that such structure is observed indicates a change in geometry in the direction of the phenyl torsion normal coordinate, i.e. a change in equilibrium value of angle ϕ. In fact DPA is known to be planar in the S_0 state but in the S_1 state the two phenyl rings twist relative to each other by about 35° in opposite directions relative to the C–N–C plane to produce a non-planar equilibrium geometry [2]. Thus in the state with lowest symmetry, the S_1 state with C_2 point group symmetry, the phenyl torsion is a totally symmetric normal mode and so there is no vibrational selection rule, i.e. Δv can take any value in the S_1–S_0 transition.

A number of conclusions can be drawn from inspection of Figure 17.3. First, the vibrational structure is dominated by a progression in a single mode, as is especially evident for photoionization from $v' = 0$. The vibrational interval is rather small, about 53 cm^{-1}, and therefore a low frequency vibration in the cation is responsible. The phenyl torsion would be expected to be a low-frequency mode, since it involves the twisting of two relatively heavy phenyl groups, and is therefore the likely candidate for the progressions. In fact it is possible to discern a small contribution from another active mode, with a frequency of ~400 cm^{-1}, which is most noticeable in the spectrum for photoionization from $v' = 2$. However, this additional active mode will be ignored since it gives only very weak features.

The vibrational numbering for the torsional mode in the ion is given at the top of Figure 17.3 and applies to all three spectra. How was this numbering arrived at? It is not always easy to establish the vibrational numbering in an electronic spectrum by simple inspection, since it is not unusual to find that the early members of a progression are too weak to observe. Consider the bottom trace in Figure 17.3, which is the spectrum for photoionization via $v'=0$ in the S_1 state. The most intense band clearly corresponds to $v_{ion} \gg 0$, which reveals that there is a substantial change in the torsion angle in moving from S_1 to the ground electronic state of the cation. It is not obvious that the very weak peak attributed to $v_{ion} = 0$, which is barely perceptible above the background noise, is correctly assigned. However, confirmation is provided by the middle and top spectra in Figure 17.3, where the observed band at lowest wavenumber is the same for these spectra. The first band becomes much stronger for excitation through $v' = 1$ and $v' = 2$ and yet no additional band appears at lower wavenumber, proving conclusively that the first band corresponds to $v_{ion} = 0$.

The different intensity distributions in the three spectra are rather interesting and can be explained by employing the quantum mechanical form of the Franck–Condon principle. This states that the transition probability for a particular member of a vibrational progression is proportional to the square of the vibrational overlap integral for the two electronic states involved in the electronic transition (see Section 7.2.2). Key to interpreting the intensity distributions is to recognize that the long progression in the torsional mode indicates a substantial change in the torsional angle ϕ on excitation from S_1 to the ground electronic state of the ion. This is represented in Figure 17.4 by a displacement of potential energy curves for these two states. Consequently, to explain the intensity profiles it is only necessary to consider those parts of the vibrational wavefunctions where significant overlap is possible. This region is marked on Figure 17.4 by the dashed vertical lines for the specific case of transitions out of $v' = 1$, and corresponds to the full spatial extent of the lower state vibrational wavefunction, $\psi_{v'=1}$.

Projecting $\psi_{v'=1}$ vertically upwards, overlap with the ion vibrational wavefunction improves from $v_{ion} = 0 \rightarrow 2$. Thereafter the overlap decreases because of cancellation of regions of positive and negative overlap. This is specifically illustrated in the inset of Figure 17.4 in the lower right corner, which brings together the wavefunctions for $v' = 1$ and $v_{ion} = 5$. The peak of the wavefunction for $v_{ion} = 5$ lies almost directly above the node for $v' = 1$. As usual in integration, a definite integral evaluated between the limits a and c

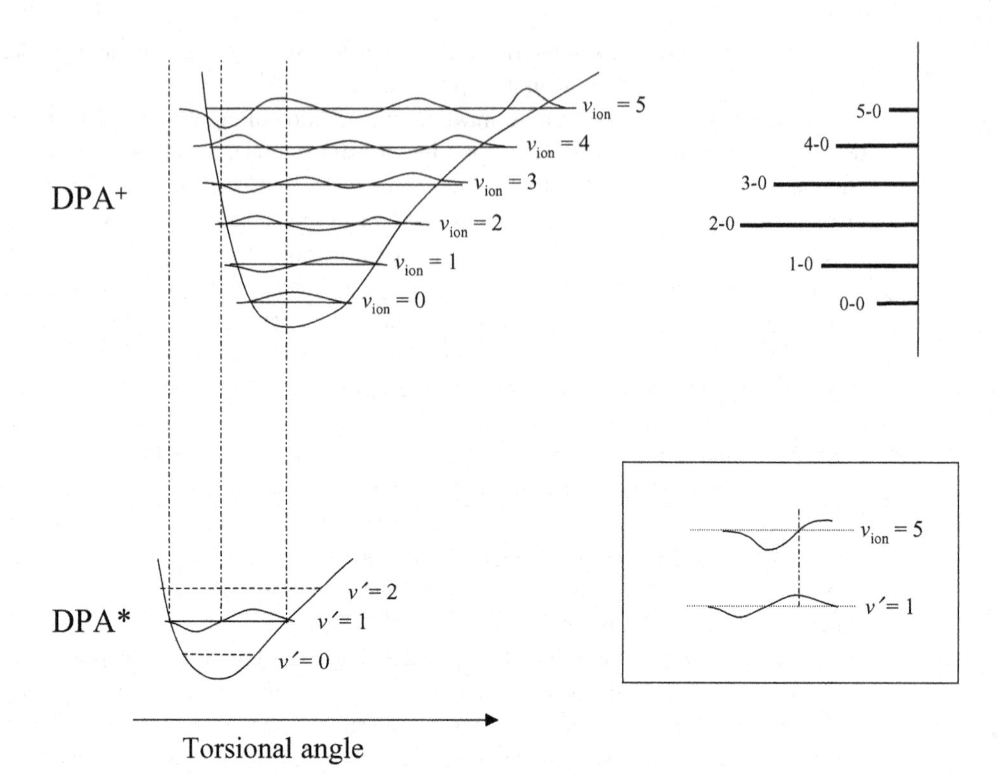

Figure 17.4 This diagram (not to scale) provides the basis for understanding the relative band intensities in the torsional progressions in Figure 17.3. The upper potential energy curve is for the DPA cation, while the lower curve is for the S_1 excited electronic state of neutral DPA. The horizontal coordinate corresponds to a change in torsional angle, ϕ. Vibrational wavefunctions in the ion up to $v_{ion} = 5$ are shown, while only that for $v' = 1$ in the S_1 state is explicitly shown. On the upper right-hand side a stick drawing of the MATI spectrum is shown with the lines coinciding with the specific ion vibrational levels accessed. The inset in the bottom right corner shows an expanded view of the $v' = 1$ and $v_{ion} = 5$ wavefunctions (see text for more details).

can be rewritten as

$$\int_a^c f(x)\,dx = \int_a^b f(x)\,dx + \int_b^c f(x)\,dx$$

where $f(x)$ is the function being integrated, and b lies between a and c. In other words, the overlap integral between $v' = 1$ and $v_{ion} = 5$ can be expressed as the sum of the overlap integrals originating either side of the node in $v' = 1$. These overlap integrals will have similar magnitudes but opposite signs, since the phase of $\psi_{v'=1}$ changes as the node is crossed whereas the vibrational wavefunction in the ion has the same sign over most of this region. Consequently, a small overlap integral will result and this explains why the transition $v_{ion} = 5 \leftarrow v' = 1$ is relatively weak.

A particularly interesting feature of the MATI spectrum via $v' = 1$ is that the intensity of the progression begins to rise again above $v_{ion} = 6$. This is because the ion torsional wavefunction becomes increasingly peaked near the inner turning point of the potential energy curve, and this turning point shifts increasingly to the left as the vibrational ladder is climbed. The overlap is now concentrated on the left-hand lobe of $\psi_{v'=1}$, reducing the effect of overlap cancellation due to the change in phase of $\psi_{v'=1}$. Eventually, above $v' = 13$ according to the spectrum in Figure 17.3, the main lobe of the ion wavefunction moves beyond the inner turning point of the potential curve for S_1, and so the Franck–Condon factor decays towards zero as v_{ion} increases further.

What is clear for the MATI transitions via $v' = 1$ is that the spectrum reflects the shape of the $v' = 1$ vibrational wavefunction. This fascinating result is mirrored for MATI transitions via $v' = 0$ and $v' = 2$, where zero and two 'nodes' are seen, respectively. (One of the nodes for the $v' = 2$ case is partly obscured by activity in an additional vibrational mode mentioned earlier.) The observation of intensity 'nodes' in long vibrational progressions is a well-known phenomenon that has been used in some cases as a means of assigning vibrational quantum numbers. Notice that in this case the original assignment of torsional vibrational quantum numbers in S_1 is confirmed by the intensity distributions.

Finally, notice that the onset of the photoionization spectra in Figure 17.3 provides a rather precise value for the *adiabatic* ionization energy of DPA.

References

1. M. G. H. Boogaarts, P. C. Hinnen, and G. Meijer, *Chem. Phys. Lett.* **223** (1994) 537.
2. J. R. Huber and J. E. Adams, *Ber. Bunsenges. Phys. Chem.* **78** (1974) 217.

18 Vibrational structure in the electronic spectrum of 1,4-benzodioxan: assignment of low frequency modes

Concepts illustrated: *low frequency vibrations in complex molecules;* ab initio *calculation of vibrational frequencies; laser-induced fluorescence (excitation and dispersed) spectroscopy; vibrational assignments and Franck–Condon principle.*

This Case Study demonstrates some of the subtle arguments that can be employed in assigning vibrational features in electronic spectra. It also provides an illustration of how important structural information on a fairly complex molecule can be extracted. The original work was carried out by Gordon and Hollas using both direct absorption spectroscopy of 1,4-benzodioxan vapour and laser-induced fluorescence (LIF) spectroscopy in a supersonic jet [1]. The direct absorption spectra were of a room temperature sample and were therefore more congested than the jet-cooled LIF spectra. Nevertheless, the direct absorption data provided important information, as will be seen shortly. For the LIF experiments, both excitation and dispersed fluorescence methods were employed (see Section 11.2 for experimental details). Only a few selected aspects of the work by Gordon and Hollas are discussed here; the interested reader should consult the original papers for a more comprehensive account [1, 2].

Possible structures of 1,4-benzodioxan are shown in Figure 18.1. Assuming planarity of the benzene ring, there are three feasible structures that differ in the conformation of the dioxan ring. One possibility is that both C—O bonds are displaced above (or equivalently below) the plane of the benzene ring yielding a folded structure with only a plane of symmetry (C_s point group symmetry). Alternatively, the dioxan ring could be in the same plane as the benzene ring (C_{2v}), or a twisted structure might occur in which the C2—C3 bond bisects the plane defined by the benzene ring by some non-zero angle (C_2). It should occur to the reader that it might be possible to distinguish between these possibilities on the basis of vibrational structure in the electronic spectrum, since the vibrational selection rules will be altered by a change of point group symmetry.

On further reflection the potential complexity of the vibrational structure might seem discouraging given that 1,4-benzodioxan has $3N - 6 = 48$ normal modes! However, three

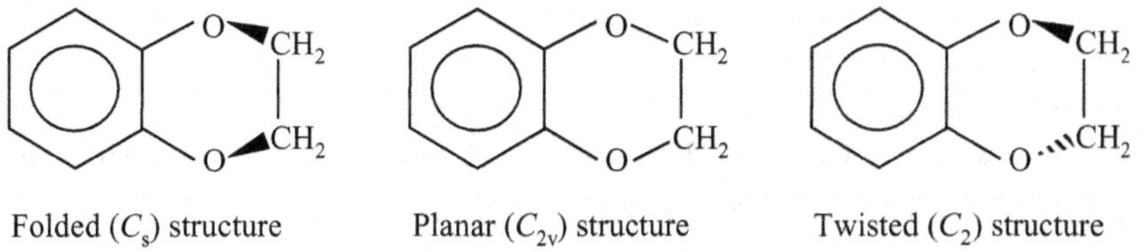

Folded (C_s) structure Planar (C_{2v}) structure Twisted (C_2) structure

Figure 18.1 Possible structures for 1,4-benzodioxan which differ in the conformation of the dioxan ring. The point group symmetries for each particular structure are in parentheses. Although not shown, the convention for numbering the framework atoms is to number them 1 through 10, starting from the upper O atom and continuing clockwise around the entire two-ring system.

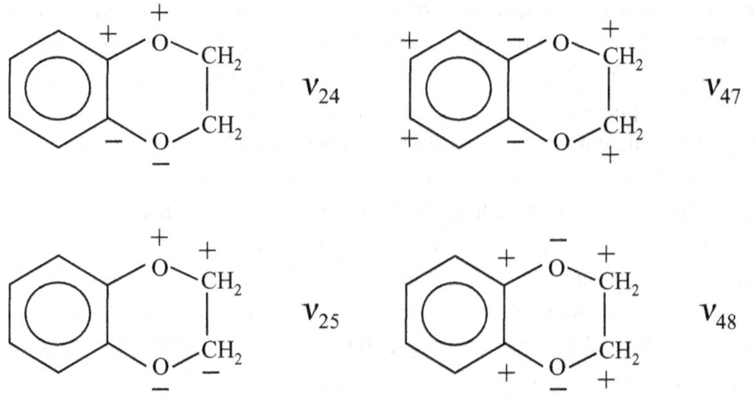

Figure 18.2 Illustration of the four low frequency skeletal vibrations ν_{24}, ν_{25}, ν_{47}, and ν_{48} for a planar (C_{2v}) structure of 1,4-benzodioxan. These vibrations take on a similar appearance for the C_s and C_2 structures, although their symmetries differ. The + and – symbols refer to out-of-plane displacements of the atoms, the – displacement being in the opposite sense to the + displacement.

factors offer some hope. First, whichever of the three conformations is the actual global equilibrium geometry, the fact that there is some symmetry in each case will limit the number of totally symmetric modes, and it is these which nearly always dominate in electronic spectra for reasons explained in Section 7.2.3 (and seen also in the previous Case Study). Second, many of the totally symmetric vibrational modes will not give rise to any detectable spectral features if there is no significant geometry change in the direction of the respective normal coordinates on electronic excitation. In other words, not all totally symmetric modes need be significantly Franck–Condon active. Finally, there are many vibrations but only the out-of-plane skeletal modes in the dioxan portion of the molecule will be significantly affected by any changes in the conformation of the dioxan ring. By comparison with similar modes in other molecules, the twisting and bending motions of the O1–C2–C3–O4 group would be expected to have relatively low frequencies, typically $<300\ \text{cm}^{-1}$. Consequently, the focus can be restricted to the region where higher frequency modes, such as skeletal stretching vibrations, cannot be observed.

The four vibrations considered by Gordon and Hollas are shown schematically in Figure 18.2. They are: the O1–C2–C3–O4 twist, designated ν_{25}; the O1–C10–C5–O4 twist,

ν_{24}; the puckering of the dioxan ring about O1–O4, ν_{48}; and the 'butterfly' bending of the two rings about C5–C10, ν_{47}. If C_s symmetry pertains, then only ν_{47} and ν_{48} will be totally symmetric, while if a twisted (C_2) structure occurs then ν_{24} and ν_{25} will be totally symmetric and ν_{47} and ν_{48} will be non-totally symmetric. All four of these modes will be non-totally symmetric if the molecule has C_{2v} symmetry.

18.1 *Ab initio* calculations

To help assign the spectrum, we will make use of the results from *ab initio* calculations. In turning to *ab initio* calculations for help the question of structure is in one sense immediately answered. However, it is always important to obtain experimental verification since the calculations may involve approximations that give misleading predictions. It is also worth noting that in the original spectroscopic work by Gordon and Hollas they did not have the luxury of being supported by *ab initio* calculations.

The *ab initio* calculations we turn to were reported by Choo and co-workers [3] several years after the studies by Gordon and Hollas. They made use of both Hartree–Fock (HF) and density functional theory (DFT), both of which are described in Appendix B. The DFT method, and especially a particular variant known as B3LYP, tends to be a significant improvement on the Hartree–Fock method for predicting both structures and harmonic vibrational frequencies without incurring much extra computational cost. The calculations by Choo *et al.* predicted that the twisted structure (C_2) is the equilibrium structure. We now use the spectra to confirm this prediction.

18.2 Assigning the spectra

Figure 18.3 shows a portion of the LIF excitation spectrum of jet-cooled benzodioxan. Two main bands are seen, the strongest at 35 563 cm^{-1} and another at 35 703 cm^{-1}. All bands in the spectrum, weak and strong, are due to transitions from the ground electronic state, which is a spin singlet and will therefore be designated S_0, to the first excited singlet electronic state, designated S_1. The strongest band corresponds to the origin transition of the S_1–S_0 system, i.e. it is due to excitation from the zero-point vibrational level in S_0 to the zero-point level in S_1. By convention, this is labelled 0^0_0. The 140 cm^{-1} separation between the origin and the other strong band, labelled A in Figure 18.3, is consistent with a vibrational progression involving excitation of a low frequency mode in the S_1 electronic state. The vibrational mode responsible will, for the moment, be left unassigned. Presumably the transition is from $v_A = 0$ in the lower electronic state (S_0) and $v_A = 1$ in the upper electronic state (S_1) and can therefore be labelled A^1_0.

Confirmation that band A is due to vibrational excitation in S_1 comes from dispersed fluorescence spectra. Dispersed fluorescence spectra obtained by laser pumping of the origin (0^0_0) and A^1_0 transitions are shown in Figures 18.4(a) and 18.4(b), respectively. A progression with an interval of ~164 cm^{-1} is obvious in the low frequency part of the spectrum in Figure 18.4(a). There are also other strong vibrational features in Figure 18.4(a) but they

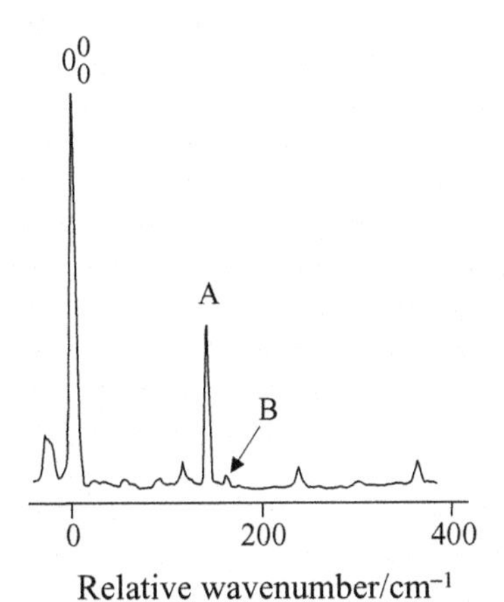

Figure 18.3 Laser-induced fluorescence excitation spectrum of 1,4-benzodioxan cooled in a supersonic jet. The wavenumber scale is relative to the position of the electronic origin band (0_0^0), which has an absolute wavenumber of 35 562.48 cm^{-1}. The assignment of bands A and B is discussed in detail in the text. (Reproduced with permission from R. D. Gordon and J. M. Hollas, *J. Chem. Phys.* **99** (1993) 3380, American Institute of Physics.)

involve modes above 400 cm^{-1} and are therefore not of interest here. The 164 cm^{-1} interval is not dissimilar to the 0_0^0–A_0^1 separation in the excitation spectrum and it is therefore tempting to suggest that mode A is also responsible for the low frequency structure in Figure 18.4(a). Proof that this is indeed the case comes from the dispersed fluorescence spectrum in Figure 18.4(b). Laser excitation of a particular mode in the excited electronic state should lead to an enhanced progression in that mode in the dispersed fluorescence spectrum. The same low frequency progression as seen in Figure 18.4(a) is clearly more prominent in Figure 18.4(b) and therefore mode A must be responsible. This use of dispersed fluorescence spectra to confirm vibrational assignments in an excitation spectrum is frequently employed by spectroscopists studying electronic spectra and is a powerful tool. The dispersed fluorescence spectrum also provides specific information on the vibrational levels in the ground electronic state, whereas the excitation spectrum provides complementary information for the excited electronic state.

Table 18.1 shows the predicted frequencies from the *ab initio* calculations by Choo and co-workers for the low frequency vibrations [3]. Comparing the measured frequency for mode A in the ground electronic state with the *ab initio* values, the only feasible assignment is to the ring twist (ν_{25}). The agreement between theory and experiment is good, the difference being only 7 cm^{-1} for the DFT calculation. Some differences would be expected due to approximations inherent in the DFT method. Furthermore, the calculations give *harmonic* vibrational frequencies whereas the experimental values are *fundamental* frequencies.[1]

[1] A transition between $v = 0$ and $v = 1$ levels for a given vibrational mode is known as the vibrational *fundamental* transition. The separation between these levels is approximately equal to the harmonic vibrational wavenumber, but an exact value would take into account the small but non-negligible contributions from vibrational anharmonicity.

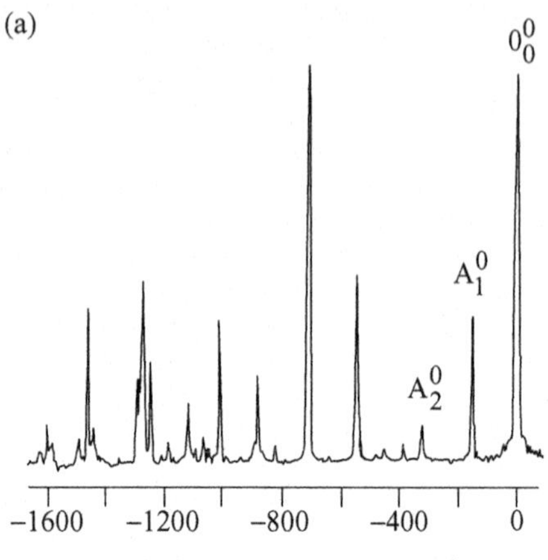

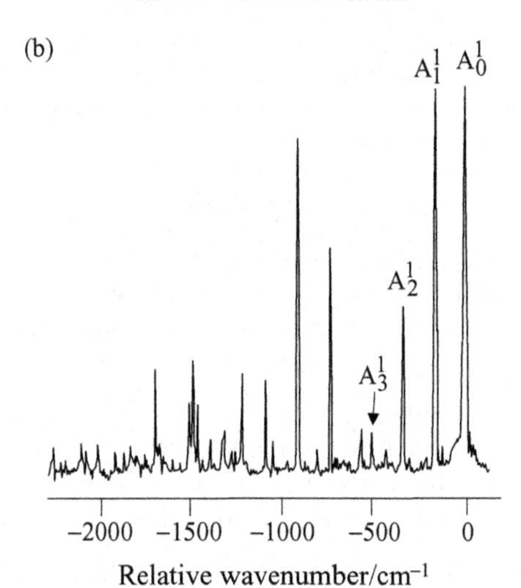

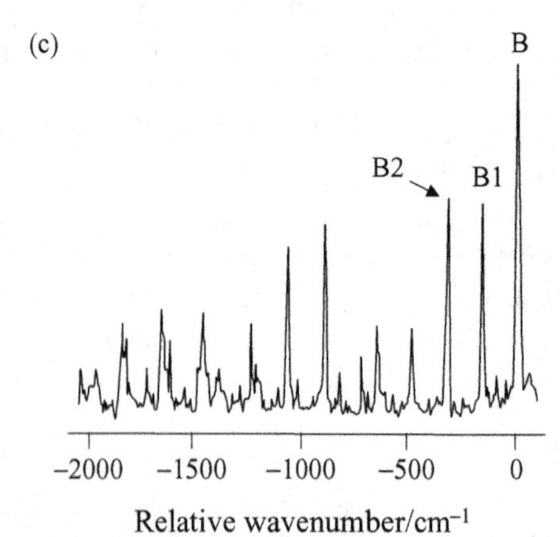

Figure 18.4 Dispersed fluorescence spectra obtained by laser excitation of the (a) 0^0_0, (b) A, and (c) B transitions shown in Figure 18.3. (Reproduced with permission from R. D. Gordon and J. M. Hollas, *J. Chem. Phys.* **99** (1993) 3380, American Institute of Physics.)

Table 18.1 Ab initio *frequencies*[a] *for out-of-plane skeletal vibrations in 1,4-benzodioxan*

Mode	Approximate description[b]	Symmetry[c]	HF/6-31G*	DFT-B3LYP/6-31G*
ν_{24}	Ring twist	A	316	334
ν_{25}	Ring deformation	A	171	171
ν_{47}	Ring flapping (butterfly)	B	295	305
ν_{48}	Ring puckering	B	91	107

[a] The HF (Hartree–Fock) values have been scaled by multiplying the original values by 0.9. This was applied by Choo and co-workers [3] because HF vibrational frequencies tend to overestimate experimental values by approximately 10%.
[b] The ring referred to in the mode descriptions is the dioxan ring.
[c] These symmetries assume a twisted (C_2 point group) equilibrium structure for the molecule.

Nevertheless, the level of agreement is such that we can have reasonable confidence in assigning other vibrational bands using the *ab initio* results.

We shall also consider one additional band, band B in Figure 18.3. This is only one of several very weak bands in the excitation spectrum but it takes on special significance when establishing the structure of 1,4-benzodioxan, as was realized by Gordon and Hollas. Band B is 159 cm^{-1} above the 0^0_0 band, and when the laser is tuned to this transition it gives the dispersed fluorescence spectrum shown in Figure 18.4(c). The resulting spectrum is similar to that obtained by laser exciting 25^1_0 (formerly designated A^1_0), but the interval between bands is larger, being ~208 cm^{-1}. It looks as if the assignment of band B is to some transition B^1_0, where B is another of the low frequency skeletal modes. However, inspection of Table 18.1 shows that there is no out-of-plane skeletal mode with a frequency close to this value. An alternative assignment must therefore be sought.

A possibility that must be considered is that the vibrational mode responsible for band B is non-totally symmetric. In the Franck–Condon limit, this would mean that only transitions with $\Delta\nu$ = even are allowed but only $\Delta\nu = \pm 2$ transitions are likely to have any significant probability. Such transitions would be expected to be very weak for reasons described in Section 7.2.3. According to Table 18.1, the ring puckering mode (ν_{48}) has just about the right frequency. Neglecting anharmonicity, the DFT calculations predict $2\nu_{48}$ at 214 cm^{-1}, compared to the observed value of 208 cm^{-1} for the first member of the progression, B1, in the dispersed fluorescence spectrum. The agreement is excellent given the approximations involved and leaves little doubt that this assignment to the 48^2_2 transition is correct. Similarly, band B2 in the dispersed fluorescence spectrum in Figure 18.4(c) is assigned to the 48^2_4 transition.

The assignment of band B in the excitation spectrum to the 48^2_0 transition is strong experimental support for the theoretical prediction that 1,4-benzodioxan adopts a twisted (C_2) structure at equilibrium. As mentioned earlier, the C_2 structure is the only one of the three shown in Figure 18.1 for which ν_{48} is non-totally symmetric. Gordon and Hollas arrived at the same assignment and the same overall conclusion about the molecular structure without the benefit of *ab initio* calculations. Their assignment of the 48^2_0 transition was derived from a careful analysis of sequence bands near the origin transitions in the absorption spectrum. Further details can be found in References [1] and [2].

References

1. R. D. Gordon and J. M. Hollas, *J. Chem. Phys.* **99** (1993) 3380.
2. R. D. Gordon and J. M. Hollas, *J. Mol. Spectrosc.* **163** (1994) 159.
3. J. Choo, S. Yoo, S. Moon, Y. Kwon, and H. Chung, *Vib. Spectrosc.* **17** (1998) 173.

19 Vibrationally resolved ultraviolet spectroscopy of propynal

Concepts illustrated: *electronic structure; symmetries of electronic states; absorption versus laser-induced fluorescence spectra; jet cooling;* ab initio *calculation of structures and vibrational frequencies; vibrational assignments and the Franck–Condon principle.*

Aldehydes and ketones have well-known electronic transitions in the ultraviolet associated with the carbonyl group. The longest wavelength (lowest energy) system is a $\pi^* \leftarrow n$ transition in which an electron from a lone pair on the oxygen atom is promoted to a C=O antibonding molecular orbital. As noted in several of the earlier Case Studies, it is common to denote the ground state singlet as S_0, and the first excited singlet state as S_1, and we talk of the $S_1 \leftarrow S_0$ transition. The exact wavelength at which absorption takes place depends on the degree of substitution and the type of substituent.

Propynal is a relatively simple aldehyde but its room temperature electronic absorption spectrum, shown in Figure 19.1, is rich in vibrational structure [1]. The presence of extensive vibrational structure is predictable if the effect of the excitation of the non-bonding electron to the π^* orbital is considered. Conjugation of the C≡C and C=O bonds is likely to result in planar (C_s point group) equilibrium geometries for both the S_1 and S_0 states:

```
              H
             /
H−C≡C−C
             \\
              O
```

However, electronic excitation should lead to a *weakening* of the C=O bond, since a non-bonding electron in the S_0 state now occupies an antibonding π^* orbital in the S_1 state: thus the S_1 state should have a longer C=O bond and a lower vibrational frequency.

Ab initio calculations would be useful to interpret the spectra, and so we have carried out calculations at the HF/6-31G* and the CIS/6-31G* levels[1] (the former for the S_0 state, the

[1] A CIS calculation on an excited electronic state is equivalent in quality to a Hartree–Fock (HF) calculation on the ground electronic state.

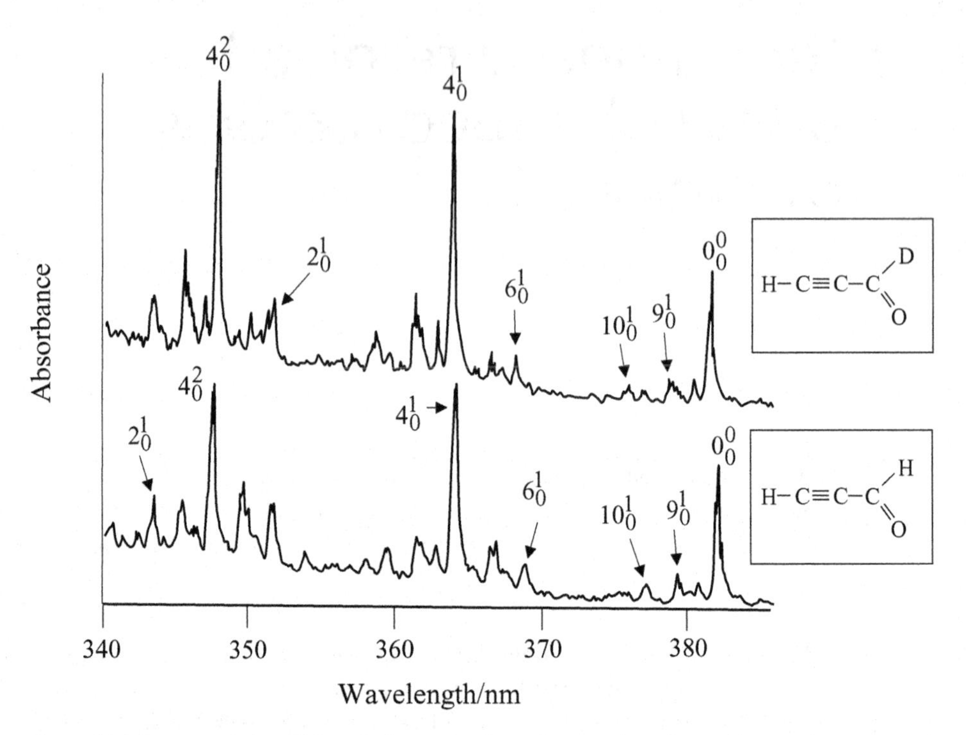

Figure 19.1 Low resolution (0.1 nm) electronic absorption spectra of room-temperature propynal (lower) and d_1-propynal (upper) vapours. (Reproduced from U. Brühlmann and J. R. Huber, *Chem. Phys.* **68** (1982) 405, with permission from Elsevier.)

latter for the S_1 state – see Appendix B for more details about these methods). Table 19.1 shows the calculated equilibrium structural parameters.

As may be seen, the major bond length change is a lengthening of the C=O bond, as expected, upon excitation. There are also other small structural changes, which result from changes in conjugation and electron repulsion brought about by electronic excitation. Application of the Franck–Condon principle suggests that the vibrationally resolved electronic spectrum will be dominated by the C=O stretch, which is a totally symmetric (a') vibration.

Propynal has twelve normal modes, and a group theoretical analysis reveals that if the molecule is planar nine vibrations have a' symmetry and three are a'' modes. It is normally safe to ignore non-totally symmetric vibrations when interpreting major vibrational features in electronic spectra. However, this statement is conditional on there being no change in equilibrium symmetry during the electronic transition. In the case of propynal, if the molecule is planar in the ground electronic state and non-planar in the excited state, then significant Franck–Condon activity in one or more out-of-plane bending modes would be expected. These bending modes would be totally symmetric in the excited state because the molecule would have only C_1 symmetry. Fortunately, this complication does not appear to arise for propynal since the *ab initio* results suggest planarity is maintained on electronic excitation.

Table 19.1 *Structural parameters (at equilibrium) for the S_0 and S_1 states of propynal from* ab initio *calculations*

Structural parameter[a,b]	S_0(HF/6-31G*)	S_1(CIS/6-31G*)
C1—H1	1.06	1.06
C1—C2	1.19	1.19
C2—C3	1.46	1.42
C3—H2	1.09	1.08
C3—O	1.19	1.27
∠C2—C3—H2	114.7	116.7
∠H2—C3—O	121.7	118.6

[a] The numbering of the carbon framework begins from the acetylenic end of the molecule.
[b] Bond lengths are in Å and bond angles are in degrees.

From the comments earlier, we therefore expect extensive Franck–Condon activity in the C=O stretch with some, but likely lesser, activity being possible in other a' vibrations.

19.1 Electronic states

The *ab initio* calculations predict a ground electronic state in which all occupied orbitals are full. This is as expected given that propynal is a relatively stable compound that can be synthesized and handled using standard laboratory techniques. The ground state will therefore be a spin singlet, as implied by the S_0 designation used earlier. However, since all orbitals are full, the overall spatial symmetry of the electronic state must be A'. It is therefore possible to dispense with the S_0 label and refer to the ground state by its full symmetry, $\tilde{X}^1A'$. The additional label $\tilde{X}$ specifies that this state is the lowest (ground) electronic state possessing $^1A'$ symmetry (there are higher energy states with this symmetry).

Excitation of an electron from an in-plane non-bonding orbital on oxygen to the carbonyl π^* orbital, which has a'' symmetry, will produce an excited electronic state of overall symmetry of A''. Singlet or triplet spin multiplicity is possible, but by far the strongest transition will be the spin-allowed $\tilde{A}^1A'' - \tilde{X}^1A'$ system.

19.2 Assigning the vibrational structure

Faced with the spectrum of propynal for the first time, there are a number of pieces of information that could be employed to assist the assignment process. Some of the structure could be assigned by a combination of chemical intuition and knowledge of the spectra of related carbonyl compounds. Since the three largest peaks are equally spaced, this structure appears to be part of a vibrational progression, and our first guess would be that it is due to the C=O stretch. The vibrational assignment process is often greatly assisted by obtaining spectra of isotopically substituted molecules, and this was done for propynal by

Table 19.2 *Harmonic vibrational frequencies of propynal obtained from* ab initio *calculations*

Mode	Approximate description	Symmetry	Harmonic frequency/cm^{-1}	
			S_0	S_1
ν_1	C—H stretch (C≡C—H)	a'	3661	3659
ν_2	C—H stretch (HCO)	a'	3237	3335
ν_3	C≡C stretch	a'	2403	2261
ν_4	C=O stretch	a'	2003	1685
ν_5	HCO bend (in-plane)	a'	1551	1327
ν_6	C—C stretch	a'	1029	1067
ν_7	CCH bend (in-plane)	a'	811	857
ν_8	CCO bend	a'	686	551
ν_9	CCC bend	a'	255	188
ν_{10}	C—H wag (HCO)	a''	1125	626
ν_{11}	CCH bend (out-of-plane)	a''	872	476
ν_{12}	CCC bend (out-of-plane)	a''	327	371

recording the absorption spectrum for d_1-propynal (see Figure 19.1). Another useful source of information, which was not used in the original studies on propynal but which is easy to generate for this molecule using modern computers, is the *ab initio* vibrational frequencies. Table 19.2 summarizes the results of a HF/6-31G* calculation on the S_0 state of propynal, together with the corresponding values for the S_1 state calculated using the CIS/6-31G* method.

We can now begin to rationalize the assignment of the absorption spectrum. The electronic origin transition, designated 0_0^0, is centred at 26 171 cm^{-1} (382.1 nm). Establishing that this band is the true electronic origin rather than a vibrationally excited feature is straightforward since it is strong, and scans to lower energy than shown in Figure 19.1 reveal no convincing alternative.

The most prominent bands in Figure 19.1 have been attributed to a progression in mode ν_4, the C=O stretch. These bands have been labelled 4_0^n where the subscript indicates the vibrational quantum number in the ground electronic state and n is the vibrational quantum number in the excited electronic state. There is very strong evidence for this assignment. First, note that adjacent members of the progression are separated by approximately 1300 cm^{-1}. Infrared spectra of aldehydes in their ground electronic states with an acetylenic CC bond between C2 and C3 show a C=O stretching band in the range 1680–1705 cm^{-1}. The much lower frequency deduced from Figure 19.1 is for the excited electronic state, and is in line both with our expectations from consideration of the bonding changes, and the results of the *ab initio* calculations in Table 19.2. Since electronic excitation *weakens* the carbon–oxygen bond by moving a non-bonding electron into the carbonyl π^* antibonding molecular orbital, the sharp fall in vibrational frequency is to be expected. Indeed it would be wrong to regard the carbon–oxygen bond as a double bond in the excited electronic state, but we will continue to retain the C=O notation for convenience.

Notice also that the spectrum of deuterated propynal is consistent with the C=O stretch assignment. Replacement of the H atom in the formyl group with a D atom should have

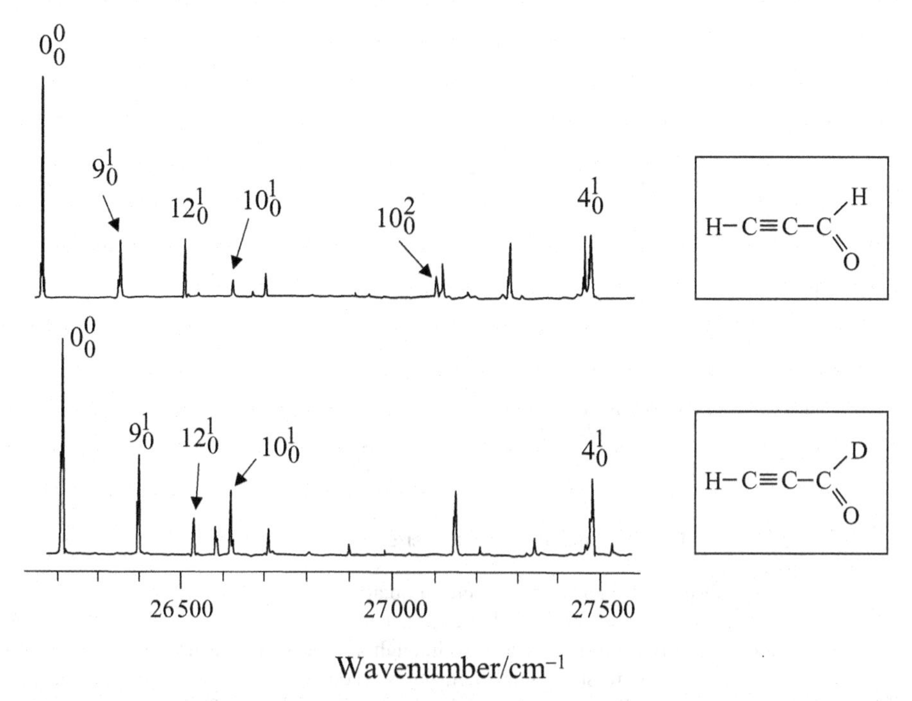

Figure 19.2 Vibrationally resolved laser excitation spectra of propynal (upper) and d_1-propynal (lower) cooled in a supersonic jet. The region shown extends to just beyond the 4_0^1 transition. (Reproduced with permission from H. Stafast, H. Bitto, and J. R. Huber, *J. Chem. Phys.* **79** (1983) 3660, American Institute of Physics.)

only a small effect on the C=O stretching frequency, and this is borne out by the similarity of the ν_4 structure in the two spectra in Figure 19.1.

Assignment of the vibrational stucture due to the C=O stretch is straightforward. It is more challenging, but possible, to assign virtually all of the remaining structure in the spectrum. However, rather than describe how the full assignment could be achieved, we will focus on *some* of the structure on the low wavenumber side of 4_0^1.

19.3 LIF spectroscopy of jet-cooled propynal

Cleaner spectra of propynal and d_1-propynal, obtained from supersonic jet expansions in argon carrier gas [2], are shown in Figure 19.2. Laser-induced fluorescence (LIF) excitation spectroscopy was used to record these spectra. Jet-cooling has lowered the rotational temperature dramatically, thereby narrowing rotational contours and thus sharpening each vibrational component. Also, although vibrational cooling is less efficient than rotational cooling, contributions from vibrational hot and sequence bands[2] are substantially reduced.

[2] Sequence bands are hot bands (transitions out of excited vibrational levels) in which the vibrational quantum does not change, e.g. 4_1^1.

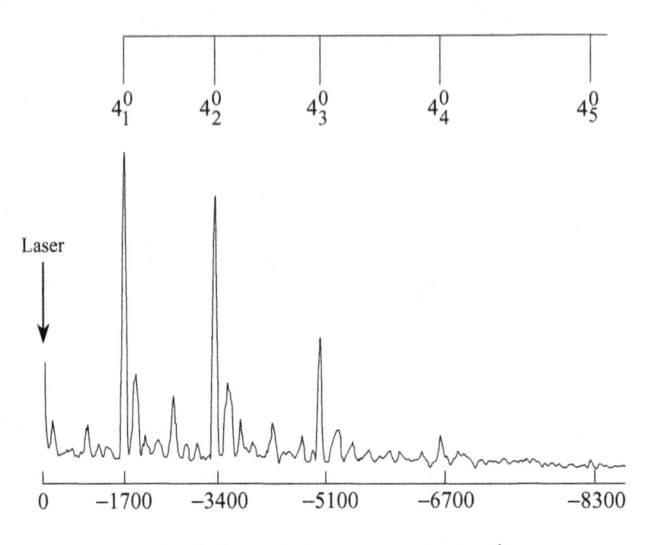

Figure 19.3 Dispersed fluorescence spectrum obtained by laser exciting the $\tilde{A}^1A''-\tilde{X}^1A'$ electronic origin (0^0_0) transition at 382.1 nm (26 171 cm^{-1}). The wavenumber scale is relative to the origin position. Note that this is a non-linear scale, so although it appears that the members of the ν_4 progression diverge, in fact the progression slightly converges, as one would normally expect. (Reproduced with permission from C. A. Rogaski and A. M. Wodtke, *J. Chem. Phys.* **100** (1994) 78, American Institute of Physics.)

Equally noticeable in comparing the room temperature absorption and jet-cooled laser excitation spectra are substantial differences in the relative intensities of some bands. This is not unusual, and arises because relative band intensities in LIF excitation spectra are affected not only by the absorbance of the molecule at a specific wavelength, but also by the fluorescence quantum yield for the excited energy level.[3] In fact the rates of fluorescence decay and non-radiative relaxation (via internal conversion) in propynal are known to be comparable. Furthermore, the rate of internal conversion depends on the vibrational mode excited, further complicating matters. Detailed studies of non-radiative decay in electronically excited propynal have been published [1, 2].

As in the absorption spectrum, the C=O stretch is active in the excitation spectrum. Dispersed fluorescence spectra show the activity in ν_4 even more clearly [3]. In Figure 19.3, which was obtained by laser excitation of the 0^0_0 transition, the dispersed fluorescence spectrum is dominated by structure in a single vibrational mode. Since this spectrum arises from emission from a single excited level, non-radiative relaxation does not affect the relative band intensities. The structure is due to emission to different vibrational levels in

[3] Other factors might also affect relative band intensities. Variation in output power of the tunable laser as a function of wavelength is one possibility, and indeed the spectra in Figure 19.2 have *not* been corrected for this variation. Similarly, the efficiency of the light detector, usually a photomultiplier tube, may also vary over the scanned wavelength range.

the S_0 state, and the separation between peaks is exactly as expected for the C=O stretch in the ground electronic state of propynal.

The first two bands above the origin in the excitation spectrum of jet-cooled propynal are medium intensity bands at $+189\ cm^{-1}$ and $+346\ cm^{-1}$ from the origin. To help assign these bands, we can draw on the *ab initio* predictions in Table 19.1. In doing so, the reader should be aware that Hartree–Fock calculations, owing to their neglect of electron correlation, are renowned for overestimating vibrational frequencies, typically by 10%. However, this overestimation is only a *trend* and it is not unusual to find some predictions outside of this range. Thus one must use the data in Table 19.1 with caution. Nevertheless, Table 19.1 reveals that only two modes have the low frequencies required, modes 9 and 12, the CCC in-plane and CCC out-of-plane bends, respectively. Assuming the predicted frequency order is correct, and for Hartree–Fock calculations this is normally far more reliable than absolute frequency predictions, then the $+189\ cm^{-1}$ band can be assigned to the 9^1_0 transition and the $+346\ cm^{-1}$ band to the 12^1_0 transition. Both of these assignments are consistent with the spectrum of d_1-propynal, which shows negligible shift of these bands relative to the 0^0_0 band.[4]

There is little doubt about the assignments of the two low-frequency bands but both show, in different ways, the limitations of the arguments we have employed to explain the presence or absence of vibrational structure. The C≡C–C framework remains linear in both ground and excited electronic states, and therefore there is no structural change in the direction of the CCC bending normal coordinate. In light of this the intensity of the 9^1_0 band is surprising.

The observation of substantial intensity in the 12^1_0 transition may seem an even bigger problem, since ν_{12} is a non-totally symmetric (a'') vibration and therefore single quantum excitation in this mode is strictly forbidden by the Franck–Condon principle. Clearly there must be a breakdown of the Franck–Condon principle, and in fact the 12^1_0 transition gains its intensity from a form of vibronic coupling known as Herzberg–Teller coupling. This is discussed in several other Case Studies and in some detail in particular in Case Study 25. Vibronic coupling amounts to a breakdown of the Born–Oppenheimer separation of electronic and vibrational motions. Its effects often manifest themselves in electronic spectra, although it is more usual for it to give rise to weak bands rather than prominent features such as the12^1_0 band of propynal.

The limitations in using *ab initio* calculations on the ground electronic state to assign vibrational frequencies in an excited state are very clearly illustrated by the one remaining assigned band in Figure 19.2, the 10^1_0 band in the HCCCHO spectrum. Mode ν_{10} is the HCO wag, another out-of-plane vibration (a'' symmetry) whose single quantum excitation requires invoking vibronic coupling. However, what is particularly noticeable in this case is the enormous difference in the observed frequency ($462\ cm^{-1}$ for the fundamental) and that estimated from the *ab initio* calculations on the ground electronic state (see Table 19.2).

4 Notice however that the absolute positions of all bands are shifted on deuteration. This is due to the fact that the zero-point energy contains contributions from modes for which deuteration at the carbonyl end of the molecule has a large effect on the vibrational frequency (ν_2, ν_5, and ν_{10}). The sum of the zero-point energies differs for the $\tilde{X}$ and $\tilde{A}$ states, giving rise to the overall shift of the d_1-propynal to higher wavenumber relative to propynal.

Fortunately, the ν_{10} frequency obtained from the CIS calculations on the S_1 state is in far better agreement with experiment, showing the value of attempting *ab initio* calculations on excited electronic states when assigning electronic spectra.

Our comments on the vibrational structure have been far from exhaustive. There is more vibrational information contained in Figures 19.1 and 19.2 than has been discussed here, and the interested reader is encouraged to consult the original references for more detailed accounts [1–4].

References

1. U. Brühlmann and J. R. Huber, *Chem. Phys.* **68** (1982) 405.
2. H. Stafast, H. Bitto, and J. R. Huber, *J. Chem. Phys.* **79** (1983) 3660.
3. C. A. Rogaski and A. M. Wodtke, *J. Chem. Phys.* **100** (1994) 78.
4. C. T. Lin and D. C. Moule, *J. Mol. Spectrosc.* **37** (1971) 280.

20 Rotationally resolved laser excitation spectrum of propynal

Concepts illustrated: *near-symmetric rotor approximation; asymmetric rotors and asymmetry splitting; parallel and perpendicular bands.*

Vibrationally resolved LIF excitation spectra of propynal were met in the previous Case Study. In the present Case Study the focus is on the rotationally resolved laser excitation spectrum of propynal. This molecule is nominally an asymmetric rotor, since the only symmetry it possesses is a reflection plane (C_s point group). However, as we will see, it is a near-prolate symmetric rotor and therefore its rotationally resolved electronic spectrum can be largely understood in terms of the properties of a prolate symmetric top.

Figure 20.1 shows the excitation spectrum for the 6_0^1 band, where mode ν_6 is dominated by C≡C stretching character. This was taken from original work by Stafast and co-workers [1] in which propynal was seeded into a pulsed supersonic jet. The origin (0_0^0) band has very similar rotational structure.

20.1 Assigning the rotational structure

The rotational structure is relatively simple to assign, although it might look quite complicated at first sight. We will attempt to interpret this spectrum by treating propynal as a prolate symmetric top, and will subsequently consider what happens when this constraint is removed.

P, *Q*, and *R* branches are readily identified in the central portion of the spectrum in Figure 20.1. The intense *Q* branches are the most obvious features, and once identified then it is relatively straightforward to see that each is flanked by fully resolved *P* and *R* branches. On both sides of the central, and most intense, *P*/*Q*/*R* system there are additional, weaker *P*/*Q*/*R* systems. With some experience, this structure suggests to the spectroscopist that transitions out of different *K* levels are being observed. For a prolate symmetric top the rotational energy levels are described by equation (6.15) and the transition selection rules are

Parallel transitions	$\Delta K = 0, \quad \Delta J = 0, \pm 1$
Perpendicular transitions	$\Delta K = \pm 1, \Delta J = 0, \pm 1$

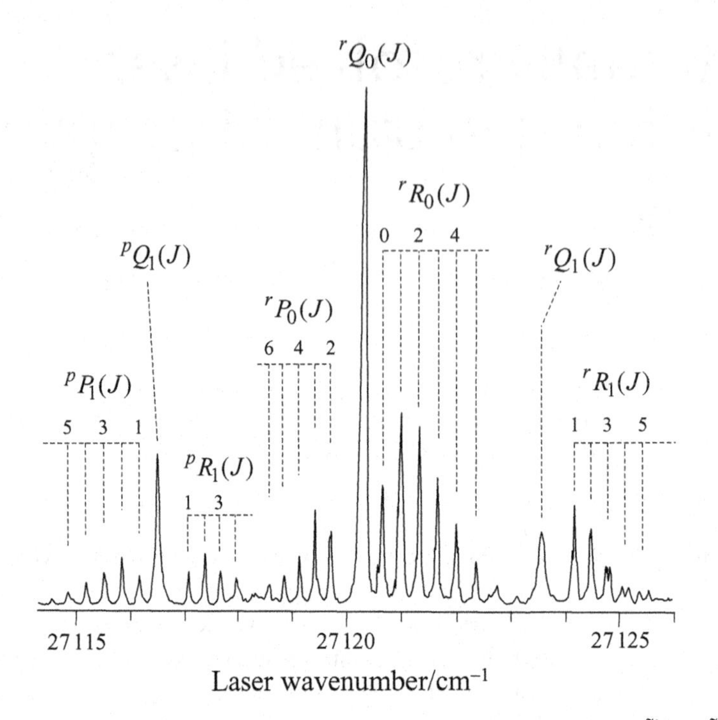

Figure 20.1 Rotationally resolved laser excitation spectrum of the $\tilde{A}^1A''-\tilde{X}^1A'$ 0^0_0 band of jet-cooled propynal. (Reproduced with permission from H. Stafast, H. Bitto, and J. R. Huber, *J. Chem. Phys.* **79** (1983) 3660, American Institute of Physics.)

A parallel transition is one in which the transition dipole moment is oriented along the *a* inertial axis. If the *A* rotational constants are similar in the upper and lower electronic states, then the second term on the right-hand side of (6.15) is irrelevant and we find that the rotational structure should consist of single *P*, *Q*, and *R* branches. This is *not* what is observed in Figure 20.1, so the transition must be dominated by perpendicular character. Using equation (6.15) transitions are expected at

$$\begin{aligned}\overline{\nu} = \overline{\nu}_0 &+ [(A' - B')K'^2 + B'J'(J' + 1)] \\ &- [(A'' - B'')K''^2 + B''J''(J'' + 1)]\end{aligned} \tag{20.1}$$

where $\overline{\nu}_0$ is the transition wavenumber for the pure 6^1_0 transition, i.e. in the absence of rotational structure. For simplification, assume that $A' = A''$ and $B' = B''$ so that the superscripts can be dropped. The above equation then simplifies to

$$\overline{\nu} = \overline{\nu}_0 + (A - B)(K'^2 - K''^2) + B[J'(J' + 1) - J''(J'' + 1)] \tag{20.2}$$

For *Q* branch transitions $J' = J''$ and therefore the second term on the right of (20.2) disappears. Consequently, for a perpendicular transition, *Q* branches are expected at the following positions:

$$\begin{aligned}
K' &= 1 \leftarrow K'' = 2 & \bar{\nu} &= \bar{\nu}_0 - 3(A - B)\\
K' &= 0 \leftarrow K'' = 1 & \bar{\nu} &= \bar{\nu}_0 - (A - B)\\
K' &= 1 \leftarrow K'' = 0 & \bar{\nu} &= \bar{\nu}_0 + (A - B)\\
K' &= 2 \leftarrow K'' = 1 & \bar{\nu} &= \bar{\nu}_0 + 3(A - B)\\
K' &= 3 \leftarrow K'' = 2 & \bar{\nu} &= \bar{\nu}_0 + 5(A - B)
\end{aligned}$$

Thus a series of Q branches are expected separated by $2(A - B)$. The most intense Q branch in a spectrum is expected to be that corresponding to $K' = 1 \leftarrow K'' = 0$, since $K'' = 0$ will be the most populated K level in the ground electronic state. Transitions from other K'' levels are possible but will be progressively weaker as K'' increases.

The features in Figure 20.1 can now be readily explained. The strong Q branch in the centre of the spectrum must be due to $K' = 1 \leftarrow K'' = 0$ transitions, with the high intensity deriving from unresolved contributions from various $Q(J)$ transitions. The weaker Q branches to higher and lower wavenumbers are, respectively, due to $K' = 2 \leftarrow K'' = 1$ and $K' = 0 \leftarrow K'' = 1$ transitions. We shall refer to these transitions with different values of K'' and K' as K sub-bands.

A specific labelling system is used for rotational transitions in electronic spectra of symmetric tops, which is an extension of that employed for linear molecules. P, Q and R branch transitions are labelled in the usual manner by specifying the rotational quantum number J in the lower state, i.e. $P(J)$ or $R(J)$. However, superscripts and subscripts are added to specify a particular sub-band. For example, in the ${}^{r}P_0(J)$ transition the pre-superscript 'r' reveals that $\Delta K = +1$, while the '0' subscript refers to the value of K in the lower state. In this way we can uniquely identify the upper and lower state rotational quantum numbers and this compact notation has been employed in Figure 20.1.

The P and R branches in each K sub-band are simple to interpret. From the second term on the right-hand side of equation (20.2), we expect P and R branch structure exactly analogous to that for linear molecules, i.e. a spacing of approximately $2B$ between adjacent members in a specific branch. However, certain members of a specific branch may be missing. For example, the first members of the R branches in both the $K' = 0 \leftarrow K'' = 1$ and $K' = 2 \leftarrow K'' = 1$ sub-bands are absent whereas that in the $K' = 1 \leftarrow K'' = 0$ sub-band is clearly seen. This is due to the fact that for $K'' = 0$ any value of J'' is possible but for $K'' = 1$ the lowest allowed value of J'' is 1, since K is the projection of J and therefore $J \geq K$. Although less obvious from the spectrum because of overlap with the stronger $K' = 1 \leftarrow K'' = 0$ sub-band, the first member of the P branch in the $K' = 2 \leftarrow K'' = 1$ sub-band is ${}^{r}P_1(3)$ since $J' \geq 2$ for $K' = 2$. Missing lines such as these make it possible to confirm the K quantum numbers in the upper and lower states, and this type of argument is commonly used to assign quantum numbers in rotationally resolved spectra.

20.2 Perpendicular versus parallel character

Why is the electronic transition perpendicular rather than parallel? In Case Study 19 it was suggested that the electronic transition involved is a $\pi^* \leftarrow n$ transition on the carbonyl group. In other words, an electron is moved from a non-bonding orbital on the C=O group to a π antibonding molecular orbital. In the non-bonding orbital the electron density is oriented in the plane of the molecule, whereas in the π^* orbital it is perpendicular to the plane. In electronic transitions the transition dipole moment reveals the direction in which the instantaneous shift in charge takes place. Since the charge shifts from an in-plane to out-of-plane orientation, the transition dipole moment must be perpendicular to the molecular plane. In other words, it is approximately perpendicular to the a inertial axis (see below), and explains why perpendicular character dominates in the rotationally resolved spectrum.

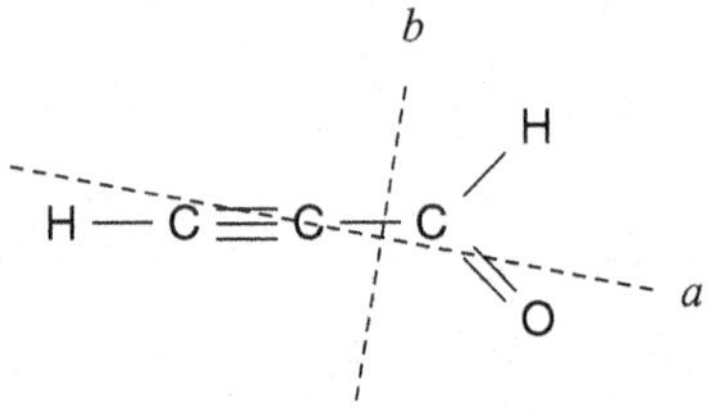

20.3 Rotational constants

In the prolate rotor limit the rotational constants can easily be determined from the spectrum in Figure 20.1. We will no longer assume that $A' = A''$ but we will continue to assume that $B' = B''$ because the $Q(J)$ transitions in a specific sub-band are unresolved (which is only possible if B' is very similar to B''). The separation between the observed Q branches can then be expressed as follows:

$$ {}^{r}Q_{1}(J) - {}^{r}Q_{0}(J) = 3A' - A'' - 2B = 3.27\,\mathrm{cm}^{-1} \tag{20.3} $$

$$ {}^{r}Q_{0}(J) - {}^{p}Q_{1}(J) = A' + A'' - 2B = 3.78\,\mathrm{cm}^{-1} \tag{20.4} $$

The wavenumbers are estimates taken from the spectrum. Similarly, $2B$ can be estimated from the average spacing between members of a particular branch, giving $B \approx 0.15\ \mathrm{cm}^{-1}$. Simultaneous equations (20.3) and (20.4) can then be solved to yield $A' = 1.91$ and $A'' = 2.17\ \mathrm{cm}^{-1}$.

Clearly these are only estimates of the rotational constants. In reality propynal is an asymmetric top so there is little point in pushing the analysis in terms of a symmetric rotor too far. Instead, it is better to consider the effect that asymmetry has on the rotational energy levels.

20.4 Effects of asymmetry

The impact of asymmetry is difficult to discern in Figure 20.1. This is partly because propynal is a good approximation to a prolate symmetric rotor, but also because of the modest resolution in the spectrum. Nevertheless, the keen-eyed reader may have noticed

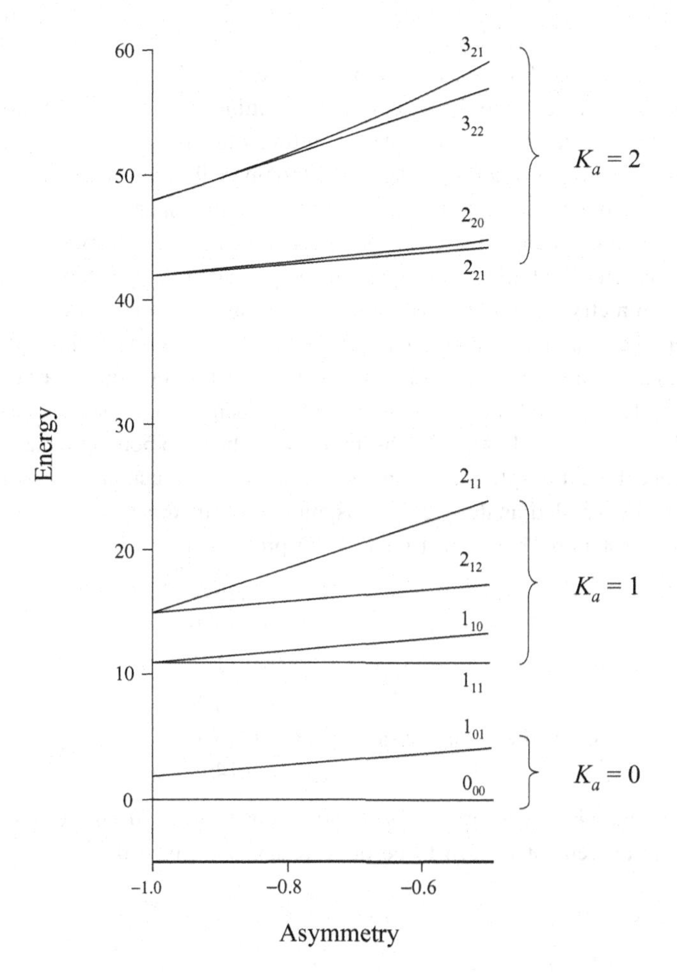

Figure 20.2 Variation of rotational energies of an asymmetric top as a function of the degree of asymmetry. This diagram was calculated for $A = 10$ and $C = 1$, the energy units being arbitrary. The asymmetry is expressed as an asymmetry parameter, κ, such that $\kappa = -1$ corresponds to a prolate symmetric top (for further details see, for example, Reference [4]). Propynal is a very good approximation to a prolate symmetric top – in its ground electronic state it has $\kappa = -0.99$.

that some of the lines in the ${}^rR_1$ branch are resolved into doublets. This splitting is caused by the breakdown of symmetric rotor behaviour.

In an asymmetric top the three rotational constants A, B, and C are all different. As a result K is no longer a good quantum number. In order to specify a particular rotational level a new labelling system must be introduced. The accepted notation is $J_{K_a K_b}$, where J has its usual definition. The quantities K_a and K_b are integers referring to the value of K with which a particular rotational level correlates in the prolate and oblate symmetric rotor limits, respectively.

Figure 20.2 shows how the energies of rotational levels change in moving from a symmetric rotor to an increasingly asymmetric top. At the extreme left the energies are those of a prolate symmetric top. The rotational energy is unaffected by the sense of rotation about the a axis, which contributes a two-fold degeneracy to each level with $K \neq 0$.[1] In an asymmetric top this degeneracy is removed, and the extent of the splitting increases as the molecule becomes more asymmetric. It is noticeable that the splitting, often referred to as asymmetry doubling, is largest for levels correlating with $K_a = 1$ (compare the splitting of the $2_{12}/2_{11}$ levels with the $2_{20}/2_{21}$ pair). This can be attributed to the higher speed of rotation about the a axis as K_a increases, which increases the prolate symmetric rotor character. Also, for a given K_a, the asymmetry doubling has an approximately quadratic dependence on J.

The effect of asymmetry should be most noticeable in transitions involving $K_a = 1$ in the upper or lower electronic state. In fact, for reasons beyond the scope of this book (see Reference [3] for more details), selection rules prevent the direct observation of asymmetry doubling in the $K' = 1 \leftarrow K'' = 0$ and $K' = 0 \leftarrow K'' = 1$ sub-bands. It is, however, visible in the R branch of $K' = 2 \leftarrow K'' = 1$. Although asymmetry doubling in both upper and lower rotational levels contribute, the splitting will be dominated by the much larger asymmetry splitting in $K'' = 1$. A detailed analysis of this asymmetry structure is possible and has yielded the following rotational constants (in cm^{-1}) for propynal [2]:

$$\text{Ground electronic state} \quad \begin{cases} A = 2.2694 \\ B = 0.1610 \\ C = 0.1501 \end{cases}$$

$$\text{Excited electronic state} \quad \begin{cases} A = 1.8893 \\ B = 0.1630 \\ C = 0.1498 \end{cases}$$

These can be compared with the estimates for A and B shown earlier from the symmetric rotor model and the agreement is seen to be quite reasonable given the approximation involved.

References

1. H. Stafast, H. Bitto, and J. R. Huber, *J. Chem. Phys.* **79** (1983) 3660.
2. J. C. D. Brand, W. H. Chan, D. S. Liu, J. H. Callomon, and J. K. G. Watson, *J. Mol. Spectrosc.* **50** (1974) 304.
3. *Molecular Spectra and Molecular Structure. III. Electronic Spectra and Electronic Structure of Polyatomic Molecules*, G. Herzberg, Malabar, Florida, Krieger Publishing, 1991.
4. *Angular Momentum*, R. N. Zare, New York, Wiley, 1988.

[1] There is also $2J + 1$ degeneracy for each rotational level so the overall degeneracy is $2(2J + 1)$ for $K \neq 0$ levels.

21 ZEKE spectroscopy of $Al(H_2O)$ and $Al(D_2O)$

Concepts illustrated: *atom–molecule complexes; ZEKE–PFI spectroscopy; vibrational structure and the Franck–Condon principle; dissociation energies; rotational structure of an asymmetric top; nuclear spin statistics.*

The study of molecular complexes in the gas phase provides important information on intermolecular forces and spectroscopy has played a vital role in this field. As an illustration, the complex formed between an aluminium atom and a water molecule is described here.

To obtain $Al(H_2O)$, it is necessary to bring together aluminium atoms and water molecules. Getting water into the gas phase is easy, but aluminium poses more of a problem since at ordinary temperatures the solid has a very low vapour pressure. An obvious solution is to heat the aluminium in an oven. However, the high temperature has a concomitant downside; if water is passed through (or near) the oven the high temperature will almost certainly prevent the formation of a weakly bound complex such as $Al(H_2O)$. Instead, the heat may allow the activation barriers to be exceeded for other reactions, leading to products such as the insertion species HAlOH.

A solution to this apparent quandary is to make $Al(H_2O)$ by the laser ablation–supersonic jet method, which was mentioned briefly in Chapter 8 (see Section 8.2.3). Any involatile solid, including metals, can be vaporized by focussing a high intensity pulsed laser beam onto the surface of the solid. The resulting plume of gaseous material above the surface, which includes metal atoms, can then be rapidly cooled by mixing with an excess of inert carrier gas, such as helium or argon. If a small amount of water vapour is seeded into the flowing carrier gas, formation of $Al(H_2O)$ complexes can occur. These are then rapidly cooled further by expanding the gas mixture into vacuum to form a supersonic jet (see Section 8.2.2).

In a recent study, Agreiter *et al.* formed $Al(H_2O)$ and $Al(D_2O)$ by the above procedure and obtained spectra of this complex for the first time using ZEKE spectroscopy [1]. This provided new information on both the neutral complexes and the corresponding cations, as described below.

21.1 Experimental details

$Al(H_2O)$ is unlikely to be the sole product when laser ablating solid aluminium in the presence of H_2O vapour. Despite the effort to cool the gas mixture, other reactions are almost certainly unavoidable. Furthermore, a variety of clusters and complexes might be formed involving multiple metal atoms and/or multiple water molecules. Experimental conditions can be optimized to favour production of $Al(H_2O)$, e.g. by adjusting the partial pressure of water vapour, but there will always be other species in the supersonic jet. Consequently, some means of selectively detecting the spectrum of $Al(H_2O)$ is beneficial.

Agreiter and co-workers recorded spectra using the PFI version of ZEKE, in which the laser wavelength is tuned to just below the ionization threshold and the complex is then ionized by application of a delayed pulsed electric field (see Section 12.6 for more details). The apparatus employed by Agreiter *et al.* was also equipped with a time-of-flight mass spectrometer, and so it proved possible to estimate the ionization energies of $Al(H_2O)_n$ complexes with different n by tuning the laser wavelength and looking for the onset of photoionization in a given mass channel. In this way, Agreiter *et al.* were able to confirm earlier work by Misaizu and co-workers in which the ionization energies of $Al(H_2O)_n$ complexes were found to decrease rapidly as a function of n [2]. Since pulsed field ionization is only observed close to the threshold for ionization, this information provides the means of distinguishing between the spectra of the various possible $Al(H_2O)_n$ complexes. The ionization energy of $Al(H_2O)$ is approximately 5.1 eV, much lower than that expected for chemical products such as HAlOH. It is therefore possible to be confident that the ZEKE spectra recorded in the region close to 5.1 eV (≡243 nm) originate from $Al(H_2O)$.

21.2 Assignment of the vibrationally resolved spectrum

ZEKE spectra of $Al(H_2O)$ and $Al(D_2O)$ are shown in Figure 21.1. Both spectra show an obvious vibrational progression. In addition, there is finer structure, which is particularly noticeable in the case of $Al(D_2O)$. This additional structure will be discussed later.

It is worth briefly reviewing what ZEKE spectroscopy reveals. In essence, resonant transitions between energy levels of the neutral molecule and the ion are recorded. Consequently, assuming most of the $Al(H_2O)$ and $Al(D_2O)$ complexes are initially in the zero-point vibrational level, as is reasonable given that they are entrained in a supersonic jet, then the observed vibrational structure is representative of the corresponding cations.

The fact that a single vibrational progression dominates the spectrum makes the assignment relatively easy. The active mode has a frequency of approximately 328 cm^{-1} for $Al^+(H_2O)$, as deduced from the spacing between adjacent peaks in the progression. The only challenge is to identify the specific vibrational mode responsible. $Al(H_2O)$ and its cation each have six degrees of vibrational freedom, which are illustrated in Figure 21.2.[1]

[1] The form of the six vibrations can be readily deduced. H_2O will contribute the same three vibrational modes as the free H_2O molecule, although the mode frequencies will differ from those of the free molecule. The formation of an Al—O bond will then add three further vibrations, an intermolecular (Al—O) stretching mode and two bending modes, one an in-plane and the other an out-of-plane (umbrella-like) deformation.

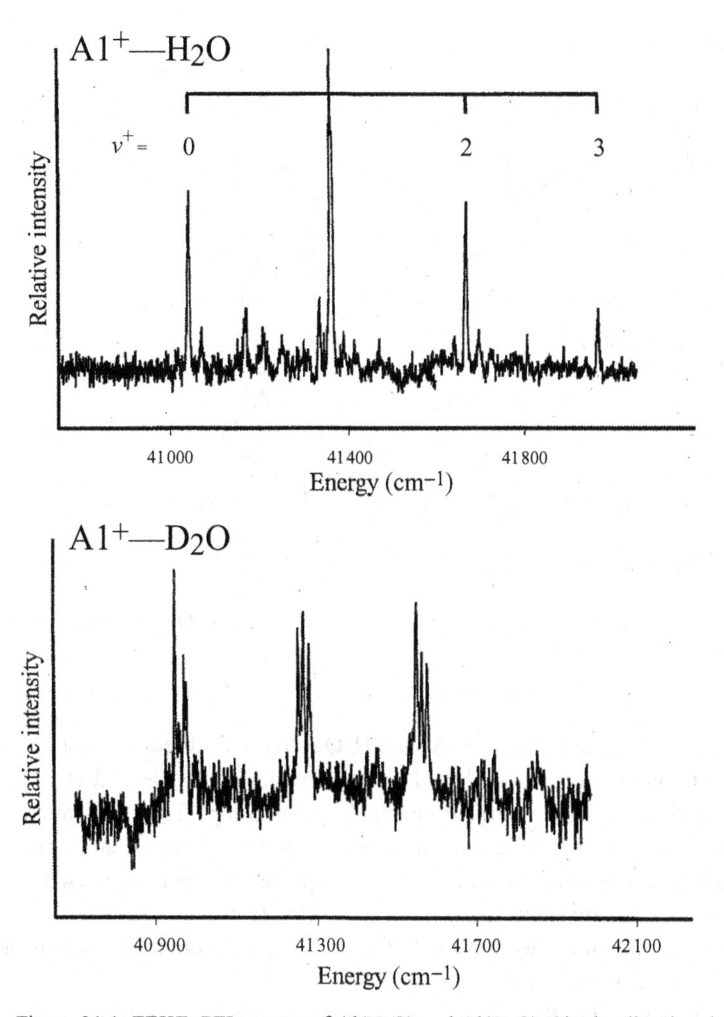

Figure 21.1 ZEKE–PFI spectra of $Al(H_2O)$ and $Al(D_2O)$. Single vibrational progressions dominate both spectra and the vibrational quantum number in the ion formed is shown above the $Al(H_2O)$ spectrum. (Reproduced from J. K. Agreiter, A. M. Knight, and M. A. Duncan, *Chem. Phys. Lett.* **313** (1999) 162, with permission from Elsevier.)

It will be assumed for the moment that the molecule has the C_{2v} structure shown in Figure 21.2, with the Al coordinated to the O atom. However, it is worth emphasizing that as yet we have presented no evidence to support this assumption.

The six vibrational modes can be divided into two groups, vibrations localized primarily on the water molecule and vibrations that are intermolecular in character. The former are essentially the same vibrations as found in a free water molecule, although with somewhat different frequencies because of the binding to an aluminium atom. We can be sure that these vibrations are not responsible for the progression in Figure 21.1 since all three water vibrations will possess far higher frequencies than 328 cm^{-1}.

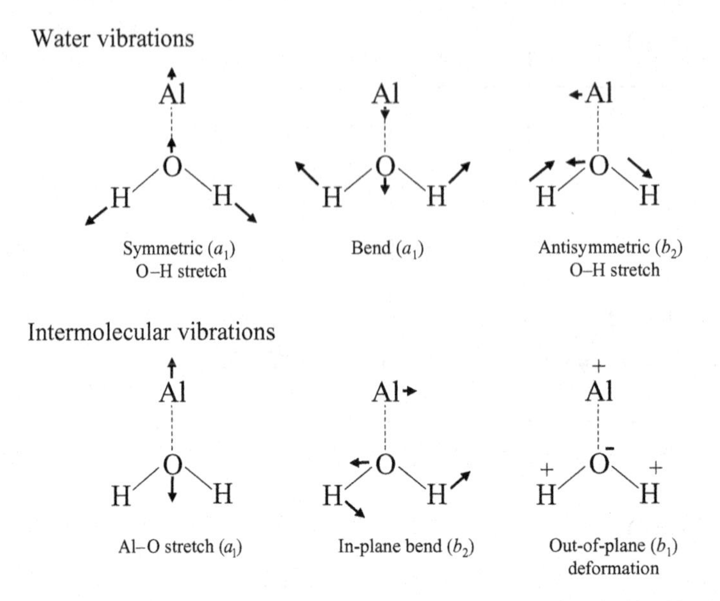

Figure 21.2 Schematic illustration of the six vibrational modes of Al(H_2O). C_{2v} point group symmetry has been assumed for the complex.

The formation of a bond between Al and H_2O introduces three additional vibrational modes, the intermolecular modes. One of these vibrations is the Al—O stretch, which is a totally symmetric motion (a_1 symmetry in the C_{2v} point group). The other two intermolecular vibrations are bending modes, one involving in-plane twisting of the water molecule relative to the Al atom, while the other is an out-of-plane deformation. These two bending modes will be non-totally symmetric in C_{2v} symmetry. If the complex has C_{2v} symmetry in both neutral and ionic states, then the Al—O stretch is the obvious assignment for the observed vibrational progression.

However, it is possible that a lower symmetry complex may be formed in either the ion or the neutral system, and in this case one or both of the bending modes may become Franck–Condon active. For example, if the complex is non-planar but the Al atom remains equidistant from the two H atoms, then the molecule will have a single plane of symmetry and will belong to the C_s point group. In this case the out-of-plane deformation would be totally symmetric and significant vibrational structure might result if there is a change in the equilibrium deformation angle on photoionization.

A comparison of the spectra for the deuterated and non-deuterated complexes establishes the assignment. The vibrational motion in the deformation mode is dominated by motion of the two hydrogen atoms. A large change in vibrational frequency would therefore be expected in switching from Al(H_2O) to Al(D_2O). The separations between adjacent peaks in the vibrational progressions show no such change, the decline in frequency being only 12 cm^{-1}. The main vibrational structure in Figure 21.1 can therefore be assigned to the Al—O stretching vibration.

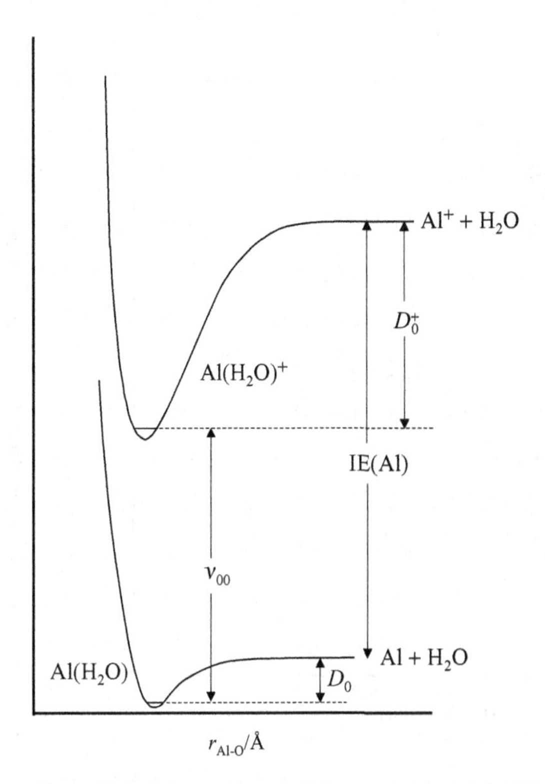

Figure 21.3 Schematic potential energy curves for $Al(H_2O)$. The potential energy is assumed to be a function of only the Al—O distance, i.e. the O—H bonds and the H—O—H angle are fixed. The quantities shown are as follows: D_0 = dissociation energy of neutral complex; D_0^+ = dissociation energy of the cation; ν_{00} is the energy of the 0_0^0 transition; IE(Al) = ionization energy of the aluminium atom.

21.3 Dissociation energies

An energy cycle, summarized in Figure 21.3, can be used to link the dissociation energies of the neutral and cationic complexes. The dissociation energy of the neutral complex, D_0, to give an Al atom and a free H_2O molecule, is related to that of the cation (D_0^+) by the expression

$$D_0 = \nu_{00} + D_0^+ - \text{IE(Al)} \tag{21.1}$$

where ν_{00} is the energy of the 0_0^0 (electronic origin) transition and IE(Al) is the ionization energy of the aluminium atom. Notice that the 0_0^0 transition energy in this case is identical to the adiabatic ionization energy of the Al-H_2O complex.

The electronic origin transition is readily identified from the ZEKE spectrum. The main vibrational progression is short, with the first member being relatively intense. There are no further members to lower energy and so the 0_0^0 transition is undoubtedly the first observed member of the progressions for $Al(H_2O)$ and $Al(D_2O)$ shown in Figure 21.1.

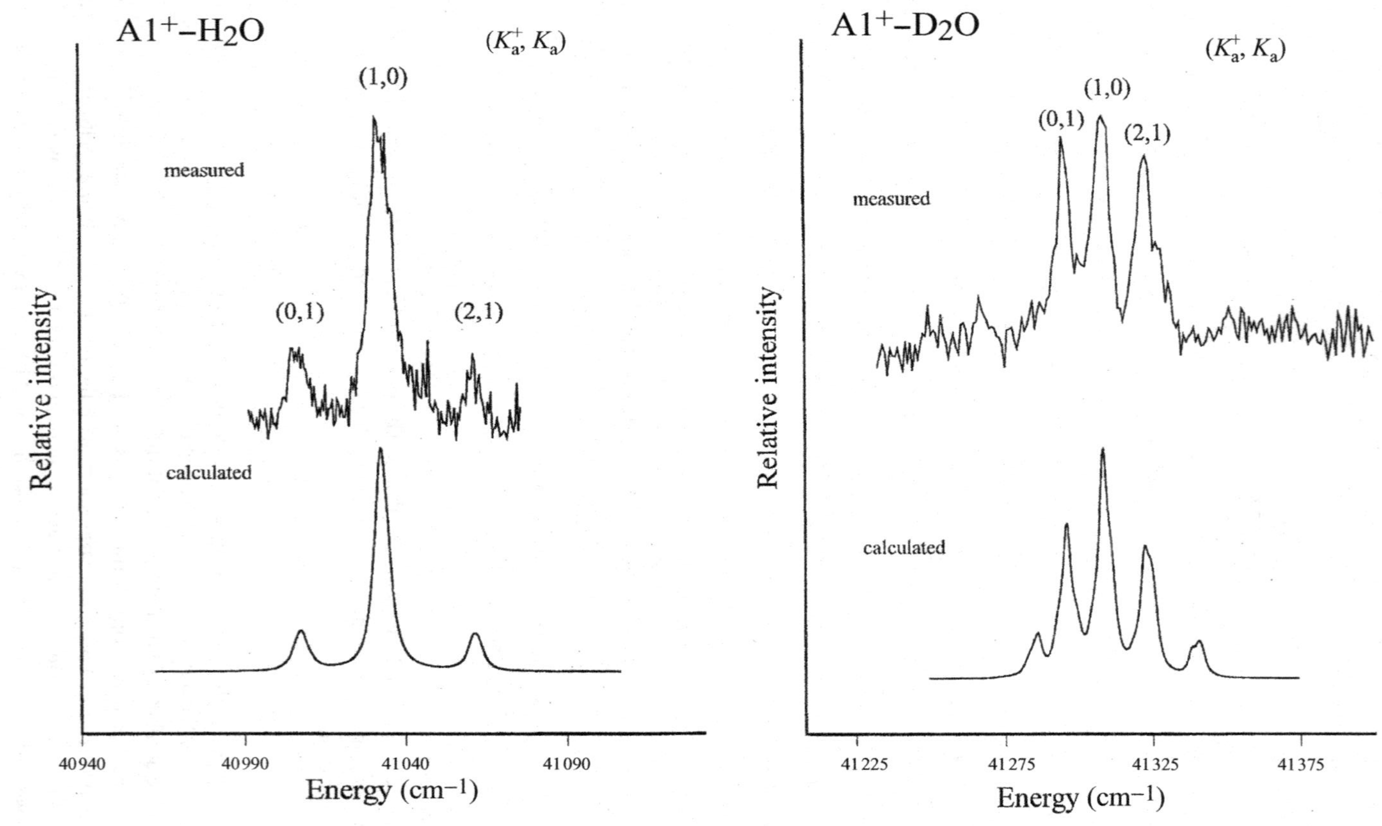

Figure 21.4 Rotational contours for Al(H_2O) and Al(D_2O). The bands shown are the electronic origin band (0_0^0) for Al(H_2O) and the 3_0^1 band for Al(D_2O). Beneath the ZEKE–PFI spectra are simulated band envelopes. The simulations assumed a rotational temperature of 10 K and the rotational constants in the neutral and ionic complexes were adjusted to achieve the best agreement with experiment. (Reproduced from J. K. Agreiter, A. M. Knight, and M. A. Duncan, *Chem. Phys. Lett.* **313** (1999) 162, with permission from Elsevier.)

Agreiter *et al.* identified the positions of these transitions as 41 018 ± 5 cm^{-1} for Al(H_2O) and 40 994 ± 5 cm^{-1} for Al(D_2O). The ionization energy of Al is known rather precisely, 48 278 cm^{-1}. This leaves the two dissociation energies as unknowns.

In principle, a Birge–Sponer extrapolation (see Case Study 23 for details) of the vibrational progression in the ZEKE spectrum could be attempted to estimate the dissociation energy of the cation. However, because the progression is relatively short this is likely to give a poor approximation to the true D_0^+. Fortunately, D_0^+ has been determined elsewhere in a mass spectrometry experiment in which the Al(H_2O)$^+$ ions were subjected to collisions with noble gas atoms [3]. The value obtained was 8700 ± 1260 cm^{-1}.

Substituting the above values into equation (21.1), we find that the dissociation energy of the neutral complex is 1440 ± 1260 cm^{-1}. The precision on this value is poor and so it is difficult to draw firm conclusions. However, as pointed out by Agreiter *et al.*, if the mean value of 1440 cm^{-1} is taken as representative, this indicates that the neutral complex might be rather strongly bound for this type of complex. A possible explanation for this is given in the next section.

21.4 Rotational structure

At higher resolution some coarse rotational structure is resolved in the ZEKE spectra of Al(H_2O) and Al(D_2O) (see Figure 21.4). At the relatively low resolution of the ZEKE data, the fine detail of the rotational structure is not revealed. Nevertheless, it is still possible to extract some useful information on the molecular structures.

If C_{2v} symmetry applies to both states, then although the neutral and ionic complexes will be asymmetric rotors, they will approximate prolate symmetric tops. In this limit the *a* inertial axis lies along the Al—O bond and therefore the *A* rotational constant is determined solely by the distance of the two H atoms from this axis. In a free water molecule the corresponding rotational constant is approximately equal to 14.5 cm^{-1}.

In a prolate symmetric top, the observed rotational structure depends on whether the transition moment is parallel or perpendicular to the *a* axis. In the parallel case, the selection rules are

$$\Delta K = 0 \quad \text{and} \quad \Delta J = 0, \pm 1$$

whereas for a perpendicular transition

$$\Delta K = \pm 1 \quad \text{and} \quad \Delta J = 0, \pm 1$$

At the relatively low resolution in the ZEKE experiments, the only structure that could possibly be resolved is the coarse structure due to $\Delta K = \pm 1$ transitions. It can therefore be concluded that the transition moment is perpendicular to the *a* axis.[2] Combining the $\Delta K = \pm 1$ selection rule with the formula for the energies of prolate symmetric rotors

[2] In near-prolate asymmetric rotors there are two 'perpendicular' inertial axes, *b* and *c*. The rotational structure for transition moments directed along these axes will differ, noticeably so if the corresponding rotational constants *B* and *C* differ substantially. It turns out that for Al(H_2O) a *b*-type transition gives the best agreement with experiment.

(equation (6.15)), a set of K *sub-bands* is expected in the perpendicular case with adjacent pairs separated by $\sim 2A$ (since $A \gg B$ in $Al(H_2O)$). The structure resolved in Figure 21.4 is consistent with this prediction. A strong central band is observed corresponding to the $K^+ = 1 \leftarrow K = 0$ sub-band. The P, Q, and R branch structure expected for this sub-band is unresolved in the ZEKE experiments. Either side of the central band are two weaker transitions originating out of the first excited K level, i.e. $K^+ = 0 \leftarrow K = 1$ and $K^+ = 2 \leftarrow K = 1$. These weaker bands are separated from the strongest band by ~ 28 cm^{-1} for $Al(H_2O)$, i.e. $\sim 2A$. In $Al(D_2O)$ the A constant will be a factor of two smaller and the actual band separations reflect this. In both isotopomers the populations of $K > 1$ levels are too small to register observable structure.

The comments above are consistent with the assumed C_{2v} symmetries for both neutral and cationic complexes. However, there is a further test that can be applied to the rotational structure to establish whether this symmetry really is applicable. In a C_{2v} geometry the two H atoms are equivalent and can be interchanged by a C_2 rotation about the a axis. It is therefore necessary to consider the effect of nuclear spin statistics (see Appendix F) in analysing the rotational structure. We will not attempt to derive the nuclear spin statistics for this particular case, but merely note the result. If the complex has C_{2v} symmetry then nuclear spin statistics introduces a 3:1 degeneracy ratio for odd:even levels of K.[3] The population of odd K levels is therefore boosted by a factor of three compared with even K levels. The effect of this is to increase the intensities of the $K^+ = 0 \leftarrow K = 1$ and $K^+ = 2 \leftarrow K = 1$ sub-bands relative to $K^+ = 1 \leftarrow K = 0$.

Simulations of the rotational structure by Agreiter *et al.* show that these nuclear spin statistics do not hold. In particular, the $K^+ = 0 \leftarrow K = 1$ and $K^+ = 2 \leftarrow K = 1$ sub-bands are far weaker than expected for a C_{2v} geometry. The simulated spectra shown beneath the actual ZEKE spectra in Figure 21.4 were generated assuming a non-planar (C_s) structure for the neutral complex. *Ab initio* calculations on $Al(H_2O)$ and $Al(H_2O)^+$ had been attempted by several groups prior to the ZEKE studies [3–5]. All agree that the cation is planar, but there is disagreement on whether the neutral complex is planar or not. The evidence from the ZEKE work suggests that the neutral complex is non-planar.

21.5 Bonding in $Al(H_2O)$

The simulated rotational structure in Figure 21.4 was obtained with a value of 11.75 cm^{-1} for the A rotational constant in the neutral complex. This is significantly smaller than the value in $Al^+(H_2O)$, which is similar to that expected for a free water molecule. A smaller value in the neutral complex could be obtained by a substantial lengthening of the O—H bonds and/or an opening out of the H—O—H bond angle; however, the changes required in these structural parameters are unreasonably large for such a weakly bound complex. The more likely explanation, already hinted at in the previous section, is that the complex is non-planar. Agreiter *et al.* were unable to suggest a unique equilibrium structure based on

[3] The deuterium nuclei are bosons and therefore a different nuclear spin degeneracy ratio of K = odd:even = 2:1 applies for $Al(D_2O)$ in C_{2v} symmetry.

the limited rotational structure in their ZEKE spectra, but they estimate a distortion from planarity of 30–40°.

This non-planarity is taken as evidence for some covalent bonding. The best estimates for the Al—O binding energy from *ab initio* calculations fall in the range 30–40 kJ mol^{-1} [6], which corresponds to 2500–3300 cm^{-1}. This is still a weak bond compared to typical covalent bond energies, but it is larger than expected on the basis of van der Waals forces alone. The *ab initio* estimates suggest that the mean value for the binding energy derived earlier, from a combination of the cation dissociation energy and the ZEKE data, underestimates the true bond energy in Al(H_2O).

References

1. J. K. Agreiter, A. M. Knight, and M. A. Duncan, *Chem. Phys. Lett.* **313** (1999) 162.
2. F. Misaizu, K. Tsukamoto, M. Sanekata, and K. Fuke, *Z. Phys. D.* **26** (1993) 177.
3. N. F. Dalleska, B. L. Tjelta, and P. B. Armentrout, *J. Phys. Chem.* **98** (1994) 4191.
4. S. Sakai, *J. Phys. Chem.* **97** (1993) 8917.
5. B. S. Jursic, *Chem. Phys. Lett.* **237** (1998) 51.
6. T. Fängström, S. Lunell, P. Kasai, and L. Eriksson, *J. Phys. Chem. A* **102** (1998) 1005.

22 Rotationally resolved electronic spectroscopy of the NO free radical

Concepts illustrated: *REMPI spectroscopy; cooling in molecular beams; rotationally resolved spectroscopy of an open-shell molecule; Hund's coupling cases.*

A rotationally resolved electronic spectrum of NO is shown in Figure 22.1. This was obtained for NO seeded into a very cold argon molecular beam. The electronic transition excited is the lowest energy allowed transition of the molecule and the spectrum was obtained using one-colour REMPI spectroscopy.

A molecular orbital diagram can easily be constructed for NO and it is readily seen that one unpaired electron resides in a $2p\pi^*$ orbital, making the ground electronic state a $^2\Pi$ state. The lowest energy transition that is observed turns out to be due to excitation of the $2p\pi^*$ electron up into a previously vacant σ orbital, leading to an excited electronic state of Σ symmetry which is 'Rydberg' in character. A Rydberg state is essentially one where the electron resides in an orbital that is large compared to the remaining core (NO^+ in this case), and the Rydberg energy levels take on a pattern rather similar to orbitals of atomic hydrogen. In the case of NO, the lowest Σ state has the electron in a $3s$-like orbital, and is denoted the $A^2\Sigma^+$ state. The A refers to the fact that this is the lowest optically accessible excited electronic state. The electronic transition therefore labelled as $A^2\Sigma^+ \leftarrow X^2\Pi$ electronic transition.

The first thing to note is that the spectrum consists of more than one line; attempts to cool the molecular beam further lead to a slightly simpler spectrum consisting of three lines, but further cooling does not significantly change the spectrum. We generally expect a single rotational line for a closed-shell molecule in the limit of zero absolute temperature, with this line corresponding to a transition from the lowest rotational level ($J'' = 0$) in the ground electronic state to the $J' = 1$ level in the upper electronic state.[1] The additional lines for NO

[1] A closed-shell diatomic molecule will always have a $^1\Sigma^+$ electronic ground state. If the excited state is also a $^1\Sigma^+$ state then the rotational selection rule is $\Delta J = \pm 1$, i.e. the Q branch is absent. If the transition is to a $^1\Pi$ excited state then the selection rule is modified to $\Delta J = 0, \pm 1$. However, the lowest possible value of J in a $^1\Pi$ state is $J = 1$ and so the first member of the Q branch is $Q(1)$. In the limit of $T = 0$ K the $J = 1$ level in a $^1\Sigma^+$ state will not be populated and therefore, despite the possibility of Q branch transitions, only the $R(0)$ transition can be observed.

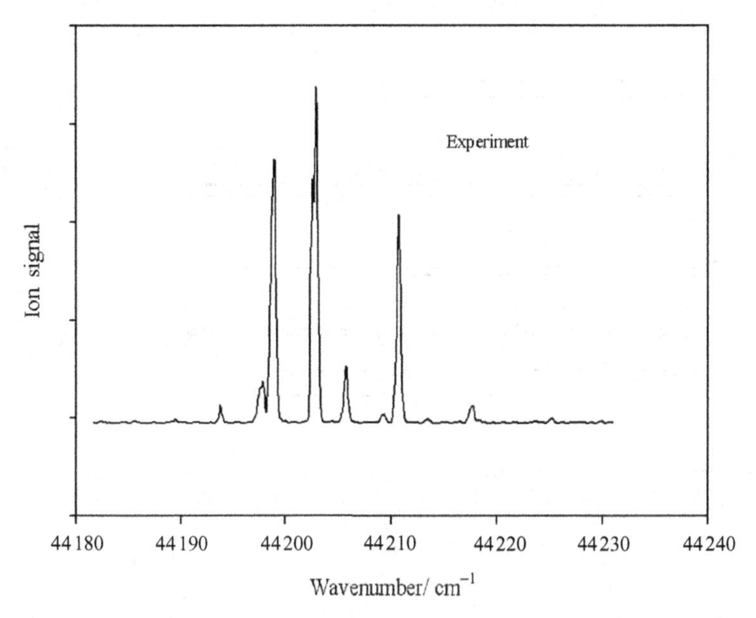

Figure 22.1 One-colour (1 + 1) REMPI spectrum of the NO $A^2\Sigma^+ \leftarrow X^2\Pi$ electronic transition recorded under molecular beam conditions.

must be the result of its open outer electronic shell, and in order to explain these lines it is necessary to consider how the rotational energy levels of open-shell molecules differ from closed-shell molecules.

The key thing to note is that open-shell molecules have spin and orbital angular momenta associated with the unpaired electron(s), and these angular momenta can couple with the rotational angular momentum of the molecule. This coupling can occur in several ways, but the two most common are outlined in Appendix G and are known as Hund's cases (a) and (b) – a fuller account of Hund's coupling cases may be found in References [1] and [2]. It is known that the $X^2\Pi$ state of NO closely matches Hund's case (a) behaviour, the reason being the large magnitude of the spin–orbit coupling (the splitting between the $^2\Pi_{1/2}$ and $^2\Pi_{3/2}$ spin–orbit sub-states is >120 cm^{-1}) relative to the rotational constant (<2 cm^{-1}). The $A^2\Sigma^+$ state has no orbital angular momentum and therefore exhibits Hund's case (b) behaviour.

In the $X^2\Pi$ state the spin and orbital angular momenta couple together to give a total electronic angular momentum along the internuclear axis, which is represented by the quantum number Ω, where $\Omega = \frac{1}{2}$ or $\frac{3}{2}$. A formal definition of Ω is given in Appendix G. It turns out that the $\Omega = \frac{1}{2}$ spin–orbit component is the lower in energy. Both spin–orbit components will have associated with them a series of rotational energy levels formed by coupling Ω with $\boldsymbol{R}$, where $\boldsymbol{R}$ is the rotational angular momentum of the molecule. The coupling together of Ω and $\boldsymbol{R}$ gives a total angular momentum denoted by quantum number J. Since the $\Omega = \frac{3}{2}$ manifold lies more than 120 cm^{-1} above the $\Omega = \frac{1}{2}$ manifold, then under efficient jet-cooling conditions only transitions from the lower spin–orbit

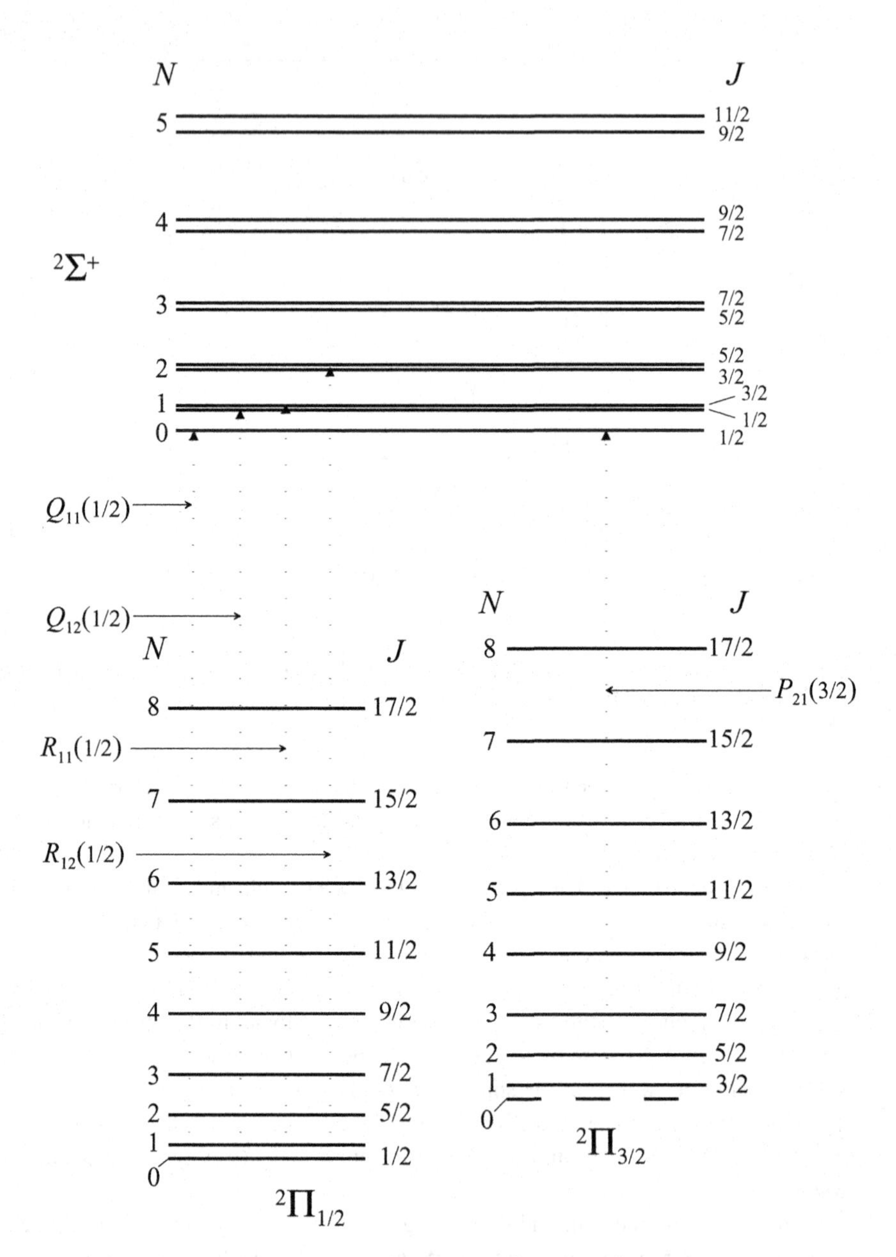

Figure 22.2 Rotational energy level scheme in the $A^2\Sigma^+$ and $X^2\Pi$ states of NO. The diagram is not to scale. The transitions responsible for the main lines seen in Figure 22.1 are also shown.

component should be observed. The energy level pattern is described by equation (G.2) in Appendix G.

As already mentioned, the $A\,^2\Sigma^+$ state follows Hund's case (b) coupling, and in this case the spin angular momentum of the unpaired electron cannot couple to the orbital angular momentum, since the latter is absent. However, to be consistent with other Hund's case (b) molecules, cases where the orbital angular momentum might not be zero, the quantum number N is used to represent the total angular momentum (orbital + rotational) minus spin. Here, the spin angular momentum couples to N to give the total angular momentum J. This has the effect of splitting each N level into two sub-levels, the splitting being known as *spin–rotation* splitting. One level in the spin–rotation pair has $J = N + \frac{1}{2}$ and the other $J = N - \frac{1}{2}$. The former levels are referred to as the F_1 manifold, and the latter as the F_2 manifold, with the energies being given by

$$F_1(N) = BN(N+1) + \tfrac{1}{2}\gamma N$$

$$F_2(N) = BN(N+1) - \tfrac{1}{2}\gamma(N+1)$$

where the quantity γ is known as the spin–rotation constant (which can be positive or negative, but is usually >0).

The arrangements of the rotational levels in the upper and lower electronic states of NO are illustrated in Figure 22.2. Note that for the lowest level in the $^2\Sigma^+$ state there is no spin–rotation splitting since the molecule is not rotating ($N = 0$) in this level, and so there is no rotational angular momentum to which $\boldsymbol{S}$ can couple.

At the very lowest temperatures, we expect that only the lowest level in the $X\,^2\Pi$ state, the $J = \frac{1}{2}$ level of the $\Omega = \frac{1}{2}$ manifold, will be populated. Since the transition involves $\Delta\Lambda \neq 0$, the selection rule for J is $\Delta J = 0, \pm 1$. Consequently, the $J = \frac{1}{2}$ and $\frac{3}{2}$ levels in the $^2\Sigma^+$ state can be accessed from the $J = \frac{1}{2}$ level in the $X^2\Pi$ state. Because of spin–rotation splitting there are actually four accessible levels, which we denote as (N, J), as follows:

$$\left(0, \tfrac{1}{2}\right), \left(1, \tfrac{1}{2}\right), \left(1, \tfrac{3}{2}\right), \quad \text{and} \quad \left(2, \tfrac{3}{2}\right)$$

The possible transitions can be labelled P, Q, and R in the usual manner where these denote $\Delta J = -1, 0$ and $+1$, respectively. The full labels used are $Q_{11}(\frac{1}{2})$, $Q_{12}(\frac{1}{2})$, $R_{11}(\frac{1}{2})$, and $R_{12}(\frac{1}{2})$, where the first subscript labels the initial F manifold and the second labels the terminating manifold. The number in parentheses is the value of J in the lower state, J''.

Figure 22.3 shows simulations of the $A^2\Sigma^+ \leftarrow X^2\Pi$ spectrum of NO at temperatures of 1, 3, and 10 K. The procedure employed to generate simulations like these is outlined in Appendix H. At the lowest temperature only three lines appear, which is consistent with the conclusion earlier based on experimental studies but which apparently contradicts the prediction above of a minimum of four rotational lines even at a temperature of absolute zero. However, the astute reader might attach significance to the fact that the middle line in the 1 K simulation is considerably more intense than the other two. Referring to the energy level diagram, and considering the transitions described above, we see that two of the transitions terminate at $N = 1$ in the $A^2\Sigma^+$ state, but with different J values. The splitting between these two levels is determined by the spin–rotation constant, γ, which is normally very small compared to the rotational constant. The resolution used in the simulation is too

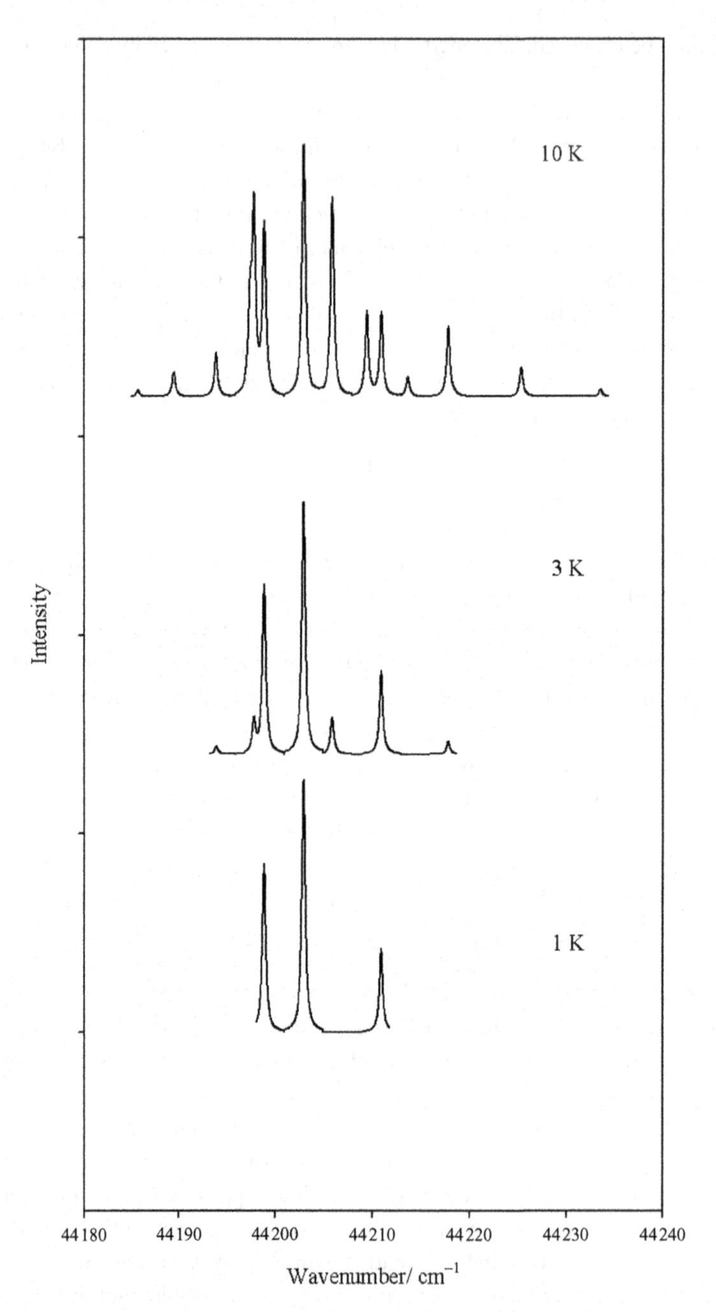

Figure 22.3 Simulations of the $A^2\Sigma^+ \leftarrow X^2\Pi$ spectrum of NO at 1, 3, and 10 K.

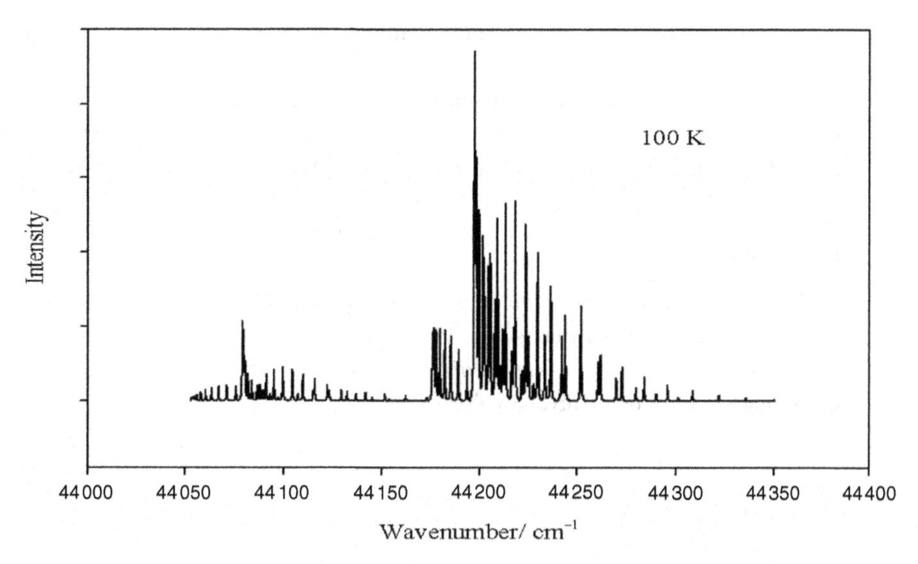

Figure 22.4 Simulation of the $A^2\Sigma^+ \leftarrow X^2\Pi$ spectrum of NO at 100 K.

low to resolve the spin–rotation splitting and so the middle line is actually a convolution of two transitions.[2] We can therefore assign the first spectral line as $Q_{11}(\frac{1}{2})$, the middle line to the unresolved $R_{11}(\frac{1}{2})$ and $Q_{12}(\frac{1}{2})$ transitions, and the highest wavenumber line as $R_{12}(\frac{1}{2})$.

If the NO sample is warmed, then additional rotational levels in the $\Omega = \frac{1}{2}$ manifold will become populated leading to a more complex spectrum. The simulations in Figure 22.3 at 3 K and 10 K begin to show these additional transitions from $J'' = \frac{3}{2}, \frac{5}{2}$, and $\frac{7}{2}$.

At even higher temperatures, the spectrum becomes rather congested and, at sufficiently high temperatures, the upper spin–orbit component ($\Omega = \frac{3}{2}$) in the $X^2\Pi$ state starts to contribute to the spectrum, as shown in the 100 K simulation in Figure 22.4. The features to the left of the spectrum are similar to (but not identical with) those on the right but are clearly much weaker.

One might initially think, given the above, that the aim of most spectroscopic experiments would be to record spectra under the coldest possible conditions. However, while it is true that this can help reduce congestion and therefore make spectral assignment simpler, it is not always an advantage. For example in the case of NO, information on the lower state could not be obtained from the spectrum recorded under the coldest conditions, and even for the upper state only the barest information can be gleaned from just three rotational features. In contrast, for a spectrum at 100 K (see Figure 22.4) a wealth of information on both the upper and lower states could be extracted because of the many rotational lines observed. This information includes bond lengths for both states (derived from the respective rotational constants), the spin–orbit splitting in the $X^2\Pi$ state, and spectroscopic constants beyond

[2] Note that the spacing between spin–rotation levels increases as a function of N, and so for high N it may be possible to resolve the two components even at modest spectral resolution.

the rigid rotor approximation. The spin–rotation parameter could also be obtained from the high-N regions of the spectrum.

Finally, we note that a comparison of the simulations in Figures 22.3 and 22.4 and the experimental spectrum in Figure 22.1 allows the temperature of the NO sample to be estimated – a temperature of ~3 K gives the best agreement.

References

1. *Molecular Spectra and Molecular Structure. I. Spectra of Diatomic Molecules*, G. Herzberg, Malabar, Florida, Krieger Publishing, 1989.
2. *Rotational Spectroscopy of Diatomic Molecules*, J. M. Brown and A. Carrington, Cambridge, Cambridge University Press, 2003.

23 Vibrationally resolved spectroscopy of Mg^+–rare gas complexes

Concepts illustrated: *ion–molecule complexes; photodissociation spectroscopy; symmetries of electronic states; spin–orbit coupling; vibrational isotope shifts; Birge–Sponer extrapolation.*

Laser-induced fluorescence, resonance-enhanced multiphoton ionization, and cavity ring-down spectroscopic techniques offer ways of detecting electronic transitions without directly measuring light absorption. An alternative approach is possible if the excitation process leads to fragmentation of the original molecule. By monitoring one of the photofragments as a function of laser wavelength, a spectrum can be recorded. This is the basic idea behind photodissociation spectroscopy.

There are limitations to this approach. If photodissociation is slow, then the absorbed energy may be dissipated by other mechanisms, making photodissociation spectroscopy ineffective. It is also possible that some rovibrational energy levels in the excited electronic state will lead to fast photofragmentation whereas others will not. In this case there will be missing or very weak lines in the spectrum which, in a conventional absorption spectrum, may have been strong. Fast photofragmentation is clearly desirable on the one hand, but it can also be a severe disadvantage if it is too fast, since it may lead to serious lifetime broadening in the spectrum (see Section 9.1).

Despite the above limitations, photodissociation spectroscopy can provide important information. This is particularly true for relatively weakly bound molecules and complexes, since these have a greater propensity for dissociating. In this and the subsequent example the capabilities of photodissociation spectroscopy are illustrated by considering weakly bound complexes formed between a metal cation, Mg^+, and rare (noble) gas (group 18) atoms. These will be referred to as Mg^+–Rg complexes.

One would expect the interaction between an Mg^+ ion and a rare gas atom to be weak, since the high ionization energies and closed electronic shells of the latter preclude the formation of ionic or covalent chemical bonds. The principal contribution to the van der Waals binding in Mg^+–Rg will be the charge-induced dipole interaction. As the name implies, the positive charge on the Mg^+ cation induces a dipole moment in the rare gas atom, and the interaction of this induced dipole moment with the charge on the cation results in a net attractive force.

In this particular Case Study some of the findings from vibrationally resolved photodissociation spectra of Mg^+–Rg complexes, obtained by M. A. Duncan's research group at the University of Georgia, will be explored. In the subsequent Case Study rotationally resolved spectra of the same complexes will be considered.

23.1 Experimental details

Duncan's group produced Mg^+ ions by pulsed laser ablation of a solid magnesium target located inside a specially designed pulsed nozzle. This technique was also briefly described in Section 8.2.3. The highly energetic ablation process leads to the formation of metal ions in the gas phase as well as neutral species. High pressure rare gas flows over the metal target and carries the mixture along to the exit aperture of the nozzle, where it expands into a vacuum chamber to form a supersonic jet. The subsequent cooling of the mixture allows the formation of weakly bound Mg^+–Rg complexes. Downstream of the nozzle the jet is skimmed to form a highly directional molecular beam,[1] and then enters a second vacuum chamber housing a time-of-flight mass spectrometer.

A tunable pulsed laser beam is directed into the second chamber to excite electronic transitions in Mg^+–Rg. Mg^+ fragment ions are then detected as a function of the laser wavelength using the mass spectrometer. In the lowest lying excited electronic states the ion complexes do not undergo dissociation when excited to bound rovibrational levels within each electronic state. This potentially renders photodissociation inoperable for these electronic transitions. However, a photodissociation spectrum *was* still observed, and this was found to be due to the absorption of a second photon from the same laser, which accesses a high lying, dissociative electronic state. This *resonance-enhanced* photodissociation technique, which only occurs with any significant probability when the first photon is resonant with a specific rovibronic transition, is directly analogous to the one-colour REMPI technique described in Section 11.4. The only difference is that in this case a photofragment ion was detected rather than a parent ion.

A potentially severe obstacle to the success of this experiment is the large background signal from those Mg^+ ions that do not form complexes with rare gas atoms in the supersonic expansion – these Mg^+ ions would clearly have the same mass as the Mg^+ arising from the photodissociation process. If not tackled, this would dramatically reduce the signal-to-noise ratio in the spectrum and, in all likelihood, make it impossible to record a satisfactory spectrum. Duncan and co-workers solved this problem by using a two-stage (tandem) time-of-flight mass spectrometer known as a *reflectron*. Ions in the molecular beam are extracted into the first stage *before* laser excitation and the instrument is set to transmit only Mg^+–Rg complexes of a specific mass. At the end of the first stage the tunable laser beam is admitted

[1] A skimmer is a cone-shaped object with the tip removed to form a small aperture. The supersonic jet flows towards the sharp end of the cone and only the central portion passes through the aperture and into the second vacuum chamber.

and interrogates the selected ion beam. The ions then enter the second stage of the mass spectrometer and the Mg^+ ion signal reaching the detector is distinguished from the Mg^+–Rg by virtue of the different flight times of these ions.

23.2 Preliminaries: electronic states

Since there is no chemical bonding between the Mg^+ and rare gas atoms, the electronic structures of these entities remain largely the same in Mg^+–Rg complexes. Rare gas atoms have full electronic shells and the energy required to excite an electron to a vacant orbital is high, requiring wavelengths far into the vacuum ultraviolet. On the other hand, Mg^+ has an unpaired electron in the $3s$ orbital in its electronic ground state and this can be excited to higher lying vacant atomic orbitals using near-ultraviolet radiation. Such transitions are therefore readily accessible with laser radiation. Consequently, the spectroscopy of Mg^+–Rg complexes in the near-ultraviolet is essentially the spectroscopy of the Mg^+ ion perturbed by the nearby rare gas atom.

The presence of a nearby rare gas atom will shift the orbital energies of the Mg^+ ion. The extent of the shift will depend on the orbital and the identity of the rare gas atom, as discussed later. At the same time the loss of spherical symmetry around the cation will change the symmetries of the orbitals and will remove some orbital degeneracies previously present in the free Mg^+ ion.

Figure 23.1 shows the basic idea. In the lowest electronic state of Mg^+ the unpaired electron resides in the $3s$ atomic orbital. Since all other occupied orbitals are full, this results in a ^2S electronic ground state. When a rare gas atom approaches, the unpaired electron remains localized almost entirely on the magnesium ion and the resulting orbital may still be viewed as a Mg $3s$ orbital. However, it is only an approximation, albeit a good one, and the use of the s label is only strictly applicable in an environment with spherical symmetry. In the complex, which has $C_{\infty v}$ point group symmetry, an s orbital becomes a σ^+ orbital. Similarly, the ^2S state of the free Mg^+ ion becomes a $^2\Sigma^+$ state in the Mg^+–Rg complex. This correlation is shown in Figure 23.1.[2]

Analogous correlations can be established for higher energy electronic states. The lowest unoccupied orbital in Mg^+ is the $3p$ orbital. Excitation of the unpaired electron from the $3s$ to the $3p$ orbital gives a ^2P excited state. This is a triply degenerate state, since there are three possible orientations of the p orbital which are energetically equivalent. However, when the rare gas atom approaches this three-fold degeneracy is removed, since the p orbital can either be oriented along the internuclear axis or perpendicular to it. This is illustrated in the orbital sketches on the right-hand side of Figure 23.1.

The energies of all the orbitals are lowered relative to free Mg^+ by the charge-induced dipole interaction. However, the lowering is greatest for the $3p_x$ and $3p_y$ orbitals. These

[2] The transformation properties of atomic orbitals in lower symmetry environments are readily deduced from inspection of the appropriate character tables. Individual s orbitals always transform as the totally symmetric irreducible representation, which for the $C_{\infty v}$ point group is σ^+. The symmetries of individual p and d orbitals can be deduced from the transformation properties of the corresponding cartesian coordinates, e.g. the np_x and np_y orbitals form a degenerate pair with π symmetry.

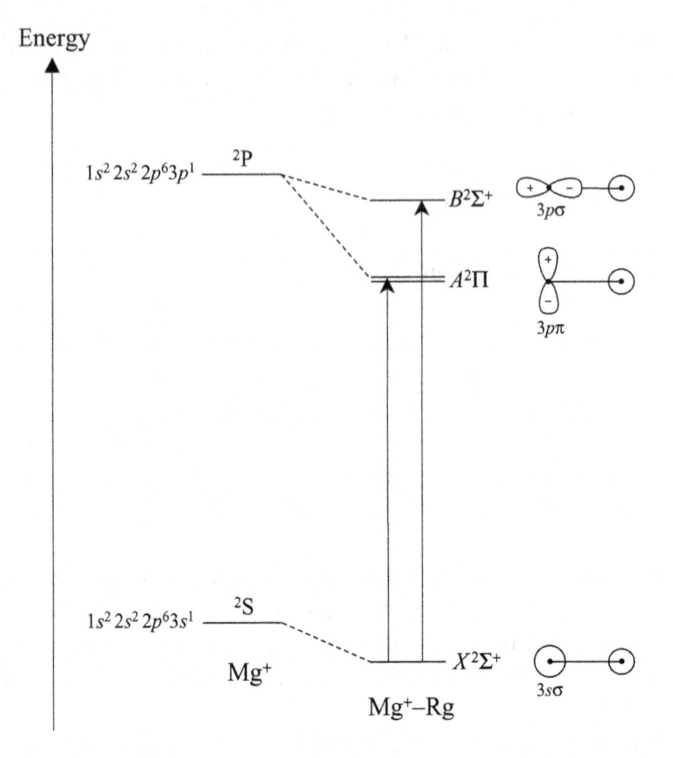

Figure 23.1 Electronic structures of the low-lying electronic states of Mg^+–Rg complexes.

form a degenerate pair of π symmetry in which there is a node along the internuclear axis. This exposes a far larger local positive charge on the metal than is the case when the $3p_z$ orbital is occupied. As a result, the charge-induced dipole interaction is particularly large for the $3p\pi$ orbitals.

23.3 Photodissociation spectra

The photodissociation spectra of Mg^+–Ne, Mg^+–Ar, Mg^+–Kr, and Mg^+–Xe in the region of the Mg^+ $3p \leftarrow 3s$ transition are compared in Figure 23.2. All four spectra are characterized by sharp bands, with the exception of Mg^+–Ne, which also has a broad, structureless feature at high wavenumber (see later). A vibrational progression can be readily identified in each spectrum. In addition, each vibrational component actually consists of a doublet due to spin–orbit coupling. Each of these points is considered in some detail below.

23.4 Spin–orbit coupling

Two electronic transitions of Mg^+–Rg in the Mg^+ $3p \leftarrow 3s$ region are expected, namely the $A\,^2\Pi$–$X\,^2\Sigma^+$ and $B^2\Sigma^+$–$X\,^2\Sigma^+$ transitions. Only the A state can give rise to spin–orbit

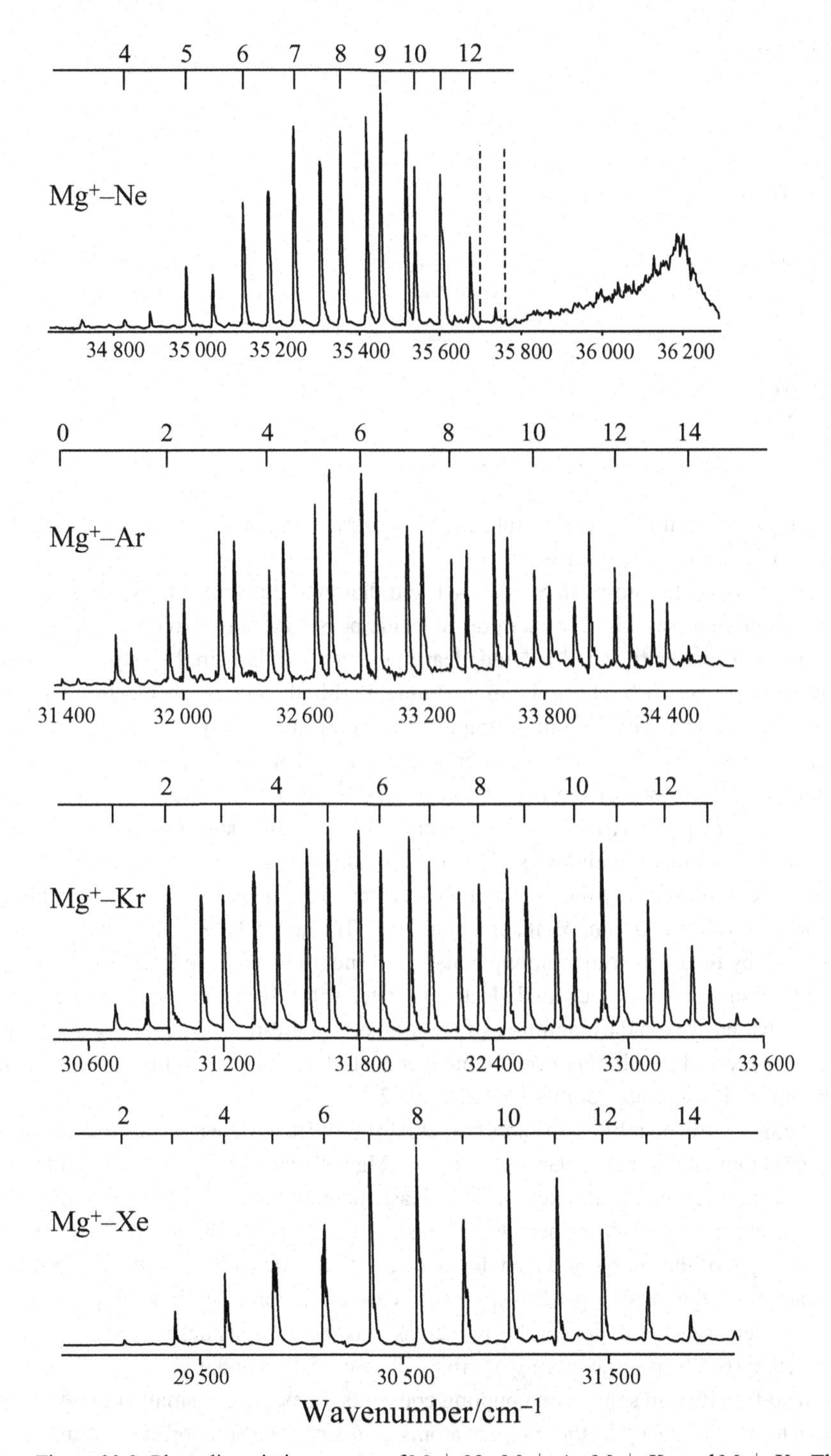

Figure 23.2 Photodissociation spectra of Mg^+–Ne, Mg^+–Ar, Mg^+–Kr, and Mg^+–Xe. The tick marks above the spectra identify vibrational structure and are aligned with the bands due to $A^2\Pi_{1/2}\ v' \leftarrow X^2\Sigma^+\ v''=0$ transitions. The corresponding spin–orbit partners from $A^2\Pi_{3/2} \leftarrow X^2\Sigma^+$ transitions are easily identified for Mg^+–Ne, Mg^+–Ar, and Mg^+–Kr. For Mg^+–Xe the vibrational frequencies and spin–orbit splittings in the excited state are very similar and hence the spin–orbit structure is hidden underneath the vibrational structure. (Adapted with permission from J. S. Pilgrim, C. S. Yeh, K. R. Berry, and M. A. Duncan, *J. Chem. Phys.* **100** (1994) 7945, and J. E. Reddic and M. A. Duncan, *J. Chem. Phys.* **110** (1999) 9948, American Institute of Physics.)

Table 23.1 *Spin–orbit coupling constants in the $A^2\Pi$ states of Mg^+–Rg complexes [1, 2]*

Complex	Spin–orbit coupling constant/cm^{-1}
Mg^+–Ne	63
Mg^+–Ar	77
Mg^+–Kr	143
Mg^+–Xe	270

coupling, and so the sharp structure in the spectra in Figure 23.2 can be assigned to the $A^2\Pi - X^2\Sigma^+$ electronic transition.

As discussed for atoms in Section 4.1 and diatomic molecules in Section 4.2.3, spin–orbit coupling arises when an atom or molecule possesses non-zero electronic orbital *and* spin angular momenta. The Mg^+ ion clearly possesses an unpaired electron, but electrons only have non-zero orbital angular momentum in orbitally *degenerate* electronic states. The ^2P excited state of Mg^+ is one example, and the orbital and spin angular momenta can couple to give $^2P_{1/2}$ and $^2P_{3/2}$ spin–orbit sub-states. The subscripts in these labels refer to the possible values of the net orbital + spin angular momenta, which for atoms are $J = L + S, L + S - 1, \ldots, |L - S|$. For an orbital less than or equal to half full, the spin–orbit component with lowest J has the lowest energy.

In the complexes only the $A^2\Pi$ state has orbital angular momentum, and coupling with the net spin yields two spin–orbit components, $A^2\Pi_{1/2}$ and $A^2\Pi_{3/2}$.[3] These will be separated in energy by the spin–orbit coupling constant, A (not to be confused with the same symbol used to designate the first excited electronic state, $A^2\Pi$). If there is little charge transfer to the rare gas atom then the magnitude of the spin–orbit splitting will depend on the properties of Mg^+ only and should therefore be independent of the identity of the rare gas atom. The experimental values are summarized in Table 23.1.

The spin–orbit coupling constants are actually found to be dependent on the identity of the rare gas atom, and in particular the values for Mg^+–Kr and Mg^+–Xe are much larger than those of the two lighter complexes. This clearly demonstrates that the assumption that the rare gas atom is largely a spectator is incorrect, especially for the heavier complexes. The strength of the charge-induced dipole interaction is dependent on the polarizability of the rare gas atom. The larger this atom, the easier it is for a nearby charge to distort the electron density, i.e. the polarizability increases as the group is descended. This increased interaction results in some mixing of orbital characteristics, and it is this that is responsible for the differences in spin–orbit coupling constants. In essence, a small amount of cationic character is introduced to the rare gas atoms, and since the spin–orbit coupling constants of the heavier rare gas atoms are large, this has a major impact on the spin–orbit coupling constant of the complex.

[3] In molecules the labels used for electronic states possessing spin–orbit coupling take the form $^{2S+1}\Lambda_\Omega$ where $\Omega = |\Lambda + \Sigma|$. See Section 4.2.3 for more details.

23.5 Vibrational assignment

It is reasonable to suppose that under supersonic beam conditions most complexes will initially be in their zero-point vibrational levels in the ground electronic state. Consequently, the dominant vibrational features will be due to excitation to different vibrational levels in the $A^2\Pi$ state. It is therefore a simple matter to estimate the vibrational frequency in the excited state from the separation of adjacent members of the vibrational progression.

However, to obtain an accurate value of the harmonic vibrational frequency, ω_e, and the anharmonicity constant, x_e, it is necessary to establish the correct vibrational numbering in the excited state. The vibrational progressions are quite long and it is clear that a substantial change in bond length must occur on electronic excitation. This makes it difficult to establish the position of the electronic origin transition, $v' = 0 \leftarrow v'' = 0$, because the Franck–Condon factor for this transition may be very small and therefore this transition may be too weak to observe.

A solution to this problem is to make use of *isotope shifts* to establish vibrational numberings. A particular vibrational component will occur at wavenumber

$$\nu = \nu_e + \omega_e'\left(v' + \tfrac{1}{2}\right) - \omega_e'x_e'\left(v' + \tfrac{1}{2}\right)^2 - \left[\omega_e''\left(v'' + \tfrac{1}{2}\right) - \omega_e''x_e''\left(v'' + \tfrac{1}{2}\right)^2\right] \tag{23.1}$$

where $'$ and $''$ refer to the upper and lower electronic states, respectively, and ν_e is the pure electronic transition wavenumber. Magnesium has three isotopes, ^{24}Mg (79%), ^{25}Mg (10%), and ^{26}Mg (11%). Assuming that equation (23.1) applies to the ^{24}Mg–Rg isotopomer, then for the heavier magnesium isotopes we can replace ω_e by $\rho\omega_e$ (see equation (5.7)) and x_e by ρx_e, where

$$\rho = \sqrt{\frac{\mu}{\mu_i}} \tag{23.2}$$

In the above expression μ is the reduced mass of the ^{24}Mg–Rg isotopomer and μ_i is the reduced mass of the heavier isotopomer (^{25}Mg–Rg or ^{26}Mg–Rg). Combining (23.1) and (23.2), and assuming that all transitions take place out of the $v'' = 0$ level, leads to the expression below for the isotope shift ΔG_{iso}:

$$\Delta G_{\text{iso}}(v') = (1-\rho)\left[\omega_e'\left(v' + \tfrac{1}{2}\right) - \tfrac{1}{2}\omega_e''\right] - (1-\rho)^2\left[\omega_e'x_e'\left(v' + \tfrac{1}{2}\right)^2 - \tfrac{1}{4}\omega_e''x_e''\right] \tag{23.3}$$

This expression is the key to determining the correct vibrational quantum numbers. It can be used to calculate isotope shifts and these are then compared with experiment. This is a trial and error process in which a particular vibrational quantum numbering is first assumed, and then approximate values of ω_e' and $\omega_e'x_e'$ are determined from the spectrum. An estimate of ω_e'' is also required (x_e'' can be neglected), which may come from observation of hot bands or must be deduced in some other manner, e.g. from *ab initio* calculations. Finally, the predictions from equation (23.3) are compared with experiment and used to determine the correct vibrational numbering. This is most easily seen graphically and an example is

Table 23.2 *Vibrational parameters for Mg^+–Rg in the $A^2\Pi$ state*

Complex	ω'_e/cm^{-1}	$\omega'_e x'_e$/cm^{-1}
Mg^+–Ne	219.4	6.7
Mg^+–Ar	271.8	3.3
Mg^+–Kr	257.7	2.3
Mg^+–Xe	258.0	1.5

These are averages over the two spin–orbit components.

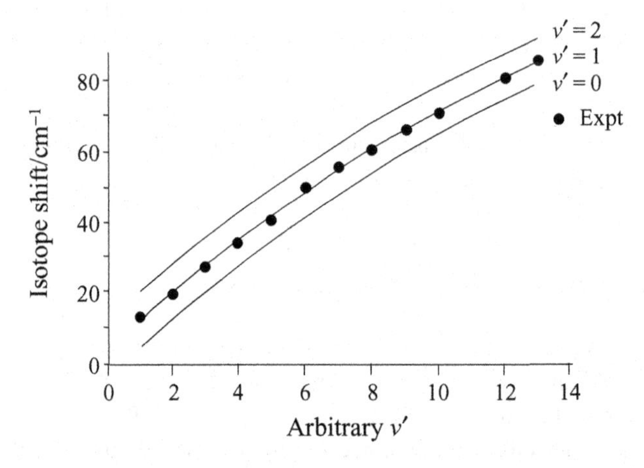

Figure 23.3 Isotope shift measurements for vibrational bands in the Mg^+–Kr spectrum. The trial assignments are for the first observable band having a vibrational quantum number of 0, 1 or 2 in the upper state. The curve for $v' = 1$ best fits the data, leading to the assignment given above the Mg^+–Kr spectrum in Figure 23.2. (Reproduced with permission from J. S. Pilgrim, C. S. Yeh, K. R. Berry, and M. A. Duncan, *J. Chem. Phys.* **100** (1994) 7945, American Institute of Physics.)

shown in Figure 23.3. This approach was used by Duncan and co-workers to firmly establish all the vibrational assignments shown in Figure 23.3.

23.6 Vibrational frequencies

The harmonic vibrational wavenumbers and anharmonicities are shown in Table 23.2 for the $^{24}Mg^+$–Rg isotopomers. The vibrational wavenumber is a function of both the bond force constant (and by implication the bond strength – see Section 5.1.2) and the reduced mass. Mg^+–Ne is the most weakly bound complex, which explains why it has the smallest vibrational frequency despite having the smallest reduced mass. For Mg^+–Kr and Mg^+–Xe the effect of the reduced mass outweighs the bond force constant contribution and therefore these complexes possess lower vibrational frequencies than Mg^+–Ar despite being more strongly bound.

Table 23.3 *Dissociation energies (D_0 in cm^{-1}) for Mg^+–Rg complexes*

Complex	$X^2\Sigma^+$	$A^2\Pi$
Mg^+–Ne	110	1700
Mg^+–Ar	1280	5550
Mg^+–Kr	1920	7130
Mg^+–Xe	4180	11030

The anharmonicities progressively decrease as the rare gas group is descended. This is because, as the bond strengthens, the potential well becomes more harmonic-like (i.e. parabolic) for the vibrational energies sampled in the photodissociation experiment.

23.7 Dissociation energies

The extensive vibrational progressions can be used to estimate the dissociation energies of the Mg^+–Rg in their $A^2\Pi$ electronic states. The dissociation energy, D_0', is simply the sum of the separations between all the vibrational energy levels, starting from $v' = 0$, i.e.

$$D_0' = \sum_{v'} \Delta G_{v'+1/2} \tag{23.4}$$

where

$$\Delta G_{v'+1/2} = G(v'+1) - G(v') \tag{23.5}$$

are vibrational term values (see Section 5.1.2). If the positions of most of the bound energy levels have been measured from the spectrum, then the area under a plot of $\Delta G_{v'+1/2}$ versus $v' + \frac{1}{2}$ extrapolated to $\Delta G_{v'+1/2} = 0$ will give an accurate dissociation energy.

In practice the vibrational structure observed in an electronic spectrum represents only a modest subset of the total set of vibrational energy levels. In this case the *Birge–Sponer extrapolation* can be employed. This extrapolation is based on the assumption that the potential energy curve is adequately described by a Morse potential, i.e. the anharmonicity constant x_e is sufficient to account for all of the anharmonicity and the vibrational term value $G(v)$ is accurately described by equation (5.14). With this approximation it is easy to show that

$$\Delta G_{v'+1/2} = \omega_e' - 2\omega_e' x_e'(v'+1) \tag{23.6}$$

and therefore a plot of $\Delta G_{v'+1/2}$ versus v' should be linear and can readily be extrapolated to $\Delta G_{v'+1/2} = 0$, allowing D_0' to be estimated.

Table 23.3 shows the dissociation energies obtained. Notice that dissociation energies for the ground electronic states are also included in the table. These can be determined from the expression

$$D_0'' = D_0' + \nu_{00} - \Delta E(^2\text{P}-^2\text{S}) \tag{23.7}$$

which follows from the conservation of energy. The quantity ν_{00} is the $A^2\Pi$ $v' = 0 \leftarrow X^2\Sigma^+$ $v'' = 0$ electronic transition energy and $\Delta E(^2\text{P}-^2\text{S})$ is the energy required to excite the unpaired electron in the free Mg^+ ion from the 3*s* to the 3*p* orbital.

The dissociation energies show the trends expected from the earlier discussion about electronic structures. Each complex is much more strongly bound in its $A^2\Pi$ state than in the $X^2\Sigma^+$ state due to the reduced shielding of the positive charge when the unpaired electron density has a π orientation. Furthermore, there is a dramatic increase in binding energy for both electronic states in moving progressively from Ne to Xe due to the increasing polarizability of the rare gas atom.

23.8 *B–X* system

In the Mg^+–Ne photodissociation spectrum in Figure 23.2 there is a prominent broad feature in addition to the sharp bands discussed above. Duncan *et al.* attribute this to the $B^2\Sigma^+$–$X^2\Sigma^+$ electronic transition. The binding energy of the $B^2\Sigma^+$ state is likely to be even less than that of the $X^2\Sigma^+$ state because of the increased electron–electron repulsion between Mg^+ and the Ne atom, a result of the orientation of the $3p_z$ orbital along the internuclear axis. There will therefore be very few if any bound vibrational levels and the separation between them will be exceedingly small, explaining why there is no evidence of any resolvable vibrational structure. In fact most of the band envelope is likely to arise from excitation to the continuum of states above the dissociation limit of the $B^2\Sigma^+$state.

References

1. J. S. Pilgrim, C. S. Yeh, K. R. Berry, and M. A. Duncan, *J. Chem. Phys.* **100** (1994) 7945.
2. J. E. Reddic and M. A. Duncan, *J. Chem. Phys.* **110** (1999) 9948.

24 Rotationally resolved spectroscopy of Mg^+–rare gas complexes

Concepts illustrated: *ion–molecule complexes; photodissociation spectroscopy; Hund's coupling cases; rotational structure in open-shell molecules; least-squares fitting of spectra.*

This Case Study follows on from the previous one. However, rotationally resolved photodissociation spectra are the focus here, specifically for Mg^+–Ne and Mg^+–Ar. Although these ions are diatomic species, their rotationally resolved spectra are not trivial to analyse. The reason for this is the presence of an unpaired electron, which gives rise to a net spin angular momentum which can interact with the overall rotation of the complex (spin–rotation coupling). In addition, in some electronic states there may also be a net orbital angular momentum, and this can interact both directly with the molecular rotation (giving rise to the phenomenon known as Λ doubling) and with the electron spin. The latter is much the strongest of these angular momentum interactions and its effect can be readily seen in the rotationally resolved spectra, as will be discussed below.

Duncan and co-workers have recorded partly rotationally resolved electronic spectra for the $A^2\Pi$–$X^2\Sigma^+$ transitions of Mg^+–Ne and Mg^+–Ar, and these form the basis of the Case Study described here [1, 2]. A photodissociation technique was employed as detailed in Chapter 23. Before describing the spectra and their analysis, the expected rotational energy level structure for the $X^2\Sigma^+$ and $A^2\Pi$ electronic states is considered. Much of this description is similar to that met for NO in Chapter 22.

24.1 $X^2\Sigma^+$ state

Figure 23.1 in the previous Case Study provides a simple and extremely useful representation of the electronic structure of Mg^+–Rg cations in their two lowest electronic states. The electrons on the rare gas atom are in tightly bound orbitals and require very high energies to excite to vacant orbitals. The remaining electrons are strongly localized on Mg^+, and all but one occupy core orbitals. Consequently, the lowest-lying electronic states in Mg^+–Rg are

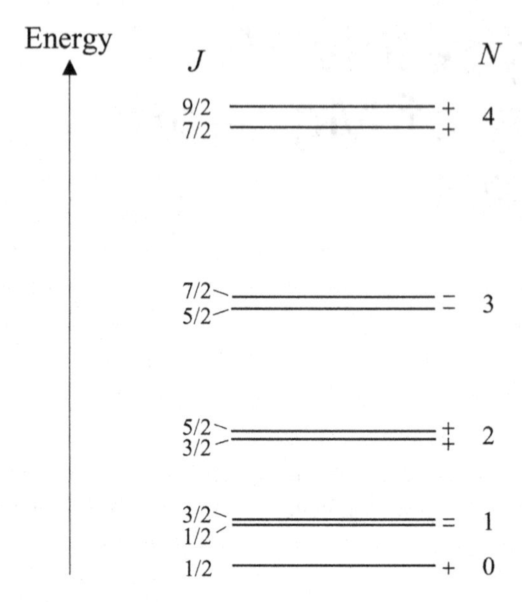

Figure 24.1 Rotational energy levels of a diatomic molecule in a $^2\Sigma^+$ electronic state satisfying Hund's case (b) coupling. The total angular momentum quantum number J is given by $J = N \pm ½$, where N is the rotational quantum number. The + or − beside each energy level refers to the parity (see text).

differentiated by the orbital occupied by the unpaired electron, and in the ground electronic state this can be described approximately as the Mg^+ 3s orbital. The resulting electronic state in the complex is a $^2\Sigma^+$ state.

The unpaired electron in this state has a non-zero spin angular momentum, which can interact with the angular momentum generated by overall rotation of the molecule. Interaction must proceed via magnetic coupling, since electron spin is a purely magnetic effect. The rotation of Mg^+–Rg will generate oscillating electric and magnetic fields and the latter can directly couple with the electron spin. However, it is important to note that indirect magnetic coupling is also possible via orbital motion of the electrons and so even in open-shell homonuclear diatomic molecules a spin–rotation interaction can occur. Regardless of the mechanism, in almost all cases spin–rotation coupling is very weak.

The coupling between spin and rotational motion in Mg^+–Rg is an example of Hund's case (b) coupling. The basic principles of Hund's coupling cases are outlined in Appendix G and were also met in Chapter 22. The total angular momentum quantum number for the molecule, J, is given by $J = N \pm ½$, where N is the rotational quantum number (= 0, 1, 2, 3, etc.). The two possible values of J, which arise for all rotational levels except for $N = 0$, are due to the two possible orientations of the electron spin (up or down). Consequently, each rotational level is actually split into two levels when spin–rotation coupling occurs, as shown in Figure 24.1. The magnitude of the splitting increases with the speed of rotation, and is given by $\gamma(N + ½)$ where γ is a quantity known as the spin–rotation coupling constant. In most molecules the effect of spin–rotation coupling is very small and can only be resolved using high resolution spectroscopy.

24.2 $A^2\Pi$ state

The first excited electronic state in Mg^+–Rg corresponds to an electron excited to the $3p\pi$ orbitals. These orbitals form a degenerate pair and as a result the unpaired electron can orbit unimpeded around the internuclear axis with an orbital angular momentum given by the quantum number $\lambda = 1$. In the resulting $^2\Pi$ electronic state, the strongest angular momentum interaction occurs between the orbital and spin angular momenta of the unpaired electron. This *spin–orbit coupling* in Mg^+–Rg was discussed in some detail in the previous Case Study. If the spin–orbit coupling constant, A, has a magnitude such that $A \gg BJ$, then Hund's case (a) coupling applies.[1] In the Hund's case (a) limit the torque provided by the electrostatic field of the nuclei locks the orbital angular momentum into precessional motion about the internuclear axis. This precession generates a concomitant magnetic field, which in turn forces the spin angular momentum to precess sympathetically about the internuclear axis. The quantum numbers describing this coupled electronic motion are Λ, S, Σ, and Ω. Λ and Σ are the quantum numbers describing the projection of orbital and spin angular momenta along the internuclear axis.[2] For Mg^+–Rg the values are $\Lambda = 1$ and $\Sigma = \frac{1}{2}$. Ω is the vector sum of $|\Lambda + \Sigma|$ and may take on the values of $\frac{1}{2}$ and $\frac{3}{2}$ in this specific example. S is a good quantum number in both Hund's cases (a) and (b).

Spin–orbit coupling splits the $^2\Pi$ electronic state into two spin–orbit components, $^2\Pi_{1/2}$ and $^2\Pi_{3/2}$, where the subscript refers to the value of Ω. Each of these sub-states has its own set of rotational levels, as shown in Figure 24.2. The rotational levels are distinguished by their total angular momentum quantum number, J, and the identity of the particular spin–orbit sub-state. One can identify a rotational quantum number with values $R = 0$, 1, 2, 3, etc., such that $J = R + \Omega$. Thus in the $^2\Pi_{1/2}$ state the smallest possible value of J is $\frac{1}{2}$ whereas in the $^2\Pi_{3/2}$ state it is $\frac{3}{2}$. This has consequences for the spectra, which will be seen later.

Additional labels, + and −, are included for the rotational levels in both Figures 24.1 and 24.2. These refer to the *parity* of the energy level. Parity is a symmetry label, but one that results from the operation in which the coordinates of all particles in the molecule (nuclei and electrons) are inverted with respect to a *space-fixed* coordinate system. This is an involved concept and will not be developed in any detail here; sophisticated treatments can be found in many books (see for example References [3] and [4]). Parity is a useful description of symmetry that aids in establishing transition selection rules, as detailed below.

1 In fact, when the spin–orbit coupling is strong there are two possible coupling cases, Hund's cases (a) and (c). See Appendix G for more details.

2 Note the potential for confusion here. As well as its use to designate electronic states in linear molecules with orbital angular momentum $\Lambda = 0$, the symbol Σ is unfortunately also used to designate the quantum number for projection of the electronic spin angular momentum on the internuclear axis.

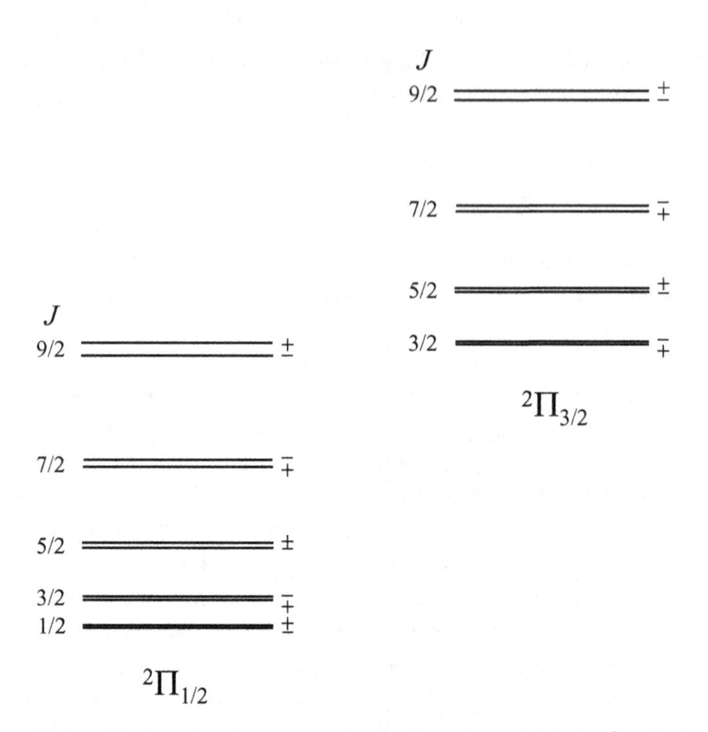

Figure 24.2 Rotational energy levels of a diatomic molecule in a $^2\Pi$ electronic state satisfying Hund's case (a) coupling. Two sets of levels are shown corresponding to the spin–orbit components $^2\Pi_{1/2}$ and $^2\Pi_{3/2}$. The overall angular momentum quantum number is given by the half integer quantum number J. The + or − beside each energy level refers to the parity (see text).

24.3 Transition energies and selection rules

The transitions must satisfy the usual single-photon selection rule for the overall angular momentum, $\Delta J = 0, \pm 1$. In addition, there is a selection rule based on parity, which derives from the fact that the dipole moment operator, $\boldsymbol{\mu}$, is a linear function of the positions of all the particles in the molecule (see equation (7.2)). Application of the parity operation switches the coordinates of the particles to their negative values and since this makes $\boldsymbol{\mu}$ change sign the dipole moment operator must possess negative parity. For an electric dipole driven transition the transition moment is given by the integral expression in equation (7.1), and this will be zero if the integrand has negative parity. If the parity in upper and lower states is the same, for example both are positive, then the parity of the integrand is $(+) \otimes (-) \otimes (+) = (-)$. Consequently, in an electric-dipole allowed transition the parity must change between the upper and lower states.

Armed with the above selection rules, it is possible to identify the allowed transitions, and these are shown in Figure 24.3. The rotational structure is more complicated than a simple three-branch *P*/*Q*/*R* structure. Focussing on the $^2\Pi_{1/2}$–$^2\Sigma^+$ sub-band, six branches can be identified. These can be divided into *P*, *Q*, and *R* branches but additional subscripts are added to the labels to designate the specific upper and lower levels. Such ideas were met in

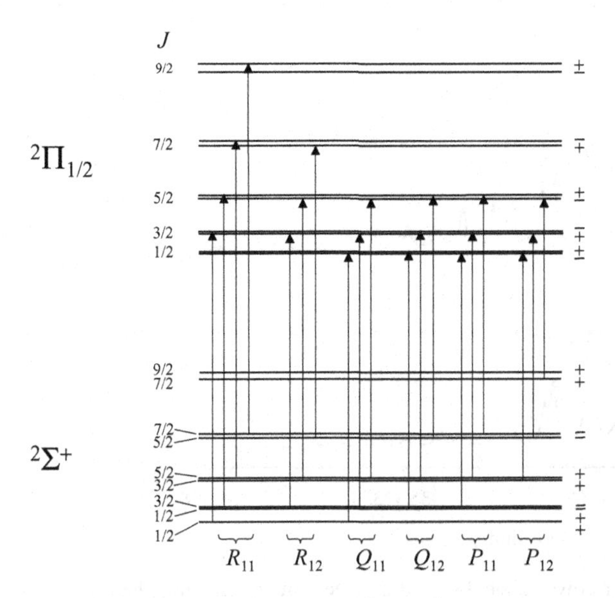

Figure 24.3 Allowed transitions in a ${}^2\Pi_{1/2}$–${}^2\Sigma^+$ electronic absorption band. Six branches occur, as shown at the bottom of the diagram. If the spin–rotation coupling in the ${}^2\Sigma^+$ state is too small to be resolved then the R_{12} and Q_{11} transitions are indistinguishable, as are the Q_{12} and P_{11} transitions, reducing the number of observable branches to four.

Case Study in Chapter 22, and the interested reader can also find out more by consulting the textbook by Herzberg [5]. The important thing to note is that if the spin–rotation splitting is too small to be resolved then the number of distinguishable branches is reduced to four. In those cases where the rotational constant is similar in the upper and lower electronic states, the branches have a structure where the spacing between adjacent transitions is roughly $3B$, B, B, and $3B$ (moving from low energy to high energy). These spacings correspond to the P_{12}, $P_{11} + Q_{12}$, $Q_{11} + R_{12}$, and R_{11} branches, respectively.

24.4 Photodissociation spectra of Mg⁺–Ne and Mg⁺–Ar

Figures 24.4 and 24.5 show rotationally resolved photodissociation spectra of Mg^+–Ne and Mg^+–Ar. In neither spectrum is the zero-point vibrational level accessed in the $A^2\Pi$ state. This is because the Franck–Condon factors for 0–0 transitions are small for these ions, and adequate signal-to-noise ratios were only obtained for rotational structure in transitions to higher vibrational levels.

The spectrum for Mg^+–Ne looks to be quite simple, but its appearance is deceptive; the limited spectral resolution means that many peaks are actually superpositions of two or more transitions. From Chapter 23 we know that the complex will be far more strongly bound in the excited electronic state than in the ground state. The rotational constant in the $A^2\Pi$ state should therefore be substantially larger than that in the $X^2\Sigma^+$ state. The marked change in rotational constants will give rise to band head formation (see also Chapter 16)

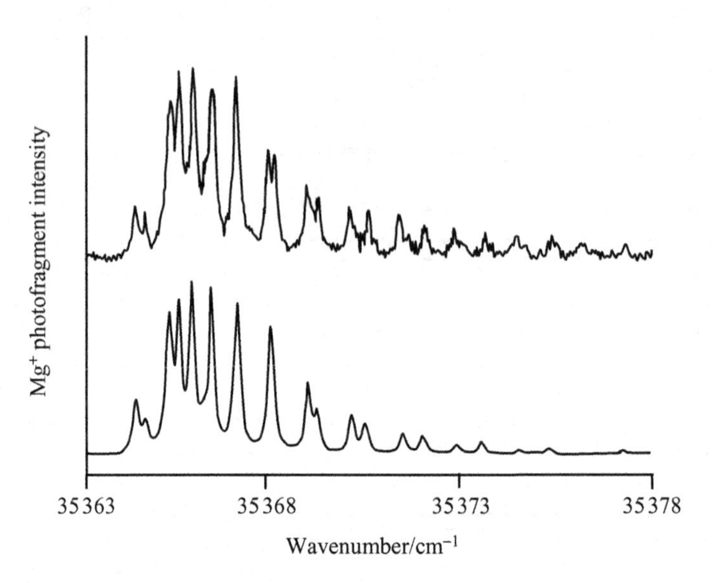

Figure 24.4 Rotationally resolved photodissociation spectrum of Mg^+–Ne. The features shown have been assigned to the $A^2\Pi_{1/2}$–$X^2\Sigma^+$ 9–0 band by Reddic and Duncan [1]. The upper trace shows the experimental spectrum while the lower trace is a simulation based on an assumed rotational temperature of 4 K. (Reproduced with permission from J. E. Reddic and M. A. Duncan, *J. Chem. Phys.* **110** (1999) 9948, American Institute of Physics.)

in some branches and rapidly divergent rotational structure in other branches. This is exactly the structure seen in Figure 24.4. It turns out that the spectrum in Figure 24.4 is due to the $A^2\Pi_{1/2}$–$X^2\Sigma^+$ transition rather than the $A^2\Pi_{3/2}$–$X^2\Sigma^+$ electronic transition, as justified later. The lowest energy feature in the spectrum is the P_{12} branch, which is relatively weak. The remaining, and much stronger features are due to the $P_{11} + Q_{12}$, $Q_{11} + R_{12}$, and R_{11} branches and all of the strong peaks contain unresolved contributions from at least two of these branches.

With so much unresolved structure it would be impossible to extract precise rotational constants from the spectrum in Figure 24.4. It is even a rather difficult task to assign the peaks to specific rotational transitions without the aid of computer simulation, but with simulations important information can be extracted readily from the spectrum. Reddic and Duncan used a program known as SpecSim to simulate the rotational structure in a $^2\Pi$–$^2\Sigma^+$ spectrum. An outline of how this and similar programs work is given in Appendix H. Most of these programs are equipped with the option of varying spectroscopic constants in a systematic (least-squares) fashion such that the best possible agreement (the best fit) between theory and experiment is obtained.

Rotational constants of 0.343 ± 0.013 and 0.238 ± 0.008 cm^{-1} were extracted for the upper and lower states of Mg^+–Ne. These can be used to estimate bond lengths of 2.59 ± 0.05 Å and 3.17 ± 0.05 Å, respectively. The spectrum in Figure 24.4 involves the $v = 9$ vibrational level in the $A^2\Pi$ state, and the larger amplitude of the vibrations in this highly excited level will yield a larger effective bond length than would be the case in the $v = 0$

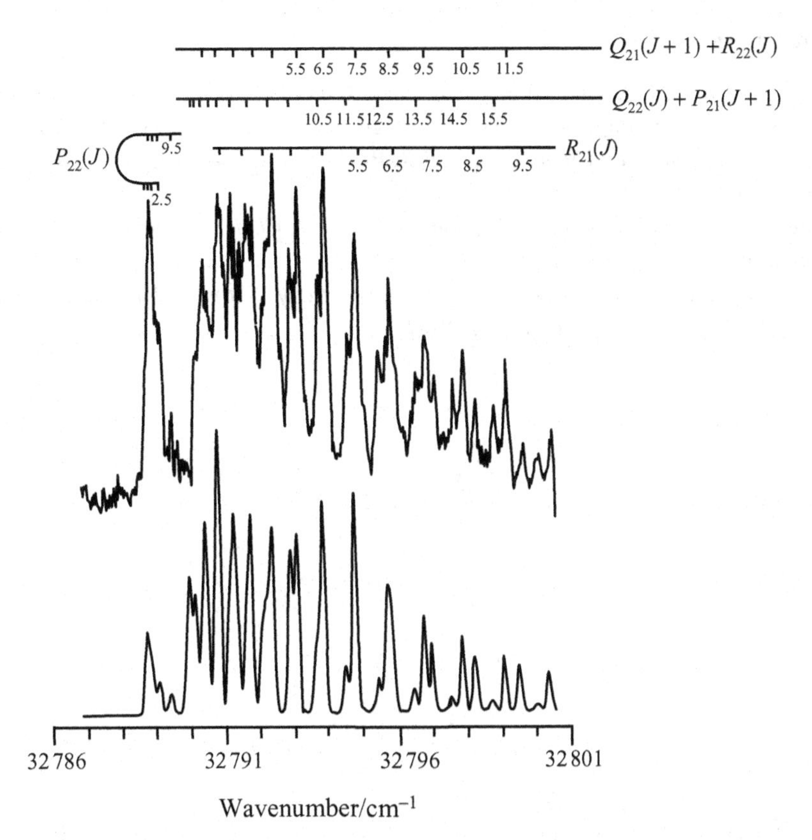

Figure 24.5 Rotationally resolved photodissociation spectrum of Mg^+–Ar. The features shown have been assigned to the $A^2\Pi_{3/2}$–$X^2\Sigma^+$ 5–0 band by Scurlock and co-workers [2]. The upper trace shows the experimental spectrum while the lower trace is a simulation based on an assumed rotational temperature of 4 K. (Reproduced with permission from C. T. Scurlock, J. S. Pilgrim, and M. A. Duncan, *J. Chem. Phys.* **103** (1995) 3293, American Institute of Physics.)

level. Nevertheless, it is clear that much shorter bond lengths are obtained in the $A^2\Pi$ state and this is consistent with the expected stronger binding in this state compared with the $X^2\Sigma^+$ state.

The findings are similar for the Mg^+–Ar spectrum in Figure 24.5. Note that here the simulations show that the band is due to the $A^2\Pi_{3/2}$–$X^2\Sigma^+$ transition. The best way to distinguish between the $^2\Pi_{3/2}$–$^2\Sigma^+$ and $^2\Pi_{1/2}$–$^2\Sigma^+$ transitions is by noting that certain transitions present in the latter are missing in the former because the lowest possible value of J in the $^2\Pi_{3/2}$ component is $J=\frac{3}{2}$. Simulations, or if the resolution is sufficient even simple inspection, should allow an assignment to $^2\Pi_{3/2}$–$^2\Sigma^+$ or $^2\Pi_{1/2}$–$^2\Sigma^+$ transitions.

A least-squares fit of the rotational structure allowed bond lengths of 2.882 ± 0.017 Å and 2.524 ± 0.014 Å to be deduced for the $X^2\Sigma^+$ and $A^2\Pi$ states of Mg^+–Ar. As with Mg^+–Ne, there is a marked shortening in bond length upon electronic excitation due to the much stronger binding in the excited electronic state.

References

1. J. E. Reddic and M. A. Duncan, *J. Chem. Phys.* **110** (1999) 9948.
2. C. T. Scurlock, J. S. Pilgrim, and M. A. Duncan, *J. Chem. Phys.* **103** (1995) 3293.
3. *Molecular Symmetry and Spectroscopy*, P. R. Bunker and P. Jensen, Ottawa, NRC Research Press, 1998.
4. *Angular Momentum: Understanding Spatial Aspects in Chemistry and Physics*, R. N. Zare, New York, Wiley, 1988.
5. *Molecular Spectra and Molecular Structure. I. Spectra of Diatomic Molecules*, G. Herzberg, Malabar, Florida, Krieger Publishing, 1989.

25 Vibronic coupling in benzene

Concepts illustrated: *Hückel molecular orbital theory; vibrational structure; vibronic coupling.*

The electronic spectroscopy of the benzene molecule has been the target of much research over the years owing to its central role in the development of the concept of aromaticity, the ubiquity of six-membered ring structures throughout organic chemistry, and the importance of these as chromophores in photochemistry.

Benzene is, of course, the prototypical aromatic molecule, and is also one of the molecules to which Hückel molecular orbital theory may be simply applied. The details of Hückel theory are not covered here and the reader is referred elsewhere for details [1] but we note that it is applicable mainly to conjugated hydrocarbons and provides a description of the π molecular orbitals formed from the overlap of carbon $2p$ atomic orbitals. This interaction causes a delocalization of the π-electron density and in cases where this leads to a lowering of energy we talk of the molecule being *resonance stabilized.*

Hückel theory is a simple model which ignores any interaction between the σ and π framework, and which makes other simplifications regarding the various integrals that arise in molecular orbital theory (see Appendix B). Since each carbon atom in benzene is sp^2 hybridized, and combinations of these hybrids give rise to the σ framework, then there is one p orbital on each carbon atom remaining for π bonding: the one perpendicular to the molecular plane. The simplifications of Hückel theory lead to the concepts of the Coulomb integral, α, and the resonance integral, β, and the energy levels in Hückel theory are expressed in terms of these two quantities. The Coulomb integral represents the energy of a C $2p\pi$ atomic orbital in the absence of any overlap with other $2p\pi$ orbitals, whereas β can be regarded as an interaction energy caused by the overlap of $2p\pi$ orbitals on adjacent atoms.

For benzene, the six carbon $2p$ orbitals give rise to six π molecular orbitals, as shown in Figure 25.1.

Each carbon atom contributes only a single electron to the π system, since the remaining electrons are employed in the σ bonding framework. If the six electrons are located in the lowest three π orbitals, all of these electrons are *lower* in energy in the resonance structure than they were before delocalization occurred (when they were at energy $= \alpha$). These orbitals are clearly bonding molecular orbitals, whereas those lying above α are antibonding. In the ground electronic state there is a net bonding effect from the π orbitals, which helps to stabilize the molecule.

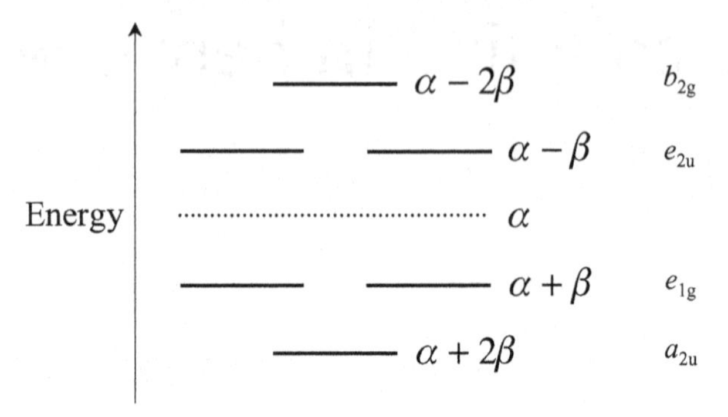

Figure 25.1 Hückel π molecular orbital energy level diagram for benzene. The quantities α and β are defined in the text. Point group symmetries of the orbitals are shown on the right-hand side of the diagram.

Hückel theory can also be used to determine the contribution of each carbon $2p\pi$ orbital to a given π molecular orbital. This is important because it reveals the symmetries of the π molecular orbitals. We note without proof that the symmetries of the bonding π orbitals are a_{2u} and e_{1g}, whereas the antibonding orbitals have e_{2u} and b_{2g} symmetries. For more details the interested reader is directed to Reference [2].

Since all molecular orbitals are full, the ground electronic state of benzene is a spin singlet and has a totally symmetric spatial symmetry in the D_{6h} point group: it is therefore a $^1A_{1g}$ state. The highest occupied molecular orbital (HOMO) has e_{1g} symmetry and the lowest unoccupied molecular orbital (LUMO) has e_{2u} symmetry. If an electron is excited from the HOMO to the LUMO, the possible excited states can be determined from the direct product $e_{1g} \otimes e_{2u}$. The result is $^{1,3}B_{1u}$, $^{1,3}B_{2u}$, and $^{1,3}E_{1u}$, but only the singlet states are of interest here because of the spin selection rule $\Delta S = 0$ in electric-dipole transitions.

It turns out that the lowest energy singlet excited electronic state is the $^1B_{2u}$ state. The lowest energy electronic transition, which can be written as $\tilde{A}\,^1B_{2u} \leftarrow \tilde{X}\,^1A_{1g}$, is symmetry forbidden, since $A_{1g} \otimes B_{2u} = B_{2u}$, and none of the x, y, or z vectors transform as this symmetry in the D_{6h} point group. Nevertheless, this nominally forbidden transition is observed in the electronic spectrum of benzene and so some explanation is required.

The relevant region of the ultraviolet absorption spectrum of benzene is shown in Figure 25.2, and was reported by Callomon *et al.* [3]. The spectrum in Figure 25.2 is an absorption spectrum recorded for the vapour above cooled liquid benzene. The spectrum was recorded at low resolution, and in fact a number of higher resolution spectra are shown in Reference [3], where some partially resolved rotational structure was obtained.

Considerable vibrational structure is seen in Figure 25.2, but all of the strong bands are built upon the single quantum excitation of the ν_6 vibrational mode in combination with quanta of the ν_1 vibration. The ν_6 vibration is actually a pair of degenerate vibrations, which cause distortions of the benzene ring, and have e_{2g} symmetry; the ν_1 vibration in benzene is the totally symmetric (a_{1g}) C—C ring breathing vibration. Approximate forms of the vibrations are shown in Figure 25.3.

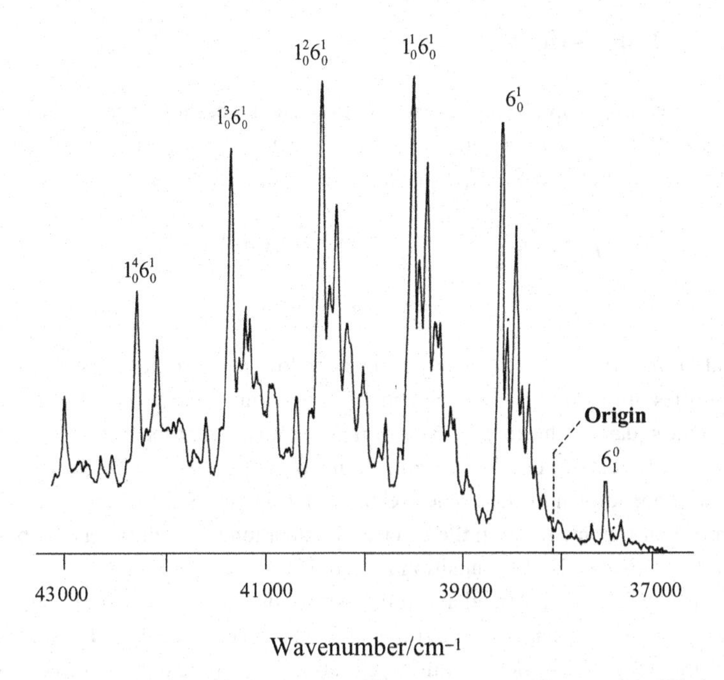

Figure 25.2 Absorption spectrum of benzene vapour. The notation N_p^q above each band refers to a transition from the $v = p$ level for vibration N in the ground electronic state to level $v = q$ in the excited electronic state.

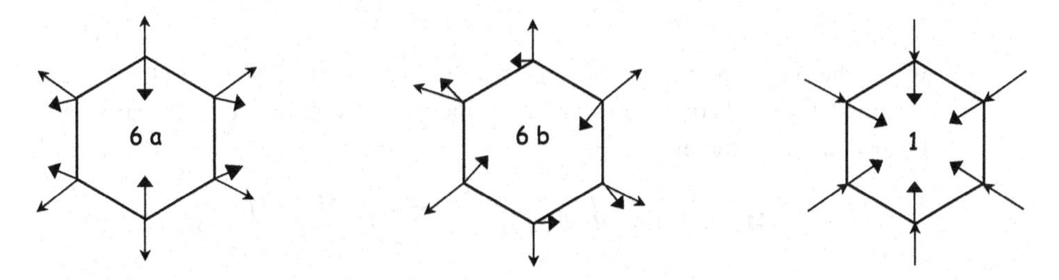

Figure 25.3 Illustration of the atomic motions for the ν_1 and ν_6 (doubly degenerate) vibrations of benzene.

Notice that the electronic origin transition (0_0^0) is not observed in the absorption spectrum. One might question how it is known that the first band is the 6_0^1 rather than the 0_0^0 transition. In fact the evidence comes from several sources, including the study of rotational structure. Also, notice the band assigned as 6_1^0 in Figure 25.2. This is a hot band transition, as shown by varying the temperature of the benzene sample. If the band assigned as 6_0^1 was really the origin transition 0_0^0 then the separation between the 0_0^0 and 6_0^1 bands would be too large to be feasible. We can therefore be certain that the 0_0^0 band is absent.

The fact that all strong bands are built upon the 6_0^1 rather than the 0_0^0 transition is an important clue as to why a nominally forbidden electronic transition is seen. The explanation is due to Herzberg and Teller [4], and is an example of a *vibronic interaction*.

25.1 The Herzberg–Teller effect

In the Born–Oppenheimer approximation, the electronic and vibrational motion is separated on the grounds that electrons move much faster than nuclei. Consequently, as discussed in Section 7.2, the transition moment, $\boldsymbol{M}_{\mathrm{ev}}$, may be expressed as follows:

$$\boldsymbol{M}_{\mathrm{ev}} = \int \Psi'_{\mathrm{ev}}\mu\Psi''_{\mathrm{ev}}\,\mathrm{d}\tau_{\mathrm{ev}} = \int \Psi'_{\mathrm{e}}\mu\Psi''_{\mathrm{e}}\,\mathrm{d}\tau_{\mathrm{e}} \int \Psi'_{\mathrm{v}}\mu\Psi''_{\mathrm{v}}\,\mathrm{d}\tau_{\mathrm{v}}$$
$$= \boldsymbol{M}_{\mathrm{e}} \int \Psi'_{\mathrm{v}}\Psi''_{\mathrm{v}}\,\mathrm{d}\tau_{\mathrm{v}}$$

where the $'$ and $''$ refer to wavefunctions in the upper and lower electronic states, respectively. All other quantities are as described in Section 7.2. The transition probability is directly proportional to the square of the above expression. The transition probability may therefore be separated into a product of a purely electronic term, $\boldsymbol{M}_{\mathrm{e}}$, and a vibrational overlap integral, the square of which is known as the Franck–Condon factor (see Sections 7.2.2 and 7.2.3). It is the symmetry of the integrand in the electronic transition moment that is the basis for deducing that the $\tilde{\mathrm{A}}^1B_{2\mathrm{u}} \leftarrow \tilde{X}^1A_{1\mathrm{g}}$ transition is forbidden.

However, this conclusion is dependent on the assumption that the electronic and vibrational motions can be fully separated. Herzberg and Teller recognized that $\boldsymbol{M}_{\mathrm{e}}$ is not strictly constant, but rather may vary somewhat during vibration. Assuming that this effect is small, then a satisfactory description can be obtained by expanding $\boldsymbol{M}_{\mathrm{e}}$ about the equilibrium structure to yield[1]

$$\boldsymbol{M}_{\mathrm{e}} = (\boldsymbol{M}_{\mathrm{e}})_{\mathrm{eq}} + \sum_{i=1}^{3N-6} \left(\frac{\partial \boldsymbol{M}_{\mathrm{e}}}{\partial Q_i}\right)_{\mathrm{eq}} Q_i$$

where the 'eq' subscript denotes the equilibrium structure and the Q_i are the individual vibrational normal coordinates. Inserting the above expression into the earlier equation for the overall transition moment gives

$$\boldsymbol{M}_{\mathrm{ev}} = (\boldsymbol{M}_{\mathrm{e}})_{\mathrm{eq}} \int \Psi'_{\mathrm{v}}\Psi''_{\mathrm{v}}\,\mathrm{d}\tau_{\mathrm{v}} + \sum_{i=1}^{3N-6} \left(\frac{\partial \boldsymbol{M}_{\mathrm{e}}}{\partial Q_i}\right)_{\mathrm{eq}} \int \Psi'_{\mathrm{v}} Q_i \Psi''_{\mathrm{v}}\,\mathrm{d}\tau_{\mathrm{v}}$$

The first term makes no contribution to the observed transition because we have already established that $(\boldsymbol{M}_{\mathrm{e}})_{\mathrm{eq}}$ is zero for a ${}^1B_{2\mathrm{u}} \leftarrow {}^1A_{1\mathrm{g}}$ transition. However, the second term may be non-zero for non-totally symmetric vibrations. This new term accounts for the weak coupling between electronic and vibrational motions, a coupling that is referred to as *vibronic coupling*. Although the separation of electronic and vibrational motions is still a reasonable description, it is no longer perfect and it is sometimes useful to think in terms of a vibronic state with a symmetry that is the direct product of the symmetries of the constituent electronic and vibrational states.

Herzberg and Teller proposed that nominally forbidden electronic transitions could gain considerable intensity by 'stealing' intensity from a nearby fully allowed electronic

---

1 This expansion is known as a *Taylor expansion* and is a well-known method in mathematics for expanding functions about a fixed point as a convergent power series.

transition. This can be achieved if there exists a vibration in the excited state of the forbidden transition which yields a vibronic symmetry (the direct product of the vibrational and the electronic symmetries) that is the same as the symmetry of the upper state in the allowed transition. These states can then mix to some extent, and the result is that the forbidden transition acquires intensity from the fully allowed transition; this is believed to be the source of the spectrum in Figure 25.2. Notice that the molecule must be vibrationally excited in order for vibronic interaction to occur, and this explains why the 0^0_0 transition is not observed.

For benzene, the $\tilde{C}\,^1E_{1u}$ state is close in energy to the $\tilde{A}\,^1B_{2u}$ state and is therefore the likely candidate for vibronic coupling and intensity stealing. In the $\nu_6 = 1$ vibrational level the vibronic symmetry becomes $B_{2u} \otimes e_{2g} = E_{1u}$, and this is the same symmetry as the $\tilde{C}$ electronic state and therefore suitable for vibronic coupling.

From the Hückel MO diagram shown earlier, the LUMO ← HOMO transition will result in a significant weakening of the π bonding and therefore a change in the C—C bond lengths. Consequently, the appearance of a substantial progression in mode ν_1 would be expected and indeed is observed (in combination with the ν_6 vibration). The excitation of totally symmetric vibrations such as mode ν_1 in combination with the non-totally symmetric ν_6 vibration does not change the excited state vibronic symmetry.

References

1. *Quantum Chemistry*, 4th edn., I. N. Levine, Englewood Cliffs, New Jersey, Prentice Hall, 1991.
2. *Molecular Symmetry: An Introduction to Group Theory and Its Uses in Chemistry*, D. S. Schonland, New York, van Nostrand, 1965.
3. J. J. Callomon, T. M. Dunn, and I. M. Mills, *Philos. Trans. Roy. Soc.* **259** (1966) 499.
4. G. Herzberg and E. Teller, *Z. Phys. Chem. B* **21** (1933) 410.

26 REMPI spectroscopy of chlorobenzene

Concepts illustrated: *REMPI spectroscopy; vibrational structure and assignments; Franck–Condon principle; vibronic coupling; Fermi resonance.*

There has been much work performed on the electronic spectroscopy of the benzene molecule, and some of this was included in the previous Case Study. As was noted in that earlier Case Study, benzene is an interesting molecule because:

(i) it has high symmetry, and this has implications for selection rules and therefore the appearance of the spectra;
(ii) vibronic coupling occurs;
(iii) it is a prototypical aromatic molecule, and the observed spectroscopy can be compared with predictions from quantum chemical calculations, ranging from simple Hückel theory through to state-of-the-art *ab initio* methods.

Substituted benzenes are also interesting molecules to spectroscopists. The simplest substitution is to replace one of the hydrogen atoms with a different atom. This can directly affect the electronic structure of the ring through donation or withdrawal of electron density by the substituent through inductive and mesomeric effects – an interesting phenomenon in its own right, although not of direct interest here.

Chlorobenzene is chosen for investigation here. Figure 26.1 shows the chlorobenzene molecule, indicating the axis system employed.

The outermost occupied orbital of benzene is a π molecular orbital with e_{1g} symmetry. In the lower symmetry (C_{2v}) environment of chlorobenzene this splits into two orbitals with b_1 and a_2 symmetries, with the HOMO being the b_1 orbital. Below these two orbitals lie two others which arise from the lone pairs on the Cl atom. These are non-bonding orbitals with b_1 and b_2 symmetries, the b_1 orbital lying lower in energy. The LUMO of benzene is a π* orbital with e_{2u} symmetry, which splits into $a_2 + b_1$ symmetry, with the a_2 being the lower. Consequently, the lowest energy electronic transition (LUMO $\leftarrow$ HOMO) in chlorobenzene is an $a_2 \leftarrow b_1$ transition. The first excited electronic state therefore has the outer electronic configuration $(b_1)^1(a_2)^1$, giving a symmetry $b_1 \otimes a_2 = b_2$, and so the first excited state is a $^1B_2$ state.

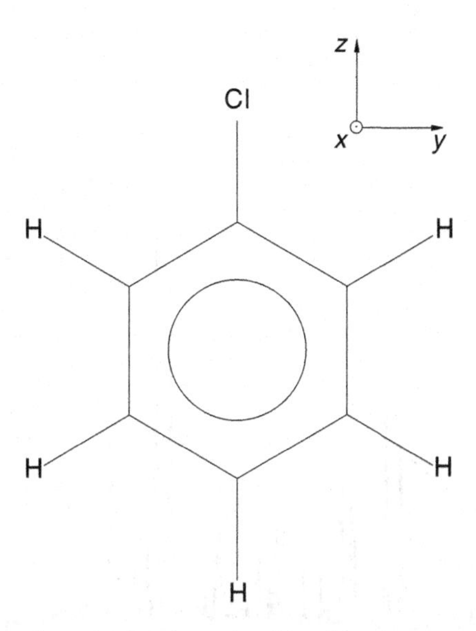

Figure 26.1 Schematic of the chlorobenzene molecule, indicating the axis system used in this work. The choice of axis system affects the symmetry labels used to specify the symmetries of the electronic and vibrational states.

By analogy with benzene, all occupied orbitals in the ground electronic state of chlorobenzene will be full and so the ground state is a $^1A_1$ state. The lowest energy singlet–singlet transition (often denoted $S_1 \leftarrow S_0$ as a shorthand and general notation for closed-shell molecules) therefore corresponds to the $\tilde{A}^1B_2 \leftarrow \tilde{X}^1A_1$ transition. This is an allowed transition, but note that it corresponds to the electric dipole-forbidden $\tilde{A}^1B_{2u} \leftarrow \tilde{X}^1A_{1g}$ transition of benzene (Chapter 25), where the symmetry labels have changed owing to the change in point group, particularly the loss of the centre of inversion. This transition has been studied by several research groups using various forms of electronic spectroscopy, with one of the earliest studies being reported in 1905 [1]: we shall concentrate on much more recent studies here [2–4].

26.1 Experimental details and spectrum

Electronic spectra of the $\tilde{A} \leftarrow \tilde{X}$ transition for chlorobenzene are shown in Figure 26.2 and have been taken from Reference [4]. A molecular beam of chlorobenzene seeded in argon was obtained by co-expanding the vapour from a room temperature sample of chlorobenzene with argon gas at a pressure of $\sim$5 bar. The supersonic expansion was then skimmed to form a molecular beam. One-colour REMPI spectroscopy was employed to record spectra. This was achieved by crossing the molecular beam with the beam from a tuneable dye laser. Ions produced were detected in a time-of-flight mass spectrometer and REMPI spectra were obtained by scanning the laser wavelength across the region of interest and recording the chlorobenzene cation current as a function of the laser wavelength.

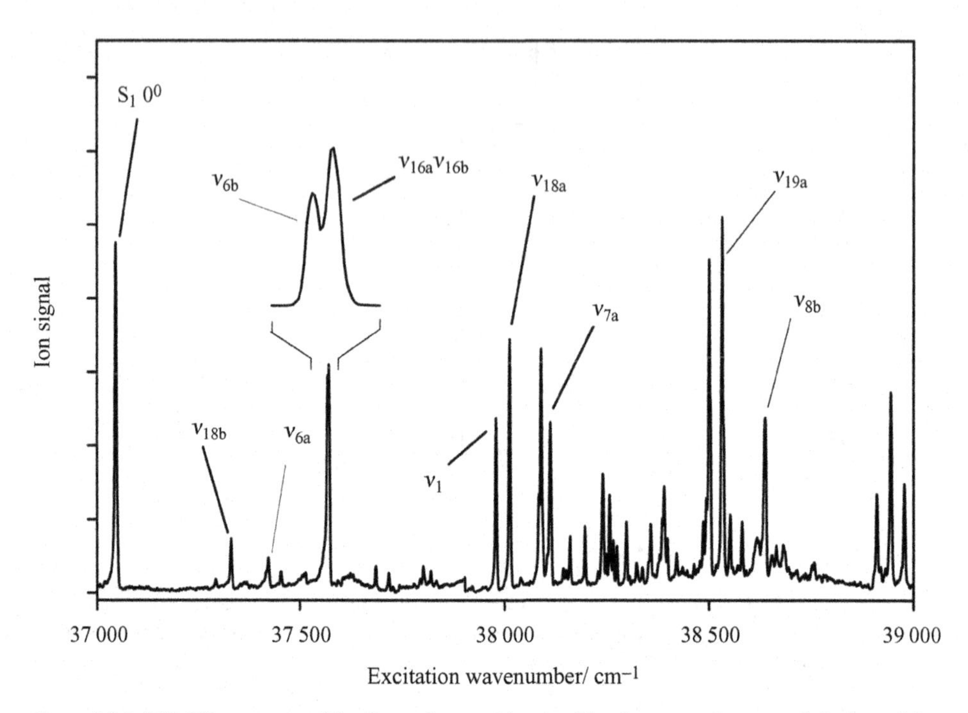

Figure 26.2 REMPI spectrum of the $S_1 \leftarrow S_0$ transition in chlorobenzene. An expanded view of the feature at 520–525 cm^{-1} above the origin band (0^0) is also shown. (Reproduced with permission from T. G. Wright, S. I. Panov, and T. A. Miller, *J. Chem. Phys.* **102** (1995) 4793, American Institute of Physics.)

26.2 Assignment

Before the assignment of specific peaks is attempted, it is necessary to establish that chlorobenzene is the molecule responsible for the spectrum. REMPI is normally excellent for this purpose, since the combination with mass spectrometry allows the mass of the spectral carrier to be determined. This is in contrast to methods such as LIF and cavity ringdown spectroscopies, where other arguments must be presented to prove that a spectrum does indeed arise from a particular molecule. However, it is worth noting that identification of the spectral carrier is not always straightforward in REMPI work, particularly when dealing with molecular complexes. This is because excess energy can be deposited into the ion in the ionization step and this can lead to fragmentation. A two-colour REMPI scheme can help to minimize fragmentation, since the wavelength of the laser used in the ionization step can be specifically chosen such that the ionization limit is only just exceeded.

The identification of the transition between the zero-point vibrational levels of each electronic state (termed the electronic *origin* transition and usually labelled as 0_0^0) is not always straightforward. Spectral features at energies below the origin can occur when the lower state is vibrationally excited – these are termed *hot bands*. Significant population of excited vibrational levels in the lower electronic state can persist even under fairly stringent supersonic cooling conditions. This is the result of the low efficiency of vibrational → translational energy transfer during the finite number of collisions that take place in the early stages of the supersonic expansion. Thus care must always be taken to identify contributions from hot bands before the origin transition is firmly assigned.

In Figure 26.2 a range of 2000 cm^{-1} is covered showing the origin (denoted 0^0 rather than the more usual 0^0_0) and a large number of additional bands. The various bands must be due to vibrational structure, and the resolution is too low to pick up the underlying rotational structure in each band.

Now consider what vibrational structure might be expected. In the cold conditions expected in a supersonic molecular beam, most of the chlorobenzene molecules will occupy their zero-point vibrational energy level. Application of the Franck–Condon principle (see Section 7.2.3) shows that the dominant vibrational structure should be due to excitation of totally symmetric (a_1) vibrations in the excited electronic state. Inspection of the known vibrational frequencies of chlorobenzene in the electronic ground state (obtained, for example, from infrared or Raman spectroscopy) quickly establishes that some of the low-frequency bands shown in Figure 26.2 cannot be due to modes with a_1 symmetry. Consequently, there must be vibrational structure that defies the Franck–Condon principle. Again, comparison with known vibrational frequencies indicates that these 'forbidden' features correspond to vibrational levels with b_2 symmetry, and so we need to explain how they gain their unexpectedly high intensities. Also of interest is the fairly strong band at approximately 37 560 cm^{-1}, which has been expanded in Figure 26.2 and is seen to consist of a closely spaced pair of peaks. Specific assignments will be proposed for these low-energy features, and then some briefer comments will be made regarding the remaining bands shown in Figure 26.2.

In Reference [4], vibrational frequencies calculated at the RHF/6–31G* level of *ab initio* theory were presented. This is a relatively low level of theory, but there is a well-established scaling factor for such calculations, which normally leads to fairly reliable predicted vibrational frequencies. We have performed additional calculations here. In particular we have obtained vibrational frequencies for the S_1 state, which are more appropriate for comparison with the REMPI spectra since the observed vibrational intervals are those exhibited by the S_1 state. Table 26.1 shows a list of calculated, scaled vibrational frequencies for the S_0 and S_1 states of chlorobenzene, together with the symmetry of each normal coordinate. Note that the labelling in Table 26.1 has been given in terms of both the Mulliken and the Wilson notations. The Mulliken notation lists the vibrations in order of symmetry, and within each symmetry block in order of descending frequency. This is the more usual and systematic way of numbering vibrational modes in polyatomic molecules. However, the Wilson nomenclature is based upon the mode numbering employed for benzene and makes the comparison with that molecule somewhat easier; we will use it in the discussion below. However, note that the comparison of vibrations in benzene with those in substituted benzenes can be misleading because the form of some vibrational modes can change significantly on substitution. The level of complexity is perhaps indicated by the fact that there is an entire book devoted to the vibrational spectroscopy of substituted benzenes [5].

The vibrational frequencies predicted by the *ab initio* calculations greatly aid the assignment of vibrational structure in Figure 26.2. The band at 378 cm^{-1} above the origin transition may be straightforwardly assigned to single quantum excitation of vibration ν_{6a}, which has

Table 26.1 *Calculated vibrational frequencies for the S_0 and S_1 states of chlorobenzene*

			Vibrational frequency/cm^{-1}	
Mode (Mulliken)	Mode (Wilson)	Symmetry	S_0[a]	S_1[b]
1	2	a_1	3030	3038
2	20a	a_1	3016	3024
3	13	a_1	2994	3006
4	8a	a_1	1594	1525
5	19a	a_1	1473	1410
6	9a	a_1	1154	1128
7	7a	a_1	1071	1047
8	18a	a_1	999	959
9	1	a_1	970	934
10	12	a_1	681	652
11	6a	a_1	361	366
12	17a	a_2	980	736
13	10a	a_2	970	618
14	16a	a_2	407	143
15	5	b_1	1002	818
16	17b	b_1	922	726
17	10b	b_1	748	603
18	4	b_1	679	397
19	16b	b_1	470	303
20	11	b_1	187	129
21	20b	b_2	3027	3035
22	7b	b_2	3004	3015
23	8b	b_2	1590	1637
24	19b	b_2	1435	1461
25	3	b_2	1303	1376
26	14	b_2	1184	1265
27	9b	b_2	1077	1125
28	15	b_2	1049	994
29	6b	b_2	601	513
30	18b	b_2	286	282

[a] Harmonic vibrational frequencies obtained using DFT calculations (B3LYP/6–31++G** level of theory).
[b] Obtained using CIS calculations with a 6–31++G** basis set. CIS calculations on excited electronic states are roughly equivalent to Hartree–Fock calculations on ground electronic states. Since vibrational frequencies in the latter are normally scaled by 0.89 to bring them into agreement with observed vibrational fundamental frequencies, the same scaling factor has been used here.

a_1 symmetry, while the lower energy feature at 288 cm^{-1} is assigned to the ν_{18b} mode, which has b_2 symmetry.

Of additional interest is the feature between 520 and 525 cm^{-1}, which in the expanded view can be seen to be a doublet. This is in the correct region for single quantum excitation of the ν_{6b} vibration (b_2 symmetry), but Table 26.1 reveals no other obvious candidate for the second peak. Two assignments have been put forward in the research literature for the second peak, between which it is difficult to differentiate, and both are based upon a combination band: $\nu_{16a} + \nu_{16b}$ or $\nu_{11} + \nu_{16a}$. Both of these have b_2 symmetry, since each consists of single quantum ($v = 1$) excitation of both an a_1 and a b_2 vibration, and the combined symmetry is obtained from the direct product $a_1 \otimes b_2 = b_2$. We employ the former assignment here, but note that it is not definitive. The proximity of vibrational levels of the same symmetry can lead to interaction, a process known as *Fermi resonance*. Briefly, if ψ_a and ψ_b are vibrational wavefunctions in close energetic proximity, then mixing becomes possible through a mechanism derived from the anharmonicity of vibrations providing the vibrational wavefunctions have the same symmetry. New perturbed vibrational states are generated with wavefunctions $a\psi_a + b\psi_b$ and $a\psi_b - b\psi_a$, where a and b are coefficients describing the extent of mixing. The term 'resonance' is indicative of the fact that this interaction is only significant if the unperturbed energy levels are close together, and Fermi resonance then results in the levels being pushed apart. Thus the current favoured assignment for the 520–525 cm^{-1} doublet in Figure 26.2 is a Fermi doublet involving the ν_{6b} and $\nu_{16a} + \nu_{16b}$ vibrational levels.

For the remainder of the spectrum in Figure 26.2, the majority of the features are assignable to totally symmetric (a_1) vibrations, but there are other bands attributable to b_2 vibrations. It is not, at the present time, possible to assign reliably all of the features in the spectrum because of the number of combination and overtone bands possible, the effects of anharmonicity, and the possibility of coupling between modes of the same symmetry.

Finally, we need to address the issue of how the b_2 vibrations appear with such high intensities in the spectra. Referring back to the earlier example of benzene (see Chapter 25), the observation of structure due to an e_{2g} vibration was attributed to a *vibronic* interaction that led to intensity borrowing by the S_1 state. In C_{2v} symmetry, a (doubly degenerate) e_{2g} vibration in benzene will transform into two distinct vibrations of a_1 and b_2 symmetry in the lower symmetry environment of chlorobenzene. In chlorobenzene the a_1 and b_2 vibrations may have very different frequencies (see Table 26.1) and should therefore be regarded as distinct vibrations. (Vibrations with the same number but additional labels a and b for doubly degenerate vibrations in benzene.) The substantial structure due to b_2 modes in the REMPI spectrum suggests that, even though the $S_1 \leftarrow S_0$ electronic transition is allowed in chlorobenzene, whereas it was forbidden in benzene, there is still some 'memory' of the higher symmetry in the parent benzene molecule and a vibronic effect gives rise to the b_2 activity in the spectrum.

In conclusion, the majority of the features in the REMPI spectrum of chlorobenzene can be assigned once it is appreciated that both totally symmetric and certain non-totally symmetric vibrations are active.

References

1. L. Grebe, *Z. Wiss. Photogr. Photophys. Photochem.* **3** (1905) 376.
2. Y. S. Jain and H. D. Bist, *J. Mol. Spectrosc.* **47** (1973) 126.
3. T. Cvitaš and J. M. Hollas, *Mol. Phys.* **18** (1970) 101.
4. T. G. Wright, S. I. Panov and T. A. Miller, *J. Chem. Phys.* **102** (1995) 4793.
5. *Vibrational Spectra of Benzene Derivatives*, G. Varsányi, New York, Academic Press, 1969.

27 Spectroscopy of the chlorobenzene cation

Concepts illustrated: *ZEKE spectroscopy; MATI spectroscopy; vibrational structure and the Franck–Condon principle;* ab initio *calculations; vibronic coupling; Fermi resonance.*

The lowering of symmetry in moving from benzene (D_{6h}) to chlorobenzene (C_{2v}) results in the removal of molecular orbital degeneracies. A convenient way of investigating this effect is through conventional photoelectron spectroscopy, and indeed Ruščić *et al.* studied this degeneracy breaking in 1981 using both HeI and HeII photoelectron spectroscopy [1]. The spectra obtained are shown in Figure 27.1, with the upper trace being that recorded using HeI radiation and the lower trace using HeII radiation.

The first two bands have similar ionization energies (maxima at 9.07 and 9.54 eV) and almost identical intensities. These bands correlate with the two components of the e_{1g} HOMO in benzene, which is a pair of π bonding orbitals (see Chapter 25) but which have split into two distinct orbitals in chlorobenzene owing to the lowering of the symmetry. Note that these two bands, and indeed most other bands in the spectra, are relatively broad. The next highest bands again form a pair, but these have considerably sharper profiles and correspond to ionization from lone pairs on the Cl atom.

The low resolution in conventional photoelectron spectroscopy restricts the amount of information that can be extracted. In this Case Study we consider alternative techniques that provide additional information about the chlorobenzene cation. This builds upon the material encountered in the previous two Case Studies.

27.1 The $\tilde{X}\,^2B_1$ state

The REMPI spectrum of chlorobenzene was described in the preceding Case Study. Once the REMPI spectrum of chlorobenzene is known, it is possible to use the vibrational levels of the intermediate S_1 state as a stepping stone to ionization, enabling two-colour ZEKE spectra to be recorded. A two-colour ZEKE spectrum is obtained by fixing the wavelength of one laser at the position of the appropriate $S_1 \leftarrow S_0$ transition, and the wavelength of the second laser is then scanned to access the cationic states (see Section 12.5 for additional experimental details). The primary advantage ZEKE spectroscopy has over photoelectron spectroscopy is its much higher resolution. In addition, in ZEKE spectroscopy, ionization

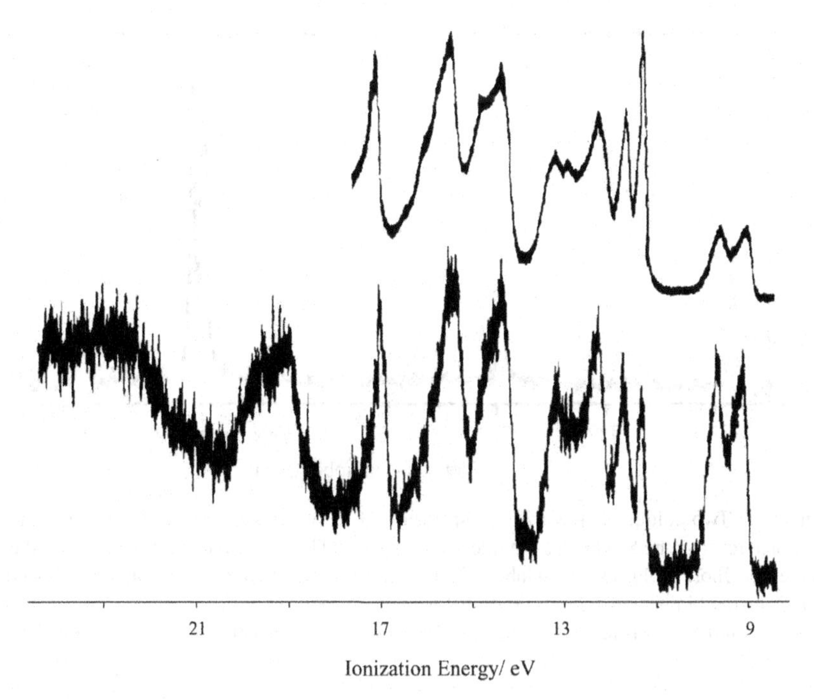

Figure 27.1 HeI (upper trace) and HeII (lower trace) photoelectron spectra of chlorobenzene. (Reproduced from B. Ruščić, L. Klasinc, A. Wolf, and J. V. Knop, *J. Phys. Chem.* **85** (1981) 1486, with permission from the American Chemical Society.)

can take place from selected vibrational levels in the intermediate electronic state by tuning the appropriate laser wavelength. Of course, it would be exceedingly time consuming simply to scan the other laser in an arbitrary search for the onset of ionization, and so some prior knowledge of the adiabatic ionization energy is very useful. Very often, a good estimate can come from conventional photoelectron studies such as that carried out by Ruščić *et al.*, and generally these are used as a first approximation of where to look for a ZEKE spectrum.

Figure 27.2 shows a two-colour ZEKE spectrum for excitation via one quantum in the totally symmetric ν_1 vibration in S_1, while Figure 27.3 shows the ZEKE spectra obtained by exciting via the $\nu_{6b}/(\nu_{16a} + \nu_{16b})$ Fermi resonance duet (see previous Case Study); these spectra were originally reported in Reference [2]. The ionization laser was tuned over a region that accesses the lowest electronic state of the cation, which corresponds to the lowest energy band in the photoelectron spectrum in Figure 27.1. The e_{1g} HOMO in benzene splits into a_2 and b_1 orbitals in chlorobenzene and it turns out that the b_1 orbital has the higher energy. Removal of an electron from this orbital therefore leads to the ground electronic state of the cation, which is a $^2B_1$ state.

The assignment of the vibrational structure in each spectrum was achieved in part by comparison with the results from *ab initio* calculations. Vibrational frequencies obtained with density functional theory (B3LYP/6–31++G**) are summarized in Table 27.1. It is also possible to excite other assigned vibrational levels in the S_1 state, and then

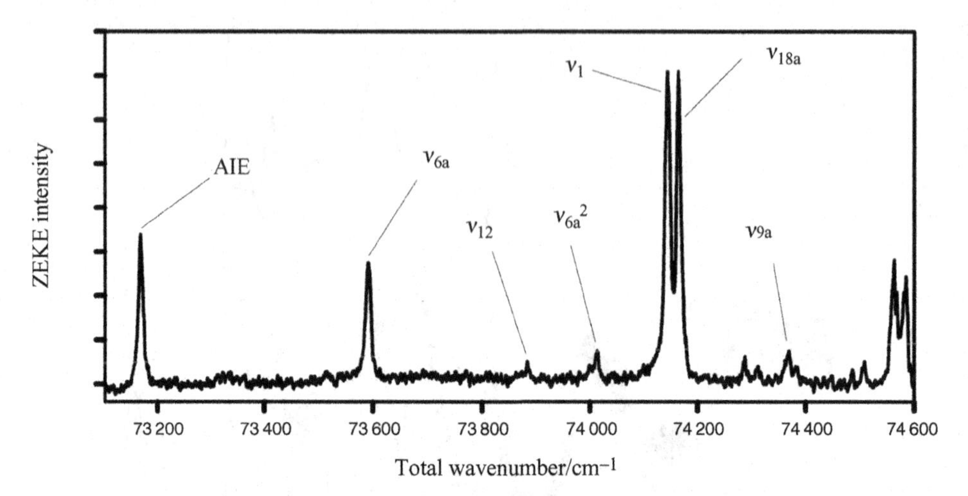

Figure 27.2 Two-colour (1 + 1′) ZEKE spectrum of chlorobenzene recorded by using the $\nu_1 = 1$ vibrational level in the S_1 state as the intermediate level. The vibrational numbering uses the Wilson scheme (see Table 27.1). The band labelled AIE refers to the adiabatic ionization process in which the cation is formed in its zero-point vibrational level. (Reproduced with permission from T. G. Wright, S. I. Panov, and T. A. Miller, *J. Chem. Phys.* **102** (1995) 4793, American Institute of Physics.)

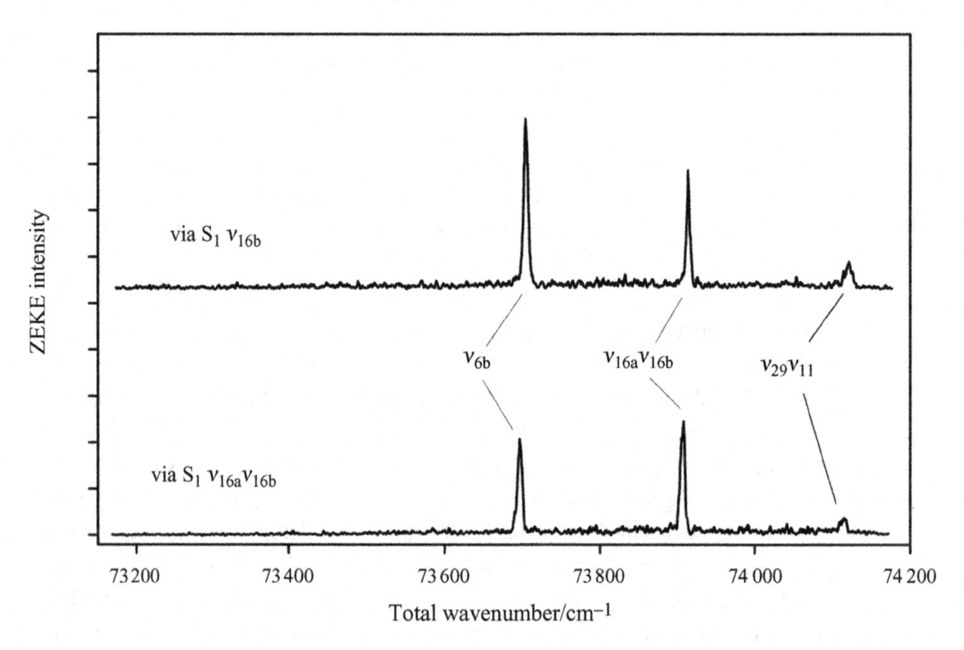

Figure 27.3 Two-colour (1 + 1′) ZEKE spectrum of chlorobenzene recorded by exciting via the ν_{6b} level (upper trace) and $\nu_{16a}\nu_{16b}$ (lower trace) vibrational levels in the S_1 state. Note that these two vibrational levels are believed to be the two components of a Fermi resonance doublet. The ZEKE spectrum is dominated by structure in vibrations with b_2 symmetry, which is consistent with the vibrational symmetry of the intermediate state. (Reproduced with permission from T. G. Wright, S. I. Panov, and T. A. Miller, *J. Chem. Phys.* **102** (1995) 4793, American Institute of Physics.)

Table 27.1 *Calculated vibrational frequencies of the chlorobenzene cation*

Mode (Mulliken)	Mode (Wilson)	Symmetry	Vibrational frequency[a]/cm^{-1}
1	2	a_1	3238
2	20a	a_1	3228
3	13	a_1	3216
4	8a	a_1	1646
5	19a	a_1	1463
6	9a	a_1	1218
7	7a	a_1	1120
8	18a	a_1	1001
9	1	a_1	989
10	12	a_1	721
11	6a	a_1	427
12	17a	a_2	1006
13	10a	a_2	801
14	16a	a_2	358
15	5	b_1	1001
16	17b	b_1	959
17	10b	b_1	773
18	4	b_1	595
19	16b	b_1	397
20	11	b_1	147
21	20b	b_2	3236
22	7b	b_2	3225
23	8b	b_2	1529
24	19b	b_2	1419
25	3	b_2	1389
26	14	b_2	1289
27	9b	b_2	1157
28	15	b_2	1103
29	6b	b_2	536
30	18b	b_2	306

[a] From DFT calculations using the B3LYP functional together with a 6–31++G** basis set.

use Franck–Condon arguments to deduce vibrational assignments in the ZEKE spectra, as has been done in Reference [2].

Notice that in the spectrum in Figure 27.2, since ionization takes place from an energy level of a totally symmetric (a_1) vibration in the S_1 state, the Franck–Condon principle leads us to expect that the main vibrational features in the ZEKE spectrum will also be due to totally symmetric vibrations in the cation. For the spectra in Figure 27.3, the S_1 vibrational levels excited have b_2 symmetry, and consequently b_2 vibrational structure should dominate in the ZEKE spectra. The Franck–Condon predictions are borne out in the spectra. Note that in Figure 27.3 the origin transition is not observed, as expected, since the wavefunction for the zero-point vibrational level of the cation has a_1 symmetry and so is not accessible from a b_2 vibrational level in the intermediate electronic state.

It is interesting to note that both Lembach and Brutschy [3] and Kwon *et al.* [4] have recorded mass analysed threshold ionization (MATI) spectra of chlorobenzene. MATI is

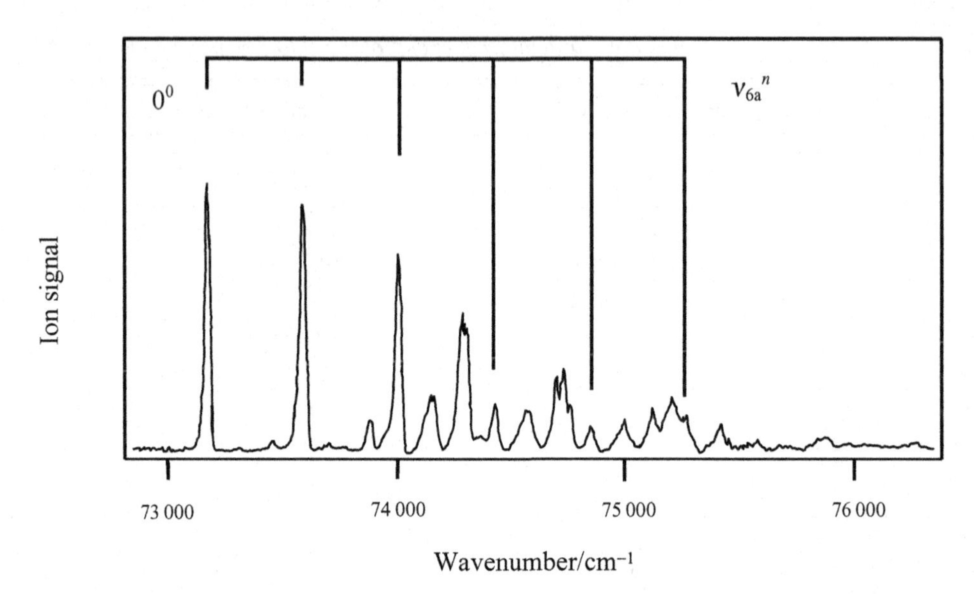

Figure 27.4 Single-photon MATI spectrum for the $\tilde{X}^2B_1 \leftarrow \tilde{X}^1A_1$ ionization process for the ^{35}Cl isotopomer of chlorobenzene. (Reproduced with permission from C. H. Kwon, H. L. Kim, and M. S. Kim, *J. Chem. Phys.* **116** (2002) 10361, American Institute of Physics.)

similar to the ZEKE technique (see Section 12.6) but in the former it is cations rather than electrons that are detected. The advantage of the MATI technique is its mass selectivity, which makes it possible to record separate spectra for the ^{35}Cl and ^{37}Cl isotopomers of chlorobenzene. Lembach and Brutschy used two-colour, two-photon ionization, whereas Kwon and co-workers employed single-photon ionization using VUV radiation. A single-photon MATI spectrum, with the excitation occurring out of the zero-point level of the S_0 state, is shown for the most prevalent isotopomer (containing ^{35}Cl) in Figure 27.4. This can be compared with the two-colour ZEKE spectrum in Figure 27.5 obtained via the zero-point level of the S_1 state.

As may be seen, the signal-to-noise (S/N) ratio is far better in the one-photon MATI spectrum, and has allowed the observation of a number of weaker features not seen in the two-colour ZEKE spectrum. The reason for the increased S/N ratio is not clear, but there is always the problem in two-colour spectroscopy of obtaining good spatial overlap of the laser beams and balancing the relative intensities of the two lasers to obtain the best signal. As noted above, Lembach and Brutschy also recorded MATI spectra of chlorobenzene, obtaining information on both isotopomers, but this time using a two-colour scheme: the spectra obtained are more similar to the two-colour ZEKE spectra than the one-colour MATI.

It is worth noting that the longer region scanned in the one-colour MATI spectrum (Figure 27.4) allows the observation of a progression in the ν_{6a} mode: this is a ring deformation mode, leading to an elongation of the ring in the direction of the C—Cl bond. Interestingly, *ab initio* calculations reported in Reference [2] revealed that the major difference in structure between the ground state of neutral chlorobenzene and the cation is a

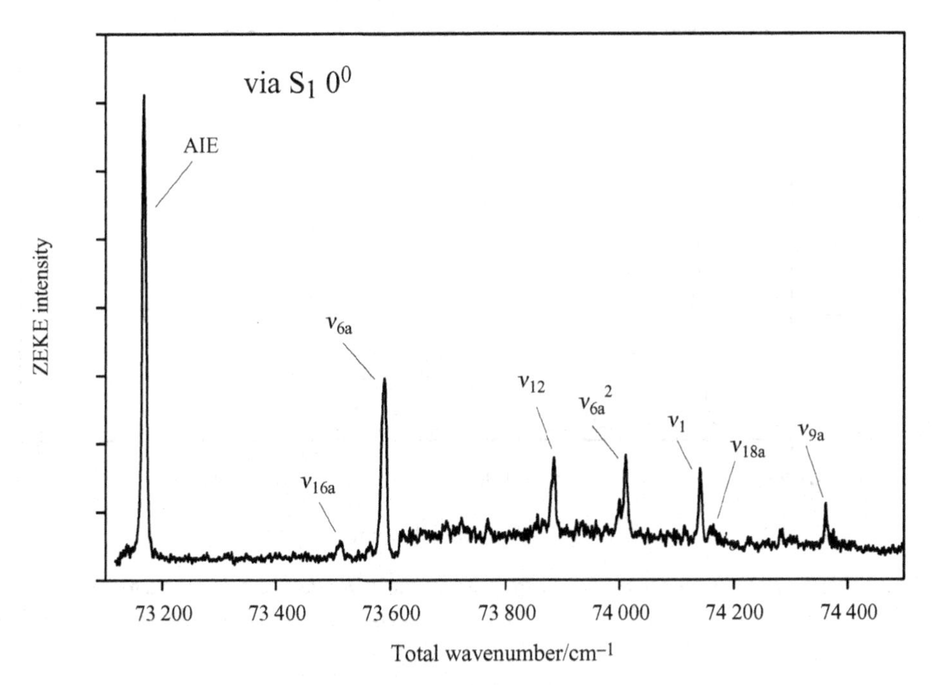

Figure 27.5 Two-colour (1 + 1′) ZEKE spectrum of chlorobenzene recorded by using the S_1 0^0 level as the intermediate state. (Reproduced with permission from T. G. Wright, S. I. Panov, and T. A. Miller, *J. Chem. Phys.* **102** (1995) 4793, American Institute of Physics.)

distortion of the latter along the ν_{6a} vibrational coordinate. The Franck–Condon principle would therefore lead us to expect the MATI and ZEKE spectra to be dominated by a vibrational progression in ν_{6a}, and this ties in nicely with the actual vibrational assignment. Note also that in the MATI spectrum there are weak features assigned that do not correspond to totally symmetric vibrations so the Franck–Condon principle is not entirely adhered to.

Returning briefly to the photoelectron spectrum, recall that the lowest energy photoelectron band is rather broad. As we have just seen from the ZEKE and MATI spectra, there is a substantial progression in the ν_{6a} vibration. This, coupled with the low resolution of conventional photoelectron spectroscopy, which is insufficient to resolve the vibrational structure, accounts for the width of the first photoelectron band in Figure 27.1.

27.2 The $\tilde{B}$ state

Since Kwon *et al.* [4] employed VUV radiation, they were also able to study excited electronic states of the cation. In particular, they concentrated on the cationic state corresponding to the photoelectron band at 11.31 eV in Figure 27.1. This corresponds to the second excited, or $\tilde{B}$ state, of the cation. The MATI spectrum obtained is shown in Figure 27.6.

As noted above, it is known from a combination of previous conventional photoelectron studies and *ab initio* calculations that this spectrum arises from removal of an electron

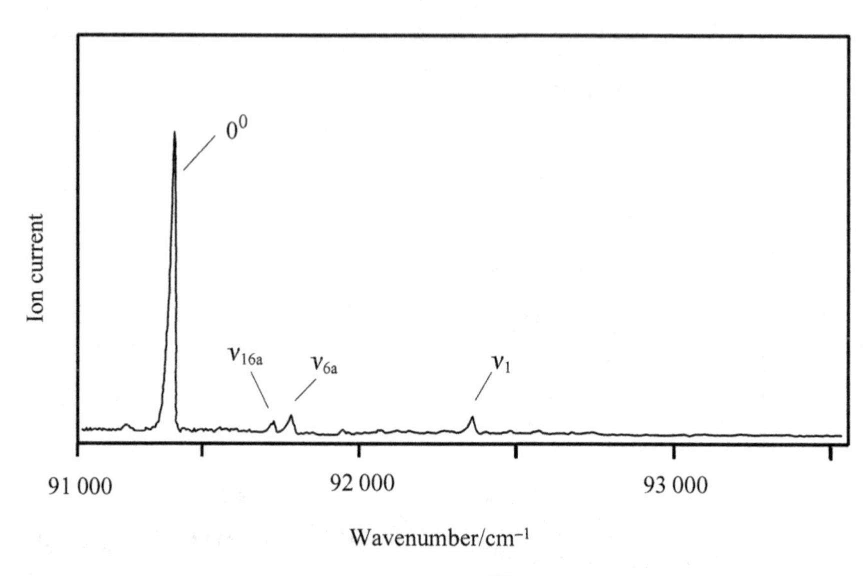

Figure 27.6 Single-photon MATI spectrum of the $\tilde{B}^2B_2$ state of chlorobenzene. (Reproduced with permission from C. H. Kwon, H. L. Kim, and M. S. Kim, *J. Chem. Phys.* **116** (2002) 10361, American Institute of Physics.)

from one of the lone pairs of the Cl atom; the lowest state of the cation that can arise from ionization of one of these electrons is the $^2B_2$ state. Since little change in molecular structure is expected for this ionization process, the dominant feature should be the origin transition in which no vibrational excitation in the ion occurs (corresponding to the adiabatic ionization energy (AIE) for the third photoelectron band). Of course, in the conventional photoelectron spectrum there was no chance to confirm this prediction, except to note that the corresponding photoelectron band was much sharper. In the MATI spectrum in Figure 27.6 it can clearly be seen that there is little vibrational structure, neatly confirming our expectations based upon prior knowledge of the ionization process.

References

1. B. Ruščić, L. Klasinc, A. Wolf, and J. V. Knop, *J. Phys. Chem.* **85** (1981) 1486.
2. T. G. Wright, S. I. Panov, and T. A. Miller, *J. Chem. Phys.* **102** (1995) 4793.
3. G. Lembach and B. Brutschy, *Chem. Phys. Lett.* **273** (1997) 421.
4. C. H. Kwon, H. L. Kim, and M. S. Kim, *J. Chem. Phys.* **116** (2002) 10361.

28 Cavity ringdown spectroscopy of the $a^1\Delta \leftarrow X^3\Sigma_g^-$ transition in O_2

Concepts illustrated: *cavity ringdown spectroscopy; Pauli principle and electronic states; Hund's coupling cases; rotational structure of an open-shell molecule; nuclear spin statistics.*

The oxygen molecule is, of course, of fundamental importance to our atmosphere and the reactions that occur in it. Oxygen is a precursor of ozone in the atmosphere, and in turn is produced when ozone is destroyed in the atmosphere. Atmospheric models of ozone concentrations depend critically upon knowing absorption coefficients for oxygen.

In this Case Study, the absorption spectrum corresponding to the $a^1\Delta_g \leftarrow X^3\Sigma_g^-$ transition is considered. This is formally a spin-forbidden electronic transition, since $\Delta S \neq 0$. It is also spatially forbidden as an electric dipole transition since the direct product $\Delta_g \otimes \Sigma_g^- = \Delta_g$, whereas the dipole moment operator has components with Σ_u^+ and Π_u symmetries. Consequently, both $\Delta\Lambda$ $(=0, \pm1)$ and $u \leftrightarrow g$ selection rules are violated, and yet remarkably the $a^1\Delta_g \leftarrow X^3\Sigma_g^-$ transition can still be experimentally observed. As one would imagine, it is an extremely weak transition and a highly sensitive spectroscopic technique is required in order to observe it.

28.1 Experimental

This Case Study is based on work by Newman *et al.* [1] using the highly sensitive absorption technique known as cavity ringdown (CRD) spectroscopy. Newman *et al.* set out to measure the spectrum and absorption coefficient data for the $a^1\Delta_g \leftarrow X^3\Sigma_g^-$ transition in order to be able to obtain accurate information for describing the absorption and emission of radiation from these electronic states.

The principles of the CRD technique have already been described in Section 11.3. Recall that this is an absorption method and therefore reliance on a second step for detecting a transition is not required (cf. LIF or REMPI). In CRD spectroscopy the decay of the intensity of a pulse of light is monitored as it bounces to and fro between two highly reflective mirrors. The rate of leakage of the light pulse out of the cavity depends on the cavity itself

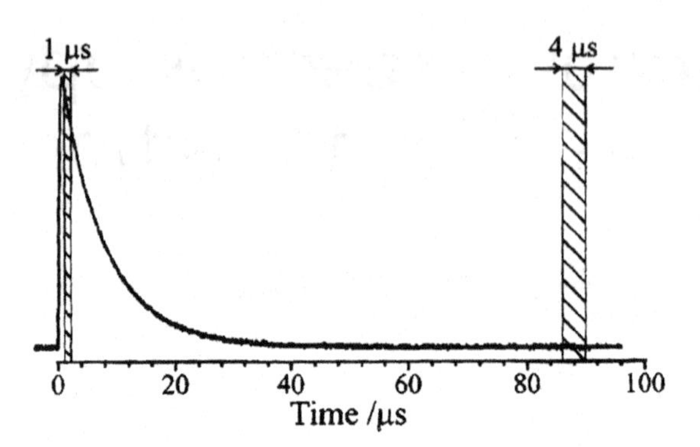

Figure 28.1 A typical cavity ringdown trace: note the exponential decay of the intensity of the light with time. (Reproduced with permission from S. M. Newman, I. C. Lane, A. Orr-Ewing, D. A. Newnham, and J. Ballard, *J. Chem. Phys.* **110** (1999) 10749, American Institute of Physics.)

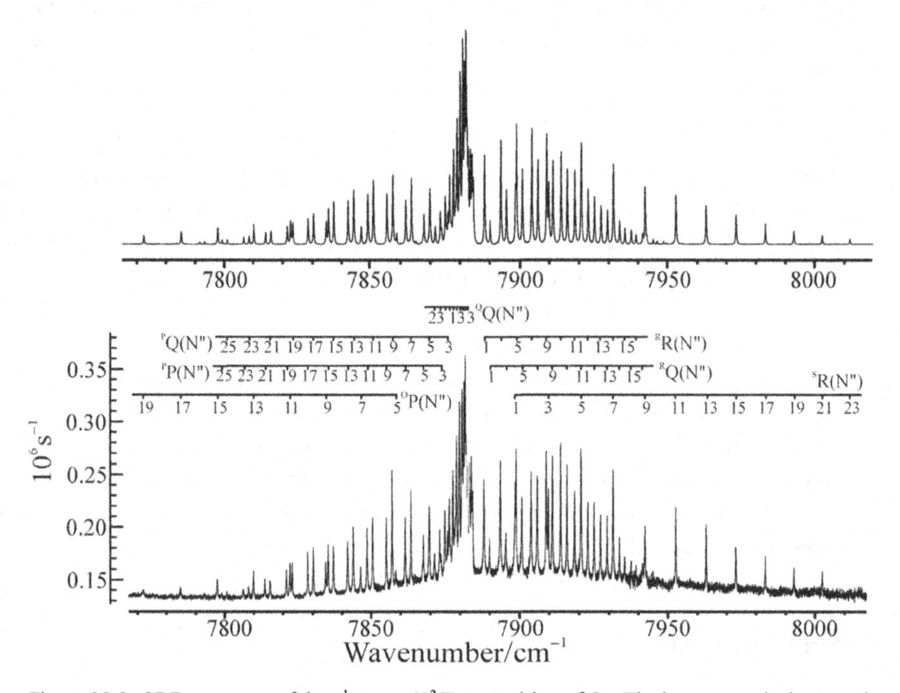

Figure 28.2 CRD spectrum of the $a^1\Delta_g \leftarrow X^3\Sigma_g^-$ transition of O_2. The lower trace is the experimental spectrum, and the upper trace is a simulation: the good agreement between experiment and theory suggests that the assignment shown is correct. The notation used for labelling the lines is discussed in the text. (Reproduced with permission from S. M. Newman, I. C. Lane, A. Orr-Ewing, D. A. Newnham, and J. Ballard, *J. Chem. Phys.* **110** (1999) 10749, American Institute of Physics.)

(specifically the reflectivity of the mirrors) and the absorption of light by molecules within the cavity. Since the separation of the a and X states of O_2 is $\sim$8000 cm^{-1}, a near-infrared light source was used by Newman and co-workers. This light source was the idler output of an optical parametric oscillator (see Section 10.8), which was pumped by the frequency-tripled output (355 nm) of a Nd:YAG laser. The wavelength of the light was varied over the range 1.25–1.29 μm. A typical ringdown trace is shown in Figure 28.1.

By accounting for the losses that are associated with the empty cavity, it is possible to deconvolute the ringdown signal so that the losses attributable *only* to sample absorption can be obtained. By scanning the laser wavelength, the ringdown data can be transformed into an absorption spectrum, and the one reported in Reference [1] is shown in Figure 28.2. This spectrum was obtained for a room temperature O_2 sample at a pressure of 1 atmosphere.

It is important to emphasize that the transition being observed is exceedingly weak by the standards of normal electronic transitions (see below) and yet a remarkably good signal-to-noise ratio is obtained because of the high sensitivity of CRD spectroscopy. The assignment of this spectrum will be discussed later after considering the low-lying electronic states of O_2 and the rotational energy levels of these states.

28.2 Electronic states of O_2

Molecular orbital theory shows that O_2 has the valence electronic configuration $(2s\sigma_g)^2(2s\sigma_u^*)^2(2p\sigma_g)^2(2p\pi_u)^4(2p\pi_g^*)^2$. This configuration can actually give rise to three electronic states and their symmetries can be determined by application of group theoretical considerations. O_2 is an example where it is necessary to take care over the Pauli principle, since there are two electrons to be distributed amongst two degenerate orbitals (see Appendix E). The spatial symmetries of the electronic states can be deduced by considering only the outer configuration $(2p\pi_g^*)^2$, since all other occupied orbitals are full and therefore make only a totally symmetric contribution to the overall electronic state spatial symmetry. The direct product $\pi_g \otimes \pi_g$ may be evaluated as $\Sigma_g^+ + [\Sigma_g^-] + \Delta_g$, where upper case symbols have been used to indicate the symmetries of electronic states. The square brackets around the Σ_g^- label indicate that an electronic state with this symmetry is antisymmetric with respect to electron exchange, whereas Σ_g^+ and Δ_g are totally symmetric. The Pauli principle requires that the overall product of the spatial and spin symmetries must be antisymmetric, since we are allowing for the exchange of equivalent fermions (electrons). The possible spin states for a two-electron case are singlet ($S = 0$) and triplet ($S = 1$). The corresponding spin wavefunctions are summarized in equations (E.1)–(E.4) in Appendix E. The triplet wavefunctions are totally symmetric with respect to electron exchange, and so can only be combined with Σ_g^- spatial symmetry to give a ${}^3\Sigma_g^-$ electronic state. In contrast the singlet spin wavefunction is antisymmetric leading to ${}^1\Sigma_g^+$ and ${}^1\Delta_g$ electronic states.

The order of these electronic states can be deduced using *Hund's rules.*[1] These predict that the lowest electronic state from a given electronic configuration will be the one with the highest spin. For states with the same spin, the one with the highest orbital angular momentum is normally the lowest in energy. These rules suggest that the energies of the electronic states lie in the order ${}^3\Sigma_g^- < {}^1\Delta_g < {}^1\Sigma_g^+$, and this is confirmed by both theory and experiment. Figure 28.3 shows potential energy curves derived from *ab initio* calculations for some of the low-lying electronic states.

1 Hund's rules are based on sound physical principles but should be used with caution. The proximity of electronic states can sometimes lead to interactions between these states that yield a different energy ordering from that predicted by Hund's rules.

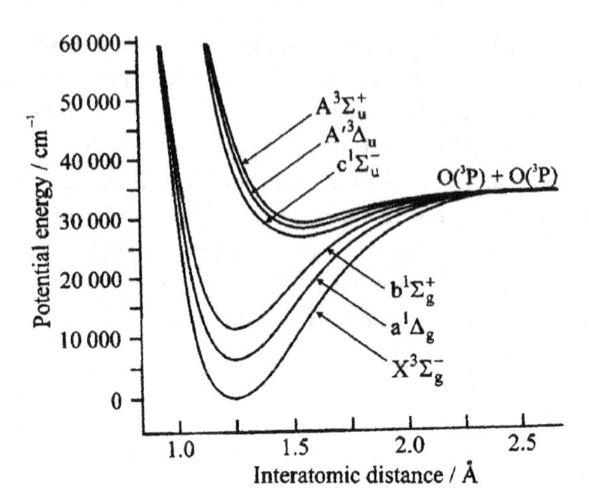

Figure 28.3 Potential energy curves for the lowest states of O_2 obtained from *ab initio* calculations. (Reproduced with permission from S. M. Newman, I. C. Lane, A. Orr-Ewing, D. A. Newnham, and J. Ballard, *J. Chem. Phys.* **110** (1999) 10749, American Institute of Physics.)

28.3 Rotational energy levels

The spectrum shown in Figure 28.2 is a rotationally resolved spectrum in the region of the electronic origin ($v' = 0 \leftarrow v'' = 0$) transition. In order to be able to assign the various lines in the spectrum, it is first necessary to understand the relevant rotational energy levels. If O_2 was a closed-shell molecule, then the simple expressions for the rotational energy levels of closed-shell diatomic molecules could be employed and the rotational analysis would be relatively simple. However, O_2 is an open-shell molecule possessing electronic angular momentum as well as rotational angular momentum, and therefore a more sophisticated approach is required. In particular, the coupling of the electronic and rotational angular momenta must be accounted for.

Considering the $^1\Delta_g$ state first, the angular momenta present are the electronic orbital angular momentum, $\boldsymbol{L}$, and the rotational angular momentum, $\boldsymbol{R}$. This state follows Hund's case (b) coupling (see Appendix G), and so the vector $\boldsymbol{L}$ will precess rapidly about the internuclear axis to give a projection described by the quantum number Λ, where $\Lambda = 2$ for a Δ state. The total angular momentum $\boldsymbol{J}$ is formed by the vector sum of Λ and $\boldsymbol{R}$. Strictly speaking, a more detailed model is required. All electronic states for which $\Lambda \neq 0$ are doubly degenerate, and coupling with the rotational angular momentum removes this degeneracy to give a pair of energy levels corresponding to each rotational level [2]. However, this so-called Λ-*doubling* normally gives rise to a very small splitting, particularly for low rotational levels, and unless working with high resolution spectra it can be safely ignored. The rotational energy levels for a $^1\Delta_g$ electronic state can therefore be satisfactorily described by the standard closed-shell expression $BJ(J+1)$, except that in this case the lowest possible value of J is 2 since the minimum angular momentum possessed by the molecule corresponds to $\Lambda = 2$.

Turning now to the $^3\Sigma_g^-$ state, this can be described satisfactorily by Hund's case (b). The spin $\boldsymbol{S}$ will couple with the rotational angular momentum resulting in a splitting of each

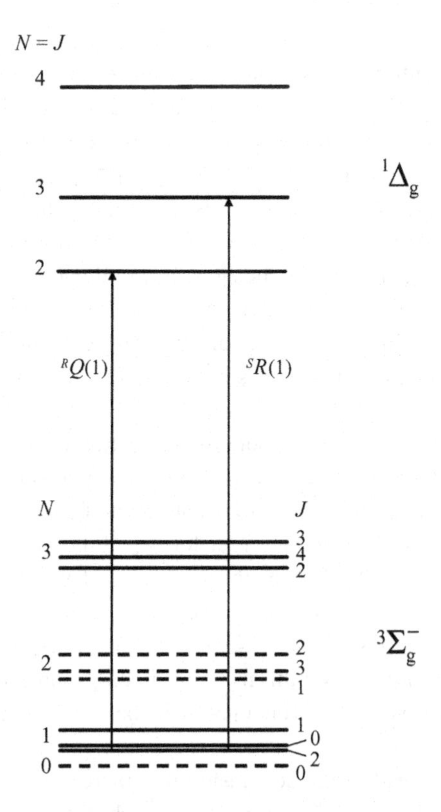

Figure 28.4 Lowest rotational energy levels of the $X^3\Sigma_g^-$ and $a^1\Delta_g$ electronic states of O_2. Note that the rotational levels in the $X^3\Sigma_g^-$ state are split due to interaction with the spin angular momentum (see text). Note also that for $^{16}O_2$ the even N levels will be absent owing to nuclear spin statistics (see text) – they are marked here as dashed lines. In addition, the two transitions expected from the lowest J'' level are indicated.

rotational level into three components corresponding to quantum numbers $J = N + 1$, $J = N$, and $J = N - 1$.[2] In fact the observed spin splitting is produced by two effects, (i) a spin–rotation interaction (see also Chapter 22) and (ii) a spin–spin interaction from the two unpaired electrons. This somewhat more complicated coupling gives rise to the energy level pattern for a $X^3\Sigma_g^-$ state shown in Figure 28.4. Note that each value of N gives rise to three values of J, except for $N = 0$.

28.4 Nuclear spin statistics

There is one further factor that must be recognized before attempting to assign the spectrum. The two atomic nuclei in O_2 are equivalent and so, as for the case of equivalent electrons,

[2] N is the quantum number conventionally employed for the combined rotational + orbital angular momentum (see Appendix G), and since there is no orbital angular momentum for a $^3\Sigma^-$ state, then N in this case can be regarded as the rotational quantum number.

the effect of the Pauli principle must be taken into account. ^{16}O has a nuclear spin of $I = 0$ and is therefore a *boson*. The Pauli principle states that the overall wavefunction must be totally symmetric with respect to the exchange of two identical bosons. The total wavefunction, Ψ_{tot}, is the product of the individual electronic, vibrational, rotational, and nuclear spin wavefunctions for a particular state, i.e. $\Psi_{tot} = \Psi_{elec}\Psi_{vib}\Psi_{rot}\Psi_{ns}$. Exchange of the two nuclei can be achieved by a 180° rotation but this also rotates the electronic wavefunction. Movement of the electronic wavefunction back to its original position while keeping the nuclei fixed is equivalent to an inversion of the electron coordinates (symmetry operation i) followed by a reflection in a plane perpendicular to the axis of 180° rotation. The point of choosing such an apparently long-winded set of symmetry operations is that the symmetry of the electronic wavefunction can then easily be established from the g/u and $\pm$ labels on the electronic state label.

For the $X^3\Sigma_g^-$ state, inversion leaves the electronic state wavefunction unchanged but reflection changes the sign, so Ψ_{elec} is antisymmetric with respect to exchange of the nuclei. The ground state vibrational wavefunction for a diatomic molecule is unaffected by nuclear exchange and hence $\Psi_{rot}\Psi_{ns}$ must be antisymmetric in order for Ψ_{tot} to be symmetric. It turns out that rotation of the molecule by 180° changes the symmetry of Ψ_{rot} by $(-1)^N$, while the fact that $I = 0$ for ^{16}O means that only a totally symmetric nuclear spin state is possible. We can therefore conclude that N must be odd to satisfy the Pauli principle, which means that the even J rotational levels *do not exist* for this molecule in the ground electronic state. Note that for $^{16}O^{18}O$ both odd and even N rotational levels do exist since in that case the nuclei are not equivalent.

For $O_2(a^1\Delta_g)$, symmetry arguments lead to the conclusion that there are no missing rotational energy levels. However, only the $\Lambda = +2$ component of each Λ-doublet occurs, and this leads to a small alternating shift in the energy of the rotational states [2]. This effect may only be observed under very high resolution.

28.5 Spectrum assignment

Owing to the nuclear spin statistics, the $N = 0$ rotational level in the $^3\Sigma_g^-$ electronic state does not exist and so the lowest occupied level corresponds to $N = 1$. Spin–rotation will split this rotational level into closely spaced $J = 0$, 1, and 2 sub-levels – note that the energy ordering of the J levels is complicated by spin–spin and spin–rotation interactions. Recall also that for the $^1\Delta_g$ electronic state the lowest rotational level corresponds to $J = N = 2$. Assuming the electric dipole selection rule $\Delta J = 0, \pm 1$, the possible transitions from the lowest J level in the ground electronic state are

$$2 \leftarrow (1, 2)$$

$$3 \leftarrow (1, 2)$$

where the two quantum numbers in the lower state refer to (N, J). These two transitions belong to Q and R branches, respectively. However, an explicit designation of the transitions also requires an indication of ΔN and so the notation employed is $^{\Delta N}\Delta J(N'')$, where ″

and $'$ are used to distinguish quantum numbers in the lower and upper electronic states, respectively. Consequently, the above transitions become $^{R}Q(1)$ and $^{S}R(1)$ in this notation – these transitions are shown in Figure 28.4. Clearly there are more than two rotational branches and so assignment of all of the transitions requires some careful consideration.

A good starting point is to recognize that the highest energy branch will be the ^{S}R branch, since both ΔN and ΔJ have their maximum values (which for ΔN is $+2$). A regular series of rotational lines can be seen in the highest wavenumber region, which can be extrapolated back to the first member, $^{S}R(1)$. A similar process at the opposite end of the spectrum can be carried out for the ^{O}P branch, and a combination of the ^{S}R and ^{O}P branch data will allow approximate rotational constants in the two states to be estimated. In fact the rotational constants (including centrifugal distortion constants) of the $X^3\Sigma_g^-$ state are already well known from earlier studies (see, for example [3]) and so the focus can be restricted to the excited state rotational constant. Assignment of lines in the other branches is more challenging because of the increased congestion but with patience the full assignment shown in Figure 28.2 can be achieved. Use of computational simulation and least-squares fitting procedures (see Appendix H) saves considerably on labour and would be the usual route to analysing a relatively complicated spectrum such as that shown here. A simulation of the spectrum is shown in the upper trace of Figure 28.2. Note that intensities as well as energies of the transitions are important for a complete understanding of a spectrum – especially in the work described in Reference [1] where the intensities were being used to derive absolute absorption coefficients.

28.6 Why is this strongly forbidden transition observed?

It was stated earlier that the $a \leftarrow X$ transition in O_2 is strongly forbidden on the basis of electric dipole selection rules. The transition intensity must therefore be carried by some other means and both electric quadrupole and magnetic dipole transitions are possibilities (these mechanisms were briefly mentioned in Section 7.1). In fact it turns out that the magnetic dipole mechanism alone is sufficient to account for the observed structure. If an electric quadrupole mechanism was also significant, the spectrum should exhibit $\Delta J = \pm 2$ transitions with appreciable intensity, which it does not. Further details can be found in Reference [1].

References

1. S. M. Newman, I. C. Lane, A. Orr-Ewing, D. A. Newnham, and J. Ballard, *J. Chem. Phys.* **110** (1999) 10749.
2. *Rotational Spectroscopy of Diatomic Molecules*, J. M. Brown and A. Carrington, Cambridge, Cambridge University Press, 2003.
3. G. Rouillé, G. Millot, R. Saint-Loup, and H. Berger, *J. Mol. Spectrosc.* **154** (1992) 372.

Appendix A
Units in spectroscopy

People working in different branches of spectroscopy tend to express spectroscopic quantities in their own preferred flavour of units. The wide variety of units in use can confuse the beginner. There is also a tendency for many practising spectroscopists to use terminology which is, strictly speaking, incorrect but which does slip by. The authors of this book have probably been guilty of this very charge on several occasions in this book.

Spectroscopic transitions involve the input or removal of energy from a molecule. The SI unit of energy is the joule (symbol J), and so the energy of a photon should be expressed in joules. If, for example, we have a blue light source with a wavelength of 450 nm, the energy of a photon ($= hc/\lambda = h\nu$) is 4.4143×10^{-19} J to five significant figures. Although correct, it is difficult to gauge the significance of such a small number. Of course, the photon energy could be expressed as 0.441 43 aJ, where the prefix a stands for atto (10^{-18}). However, this is rarely done in practice. Part of the problem is historical, but also the usable range of photon energies in spectroscopy varies over so many orders of magnitude that different types of spectroscopists have their own favourite units.

In visible and ultraviolet electronic spectroscopy, the positions of transitions are commonly expressed in terms of the photon wavelength in nanometres (nm). However, it is also quite common to employ wavenumber units, where

$$\text{Wavenumber } \bar{\nu} = \frac{1}{\lambda} = \frac{\nu}{c} \tag{A.1}$$

Although wavenumbers could be quoted in m^{-1}, they are more commonly given as cm^{-1}. For example, a wavelength of 450 nm corresponds to a wavenumber of 22 222 cm^{-1} to five significant figures. The use of frequency units is uncommon in electronic spectroscopy because the numbers obtained are so large, e.g. 450 nm corresponds to 6.6620×10^{14} Hz. Occasionally the widths of lines in high resolution electronic spectroscopy are quoted in frequency units, although in that case it normally falls in the MHz range.

In photoelectron spectroscopy, the photon energies are much larger, and therefore the transition wavenumber is also much larger and cumbersome. Consequently, in quoting ionization energies the favoured unit of photoelectron spectroscopists is the electronvolt, given the symbol eV. This is the energy required to move an electron through a potential difference of 1 V. For example, at the HeI wavelength, 58.4 nm, the photon energy is 3.40×10^{-18} J, but in electronvolts this corresponds to 21.2 eV. The conversion is easily obtained by dividing the photon energy by the elementary charge, e (1.602×10^{-19} C).

As if the above was not enough, there are other complications. It is common for energies obtained from *ab initio* and other quantum chemical calculations to be output in atomic units, known as hartrees (symbol E_h). Also, calories are also still widely used in the chemistry literature as a unit for energy despite being superseded by joules, but calories are rarely used by spectroscopists for the same reason that joules are also little used in quoting spectroscopic quantities.

A.1 Some fundamental constants and useful unit conversions

Speed of light (in a vacuum) $c = 2.997\,924\,59 \times 10^8$ m s^{-1}
Planck constant $h = 6.626\,0755\ (40) \times 10^{-34}$ J s
Elementary charge $e = 1.602\,177\,33\ (49) \times 10^{-19}$ C
Electron rest mass $m_e = 9.109\,3897\ (54) \times 10^{-31}$ kg
Proton rest mass $m_p = 1.672\,6231\ (54) \times 10^{-27}$ kg
Avogadro constant $N_A = 6.022\,1367\ (36) \times 10^{23}$ mol^{-1}
Boltzmann constant $k = 1.380\,658\ (12) \times 10^{-23}$ J K^{-1}

The numbers in parentheses represent an uncertainty of *one* standard deviation in the last two figures of each quantity.

$$
\begin{aligned}
1\ \text{hartree} &= 4.359\,75 \times 10^{-18}\ \text{J} \\
&= 2.625\,9 \times 10^{3}\ \text{kJ mol}^{-1} \\
&= 627.510\ \text{kcal mol}^{-1} \\
&= 27.211\,61\ \text{eV} \\
&= 2.194\,746 \times 10^{5}\ \text{cm}^{-1} \\
&= 6.579\,684 \times 10^{9}\ \text{MHz} \\
1\ \text{eV} &= 1.602\,177 \times 10^{-19}\ \text{J} \\
&= 96.485\,3\ \text{kJ mol}^{-1} \\
&= 23.061\ \text{kcal mol}^{-1} \\
&= 3.674\,931 \times 10^{-2}\ E_h \\
&= 8065.54\ \text{cm}^{-1} \\
&= 2.417\,988 \times 10^{8}\ \text{MHz}
\end{aligned}
$$

More details on the recommended units used in physical chemistry and spectroscopy can be found in the following book: *Quantities, Units and Symbols in Physical Chemistry*, published by Blackwell Scientific Publications (Oxford, 1993).

Appendix B
Electronic structure calculations

As was mentioned in Chapter 2, analytical solutions of the many-electron Schrödinger equation are not possible. To be able to predict properties of molecular systems, approximations are introduced and the resulting equations are solved numerically. As is usually the case with approximations, they represent a trade-off between ease of calculation and quality of prediction. It is therefore always important to bear in mind what approximations are implied because this affects both the validity and the reliability of the results.

A brief summary of some of the different kinds of calculational methods available is given in this appendix. Broadly speaking they can be divided into three groups, ***ab initio***, **semiempirical**, and **density functional** methods. Semiempirical methods are particularly important for tackling large molecules but, because of the tremendous increase in computer power over the past two decades, they have now been largely superseded by the more sophisticated *ab initio* methods for calculations on small and medium-sized molecules. Density functional calculations are now also becoming commonplace and these would seem to yield good quality results at modest computational cost. Our emphasis here is primarily on the *ab initio* approach, although we will briefly return to consider semiempirical and density functional methods later.

B.1 Preliminaries

Inherent to virtually all electronic structure calculations are two approximations, the neglect of relativistic effects and the use of the Born–Oppenheimer approximation. Neglecting the energy terms that describe relativistic effects is a rather safe thing to do if we are only interested in molecules containing first-row elements (H–Ne). For heavier atoms, especially those in the third and higher rows, relativistic effects can be highly significant and there are methods available, which will not be considered here, to deal with these [1].

The Born–Oppenheimer approximation, whose origins were briefly discussed in Section 2.12, is also satisfactory in most situations. A consequence of this approximation is that the full molecular time-independent Schrödinger equation can be divided into two separate equations

$$H_e \Psi_e = E_e \Psi_e \tag{B.1}$$

$$(T_n + E_e)\Psi_n \approx E \Psi_n \tag{B.2}$$

where H_e is the electronic Hamiltonian operator given in full in equation (2.4) and E_e is the corresponding energy (it includes the nuclear–nuclear repulsion). Equation (B.1) is the Schrödinger equation for a fixed set of nuclear positions. Equation (B.2) describes the effect of nuclear motion, with T_n being the nuclear kinetic energy operator and E being the total energy of the molecule. Notice that the potential energy 'operator' in (B.2) is the energy from solution of (B.1), so equation (B.1) must be solved before tackling (B.2).

In the case of a diatomic molecule, solution of (B.1) at various internuclear separations gives the *potential energy curve* for that molecule. In polyatomic molecules consisting of N atoms the energy E_e is a function of $3N - 6$ or $3N - 5$ internal nuclear coordinates, depending on whether the molecule is non-linear or linear, and it constitutes the *potential energy surface*. The potential energy curve or surface defines the vibrational motion of a molecule and therefore in order to predict vibrational frequencies equation (B.1) can be solved at a variety of nuclear configurations to generate the potential energy surface, and then (B.2) is subsequently solved. In fact in the majority of calculations equation (B.2) is rarely solved explicitly to extract vibrational frequencies: a quicker route, based on the evaluation of first and second derivatives of the total electronic energy with respect to the internal nuclear coordinates, is usually employed [2].

An important point is that the wavefunction must satisfy the Pauli principle. In its simplest form, this says that each electron in an atom or molecule has a unique set of quantum numbers. In formal quantum mechanics, this corresponds to the insistence that the total electronic wavefunction, Ψ_e, must be antisymmetric with respect to the exchange of any two electrons. A simple product wavefunction, one for each electron, of the type shown in equation (2.5), will *not* satisfy the Pauli principle.

Take, as an example, the case of H_2 in its ground electronic state, where the two electrons are paired up in the $1\sigma_g^+$ orbital. The wavefunctions for each electron are different, the difference being not the spatial distributions of the two electrons, which are the same, but the spins, which are opposite. We could therefore factor the wavefunction for each electron into a common spatial part, which will be written as σ_g^+, and a spin part, which is designated as either α or β depending on whether the spin is 'up' or 'down'. Notice that the spatial wavefunction represents what is commonly referred to as an *orbital*, in this case a molecular orbital. The total electronic wavefunction can therefore be written as

$$\Psi_e = \sigma_g^+(1)\alpha(1)\sigma_g^+(2)\beta(2) \tag{B.3}$$

Unfortunately, this doesn't satisfy the Pauli principle since an exchange of electrons 1 and 2 (equivalent to just switching the '1' and '2' labels in (B.3)) does not change the sign of the wavefunction. However, the following function *is* antisymmetric with respect to electron exchange:

$$\Psi_e = \sigma_g^+(1)\sigma_g^+(2)[\alpha(1)\beta(2) - \alpha(2)\beta(1)] \tag{B.4}$$

This is an acceptable form of the wavefunction for a spin singlet since it satisfies the Pauli principle and it retains, albeit in a slightly more complicated manner, the concept of molecular orbitals.

Can similar antisymmetrized electronic wavefunctions be constructed for more complicated molecules? The answer is yes, but written out in full algebraic form the expressions are extremely long even when relatively few electrons are involved. A concise and general way of writing the antisymmetrized wavefunctions is in the form of a determinant, the so-called Slater determinant

$$\Psi(1, 2, \ldots, n) = \frac{1}{\sqrt{n!}} \begin{vmatrix} \varphi_1(1) & \varphi_2(1) & \ldots & \varphi_n(1) \\ \varphi_1(2) & \varphi_2(2) & \ldots & \varphi_n(2) \\ \vdots & \vdots & & \vdots \\ \varphi_1(n) & \varphi_2(n) & \ldots & \varphi_n(n) \end{vmatrix} \tag{B.5}$$

where n is the number of electrons and φ_i represents the ith *spin-orbital*, which is a product of the spatial and spin wavefunctions. The electronic wavefunctions employed in all *ab initio* calculations are either single Slater determinants, or are linear combinations of Slater determinants.[1]

B.2 Hartree–Fock method

The Hartree–Fock (HF) method is the most common *ab initio* technique for calculating electronic structure. It is also the starting point for many of the more sophisticated methods and it is therefore worthwhile outlining the underlying philosophy. The HF method is derived from application of a well-known theorem in quantum mechanics, the *variation theorem*. We start from the proposition that (B.1) cannot be solved analytically and so we must seek approximate solutions. Suppose we make a guess at the mathematical form of the true electronic wavefunction, Ψ_e, our guess being represented by the symbol Ω (in all probability of course an arbitrary guess is likely to be a very poor one indeed!). According to the variation theorem, if the energy is calculated using this guessed, or so-called *trial* wavefunction, which can be done using the expression[2]

$$E = \frac{\int \Omega^* H_e \Omega \, d\tau}{\int \Omega^* \Omega \, d\tau} \tag{B.6}$$

then $E \geq E_e$, where E_e is the true energy of the system. This is an extraordinarily powerful and remarkable result, for it reveals that no matter how good, or bad, our guess at the wavefunction actually is, the energy calculated will *always* be above the true energy. Consequently, if a wavefunction is chosen containing adjustable parameters, then values for these parameters could be varied to give the minimum possible value of E. If the trial wavefunction is sufficiently flexible, this minimization of E may give an energy very close to the true value, E_e.

---

1 A single Slater determinant always suffices for closed-shell molecules, but for open-shell molecules more than one Slater determinant is often required for a correct representation of the electronic state within the Hartree–Fock model.

2 Equation (B.6) is obtained by replacing the wavefunction in the Schrödinger equation (B.1) with the trial wavefunction Ω. Multiplication of both sides of (B.1) by Ω^*, which is the complex conjugate of Ω, followed by integration and rearrangement, then leads to (B.6). The quantity calculated in (B.6) is known as the *expectation value* of the energy for the given trial wavefunction.

In the HF method, a trial wavefunction consisting of a product of molecular orbitals, one for each electron, is assumed. Specifically, the product wavefunction is in the form of the Slater determinant (B.5). The point was made in Section 2.1.4 that the true overall electronic wavefunction cannot be factored exactly into individual one-electron wavefunctions. Thus the imposition of molecular orbitals is only an approximation made for convenience and this should always be borne in mind when interpreting the results from Hartree–Fock calculations. Let us assume that the molecular orbitals contain variational parameters (these will be identified later). If the total electronic wavefunction and the Hamiltonian (2.4) are then inserted into (B.6) it is possible to derive, by minimizing E using differential calculus, a so-called *Hartree–Fock equation* for each molecular orbital:

$$\left\{ H_i + \sum_j (2J_j - K_j) \right\} \psi_i = \varepsilon_i \psi_i \tag{B.7}$$

where i and j label the molecular orbitals, ψ_i and ψ_j are spatial wavefunctions, i.e. the spin parts have been removed, and ε_i is the energy of the ith orbital. H_i is a one-electron operator that represents, in effect, the hypothetical energy of an electron in molecular orbital i in the absence of any other electrons. J_j and K_j are operators, called the Coulomb and exchange operators, respectively, which account for electron–electron interactions. They are given by

$$J_j \psi_i(1) = \left(\int \psi_j^2(2) \frac{1}{r_{12}} \mathrm{d}v_2 \right) \psi_i(1) \tag{B.8}$$

$$K_j \psi_i(1) = \left(\int \psi_j^*(2) \psi_i(2) \frac{1}{r_{12}} \mathrm{d}v_2 \right) \psi_j(1) \tag{B.9}$$

The Coulomb operator accounts for the repulsion between electron 1 in the ith orbital with the averaged charge cloud of electron 2 in the jth orbital. This treatment of electron–electron repulsion as being the interaction of one electron with the averaged charge cloud of another overestimates the electron–electron repulsion since it does not allow for the correlated motion of electrons, which serves to minimize the distance of closest approach. The exchange term does not have such a simple explanation, but in effect it *partially* allows for the correlated motions of electrons with identical spin, hence the minus sign preceding it in equation (B.7).

It might seem at first sight that equation (B.7) is of the same form as the Schrödinger equation (B.1), i.e. we could write a Schrödinger-like equation for each electron of the form

$$F\psi_i = \varepsilon_i \psi_i \tag{B.10}$$

It is indeed possible to write the HF equations in this abbreviated form, but it can be misleading because there is an important and vital difference between (B.1) and (B.10). The Hamiltonian in (B.1) is a mathematical operator which can be written down independently of the actual solutions of (B.1), whereas the exact form of the so-called Fock operator, F, in (B.10) is dependent on the solutions of the HF equations. This can be seen by looking at equations (B.8) and (B.9); the Coulomb and exchange operators *contain* the molecular orbitals that we wish to determine!

The way around this apparent impasse is to solve the set of equations (B.10) by using an iterative procedure referred to as the self-consistent field (SCF) method. In essence a guess is made of the mathematical form for the molecular orbitals, the HF equations are then solved using this guess to generate new orbitals and their energies, and then the new orbitals are used to solve the HF equations again. This process is continued until there is negligible change in the solutions from one cycle to the next: the calculation is then said to be *converged* and the solutions are *self-consistent*.

B.2.1 *LCAO expansions of molecular orbitals*

It is possible to solve the HF equations using a fully numerical approach on a computer. In practice this is easy to do for atoms, because of their spherical symmetry, but is very difficult for molecules. Consequently, for molecular calculations it is usual to adopt a different approach in which the molecular orbitals are expanded as linear combinations of atomic orbitals. This is the LCAO approximation and it will be familiar to any chemist who has sketched a molecular orbital diagram. Each molecular orbital, ψ_i is expanded as

$$\psi_i = \sum_p c_{ip}\phi_p \tag{B.11}$$

where the ϕ_p are atomic orbitals and the c_{ip} are the expansion coefficients for the ith molecular orbital. Substituting (B.11) into the HF equations (B.10) gives the Hartree–Fock–Roothaan equations

$$\sum_p c_{ip}(F_{pq} - \varepsilon_i S_{pq}) = 0 \tag{B.12}$$

where

$$F_{pq} = \int \phi_p F \phi_q \, \mathrm{d}V \tag{B.13}$$

$$S_{pq} = \int \phi_p \phi_q \, \mathrm{d}V \tag{B.14}$$

Equations (B.13) and (B.14) are definite integrals evaluated over all space. An equation of the type (B.12) occurs for each atomic orbital. This set of equations is particularly amenable to solution by matrix methods, and this is a great advantage over direct numerical solution of the HF equations. In effect, the SCF approach is reduced to an iterative determination of the expansion coefficients, c_{ip}, which act as the variational parameters. However, it should be noted that many integrals of the form shown in equations (B.13) and (B.14), often millions, need to be evaluated. This is clearly a massive computational task, hence the requirement for powerful computers.

Unfortunately, while the LCAO expansion is fine in principle, precise mathematical forms for the atomic orbitals are not available! The HF equations for an atom can be solved numerically, but this merely provides specific values for the amplitude of each atomic orbital at various points in space rather than an explicit mathematical function. Consequently, we make do with second best and employ one or more mathematical functions which *resemble* the actual atomic orbitals of the individual atoms. The functions most commonly chosen are Gaussian-type functions (GTFs):

$$\phi_p = N x^{\ell} y^m z^n \exp(-\alpha r^2) \tag{B.15}$$

The quantity r is the distance of the electron from the atomic nucleus (the origin), while x, y, and z are the cartesian coordinates of the electron. The exponents of x, y, and z determine the type of orbital, e.g. if $l = m = n = 0$, then we have an s-type function; if $l = 1$ and $m = n = 0$ then it represents a p_x orbital, and so on. The exponential part of (B.12) confers the behaviour expected at large r, namely as $r \to \infty$ then $\phi_p \to 0$. The parameter α is the so-called orbital exponent, which determines the 'size' of the atomic orbital (if α is small then the orbital is large and vice versa).

Since GTFs are not the actual atomic orbitals, it should come as no surprise that they are imperfect. A better approximation is to use linear combinations of several different GTFs to represent each occupied atomic orbital on an atom, e.g. three GTFs could be chosen, each with different orbital exponents, to represent a particular atomic orbital. In fact it is also quite common to include functions representing unoccupied orbitals in atoms, e.g. for molecules formed from first row atoms it is common to include d-type GTFs. These higher angular momentum functions are called *polarization functions* and they allow for the angular distortion of occupied AOs as bonds are formed. The final choice of functions employed in (B.11) is said to be the *basis set* for the calculation.

Large basis sets will generally produce more reliable results, but they will also be more costly in terms of computer time. To carry out a HF calculation on a molecule a specific basis set must be selected for each atom. In all commercial programs a list of standard basis sets is provided and in most cases one of these will suffice. These basis sets go under well-known abbreviations such as STO-3G, 6–31G, cc-pVTZ, and many others. Further information on these and other basis sets can be found elsewhere [3].

To close this section, we are now in position to see why the Hartree–Fock method is described as an *ab initio* method. *Ab initio* is Latin for 'from the beginning' and implies that an *ab initio* calculation is one carried out from first principles. This of course does not necessarily mean that there are no approximations. We have seen that the Born–Oppenheimer and orbital approximations are fundamental to the Hartree–Fock method. Furthermore, computational constraints mean that finite basis sets must be used in practice when only infinite basis sets will actually yield the 'correct' result. Neverthless the Hartree–Fock method can reasonably be described as *ab initio* because it does not make any use of empirical (experimentally determined) parameters.

B.3 Semiempirical methods

A few words on semiempirical calculations are in order here as these have been, and to some extent still continue to be, popular alternatives to *ab initio* calculations for large molecules. These lie in the middle ground between the familiar but extremely simple Hückel theory, which is based entirely on the use of empirically determined parameters, and Hartree–Fock calculations. The semiempirical methods are all based on the Hartree–Fock–Roothaan approach but many integrals are ignored and many of those not ignored are treated as empirical parameters.

An example is the so-called neglect of differential diatomic overlap (NDDO) method, in which the integrals (B.13) and (B.14) involving basis functions on different atoms are set equal to zero. The justification is that the neglected terms are relatively small and mostly compensate each other. Furthermore, the calculations are usually empirically parameterized so that good agreement with experiments is achieved for a number of test molecules before general usage. Hence the calculated results are acceptable for many purposes and the requirement in computational resources is reduced by about two orders of magnitude or more compared with HF calculations. The most commonly encountered NDDO-type semiempirical models are the MNDO, AM1, and PM3 methods. There are, in addition, many other levels of approximation, which go under abbreviations such as CNDO and INDO. Further details can be found in the books listed at the end of this appendix.

B.4 Beyond the Hartree–Fock method: allowing for electron correlation

Hartree–Fock calculations are unsuitable for the prediction of a number of physical phenomena. These include the dissociation of molecules, the non-crossing of potential energy curves of identical symmetry, and accurate predictions for excited electronic states, or open-shell states in general. Furthermore, HF calculations may in some instances yield insufficiently accurate predictions of other properties, such as bond lengths and vibrational frequencies, even for molecules in their electronic ground states. To obtain reliable information on properties such as these, one must therefore go beyond the HF method and allow for some, or ideally all, of the electron correlation (see Section 2.1.4). In other words we must go beyond the orbital approximation. In fact most of the more advanced theoretical methods begin with a HF calculation, and then subsequently apply more sophisticated procedures to recover electron correlation energy.

The post-HF method that is the easiest to understand is the *configuration interaction* (CI) method. It is really a 'sledgehammer' extension of the variational idea underlying the HF method. Suppose, instead of using just a single Slater determinant, the total electronic state wavefunction is expanded as a linear combination of Slater determinants

$$\Phi = a_0\Psi_0 + \sum_i a_i\Psi_i \tag{B.16}$$

where Ψ_0 is the original Hartree–Fock wavefunction. The way to construct different determinantal wavefunctions is by moving electrons from occupied orbitals into unoccupied, or so-called *virtual* orbitals, which are also generated in HF calculations. These substitutions can involve a single electron, or they might involve double, triple, or quadruple excitations. The unknown coefficients, a_i, are determined by the variational method using a matrix approach entirely analogous to that employed in the LCAO–HF method. The lowest root of the obtained equations gives the ground state energy of the system, whereas the higher roots yield the different excited state energies. If all possible excitations are taken into account a *full* CI calculation is performed: the solution obtained, assuming an infinitely large LCAO basis set had been used in the initial HF calculation, would give the exact non-relativistic electronic energy. Of course this ideal situation cannot be reached, but one can get close to it for very small molecules using a large but finite LCAO basis set. Even then, however, an enormous number of Slater determinants are needed, maybe 10^8, and therefore the calculation becomes extremely expensive.

To reduce the computational requirements, practical applications of CI methods usually restrict the number of excitations performed. The configuration interaction singles (CIS) method takes only the single substitutions into account: it leads to no improvement for the ground state, but it provides excited state energies in a simple way. More sophisticated are the CID, CISD, CISDQ, and CISDTQ methods with their single (S), double (D), triple (T), and quadruple (Q) substitutions.

In the multiconfiguration SCF (MCSCF) procedure, a linear combination of a finite, and carefully selected, set of Slater determinants is employed. A special case of selection is the so-called complete active space SCF (CASSCF) selection in which all possible excitations within a limited set of occupied and virtual orbitals are considered. In contrast to CI, however, the aim is to optimize both the CI and the LCAO expansion coefficients simultaneously. The result is an energy that incorporates a large part of the correlation energy (how large depends on the configuration selection) without needing as many Slater determinants as the CI method. The principal drawback of the method is that it is still very costly in terms of computer time, which restricts the number of excited state configurations that can be handled in practice. Furthermore, the choice of the active space is somewhat arbitrary.

Perhaps the most accurate practical method to calculate the correlation energy in both ground and excited electronic states is the multireference CI (MRCI) method. In this method all important configurations that contribute in the given state (this selection can be done, for example, by performing an MCSCF calculation) are treated as reference configurations in a CI procedure, whereby singly, doubly, etc., excited configurations are produced from each of the reference ones.

Finally, it should be borne in mind that there are other widely used post-HF methods available for electronic structure calculations. These include *Møller–Plesset perturbation theory* and *coupled cluster* methods. The former is particularly widely used for calculations on the ground states of molecules because at its lowest level, MP2, it is a relatively quick way of recovering much of the correlation energy. Coupled cluster methods are more computationally intensive but are becoming increasingly the method of choice for high quality calculations on molecules in their ground electronic states.

B.5 Density functional theory (DFT)

DFT is not a true *ab initio* method but it is a closely related technique. It has become very popular in the last few years because of the combination of modest computational cost and the good quality of prediction of many molecular properties. Its basic philosophy is slightly different from the *ab initio* approaches in that the calculations work with electron density rather than explicitly dealing with the electronic wavefunction. The basis of DFT is the Hohenberg–Kohn theorem, which states that the ground state energy, the wavefunction, and all other electronic properties of a many-electron system are completely and uniquely determined by the electron density distribution of the system [3].

The derivation of this theorem is beyond the scope of this book. The bottom line is that it is, in a sense, a post-HF method in that it does contain some allowance for electron correlation, and yet it is also very quick computationally, taking perhaps little more time than a standard HF calculation. The disadvantage is that it does not allow for electron correlation in any systematic fashion, unlike CI for example, and so it can be difficult to assess how 'good' a particular DFT calculation is likely to be. Traditional *ab initio* methods are generally preferable to the DFT method for small molecules composed of light atoms, as well as when high accuracy is required. For example, using the MRCI method mentioned above, in principle any level of accuracy can be achieved given sufficiently powerful computational resources. In contrast the accuracy of DFT depends on the form of the relationship chosen to link energy to electron density, the so-called *functional*. Although many forms for this functional have been proposed and tested, there is no known systematic way to achieve an arbitrarily high level of accuracy.

Comparison with experimental data continues to be the way forward and there have been many research studies in recent years devoted to testing the performance of DFT methods. The results obtained so far appear very promising, at least for certain molecular properties. However, dealing with excited electronic states is, like the HF method, very difficult and currently unreliable with DFT, although a number of research groups are actively investigating this aspect of DFT.

B.6 Software packages

There are many software packages available for *ab initio*, semiempirical and DFT calculations. Most offer a range of *ab initio*, semiempirical and DFT methods within a single program suite and so the user is free to choose the method that is most appropriate to their particular application. Also available within these packages are a wide variety of standard basis sets. A basis set will need to be chosen (for *ab initio* and DFT calculations only) that will describe the system under investigation while not exceeding the available computing resource. Reliable use of these software packages is only achieved through experience. In particular, it is important to recognize that no calculation is perfect, and some idea of the likely error range for predictions from a particular level of theory is always a useful skill.

Some software packages can be downloaded free from the web (e.g. GAMESS) but others, such as GAUSSIAN, MOLPRO, and SPARTAN, must be purchased from a registered supplier.

B.7 Calculation of molecular properties

Finally, we summarize some of the molecular properties, especially those pertinent to electronic spectroscopy, that can be obtained from electronic structure calculations.

- *Total electronic energy of a given state*. Although this value has no direct experimental relevance, it is crucial in the calculation of several of the properties listed below. Note that a comparison of total energies is only meaningful if they were obtained by the same method using identical parameters (e.g. the same basis sets).
- *Potential energy surfaces*. These define the total electronic energy (including the nuclear–nuclear repulsion) as a function of nuclear positions. They are obtained by calculating the total energy at a variety of nuclear positions and this grid of points is then usually fitted to some suitably flexible analytical function to enable the potential energy to be determined at any point in space.
- *Equilibrium geometries*. These are determined by searching for the global minimum in the potential energy surface. This can be done from an explicit calculation of the potential energy surface, as indicated above. If only the minimum is required a quicker route involves calculating and minimizing the gradient of the total electronic energy using an analytical procedure [2].
- *Vibrational frequencies*. The calculation of harmonic vibrational frequencies first requires determination of the force constants of the molecule. These are the second derivatives of the total energy with respect to the nuclear coordinates and the various components can be collected together to form the so-called Hessian matrix. Harmonic frequencies are readily determined from the Hessian using standard procedures. This method is also applicable for the calculation of vibrational frequencies in excited electronic states or for ions. By calculating higher derivatives of the energy, it is also possible to determine anharmonicity constants, although this is rarely used.
- *Ionization energies*. These can be obtained from a HF calculation on the ground electronic state of a closed-shell molecule using Koopmans' theorem, which states that the negative of the orbital energy is equal to the ionization energy of that orbital. Koopmans' theorem is an approximation, although it often yields quite accurate ionization energies. It does not apply to open-shell molecules. A computationally more expensive, but more general, way to obtain ionization energies is by calculating the difference in the total energy between the ion and the neutral molecule.
- *Electron affinities*. These are calculated as the difference in total energy between the neutral molecule and the corresponding negative ion. The quality of the calculation must be high because the outermost electron in the anion is usually very weakly bound.
- *Electronic excitation energy*. This can, in principle, be obtained as the difference in the total energies between the two electronic states in question. However, the Hartree–Fock approach is usually inappropriate for the calculation of excited state energies. The simplest (but least accurate) *ab initio* procedure to calculate excited state energies is the CIS method, providing results for excited states of similar quality to the HF method in the ground state. More sophisticated methods, notably MRCI, can predict electronic transition energies to an accuracy of better than 6 kJ mol^{-1} (500 cm^{-1}) for small molecules.
- *Franck–Condon factors*. These are important for estimating intensities of vibrational components in electronic or photoelectron spectra and require evaluation of the overlap integrals of the vibrational wavefunctions in the upper and lower electronic states. To obtain the vibrational wavefunctions, the vibrational Schrödinger equation (B.2) is

solved for both states. This can be achieved once the potential energy surfaces have been calculated.

- *Dissociation energy*. The HF method is unsuitable for deducing dissociation energies. It is essential to use methods which incorporate electron correlation in order to make reasonable predictions of dissociation energies.
- *Intermolecular forces*. Weak intermolecular forces such as hydrogen bonding, and even very weak forces such as dispersion, can be determined. Generally, this requires a fully *ab initio* method and allowance for much of the electron correlation is essential for meaningful results.
- *Other properties*. Properties such as dipole and quadrupole moments are easy to calculate from the electronic wavefunction. Polarizabilities and hyperpolarizabilities can also be calculated without too much difficulty.

References

1. P. Pyykkö, *Chem. Rev.* **88** (1988) 563.
2. *A New Dimension to Quantum Chemistry. Analytical Derivative Methods in Ab Initio Molecular Electronic Structure Theory*, Y. Yamaguchi, Y. Osamura, and H. F. Schaefer III, Oxford, Oxford University Press, 1994.
3. J. Simons, *J. Phys. Chem.* **95** (1991) 1017.
4. P. Hohenberg and W. Kohn, *Phys. Rev. B* **136** (1964) 864.

Further reading

A more detailed account of the theoretical foundations presented in this appendix can be found in several books including the following:

Ab Initio Molecular Orbital Calculations for Chemists, W. G. Richards and D. L. Cooper, Oxford, Oxford University Press, 1985.

A Computational Approach to Chemistry. D. M. Hirst, Oxford, Blackwell Scientific, 1990.

Modern Quantum Chemistry: Introduction to Advanced Electronic Structure Theory, A. Szabo and N. S. Ostlund, New York, Dover Publications, 1996.

Quantum Chemistry, 5th edn., I. N. Levine, New Jersey, Prentice Hall, 1999.

Quantum Chemistry: Fundamentals to Applications, T. Veszpremi and M. Feher, Dordrecht, Kluwer, 1999.

Essentials of Computational Chemistry, C. J. Cramer, Chichester, Wiley, 2002.

The following volume provides a modern and advanced account of how electronic structure calculations can be used to obtain spectroscopic information:

Computational Molecular Spectroscopy, P. Jensen and P. R. Bunker (eds.), Chichester, Wiley, 2000.

Appendix C
Coupling of angular momenta: electronic states

Molecules can possess several types of angular momentum. Electrons possess orbital and spin angular momenta, rotation of the molecule generates angular momentum, degenerate vibrations may give rise to vibrational angular momentum, and nuclei may also have spin angular momentum. Interactions (coupling) between these various types of angular momentum can have important implications for interpreting spectra, particularly high resolution spectra, and so it is important to be familiar with some basic results from the quantum theory of angular momentum.

As discussed briefly in Chapter 3, angular momentum is a vector quantity and a simple vector model provides a useful visual recipe for assessing how different angular momenta interact. There are more powerful and rigorous mathematical approaches to angular momentum coupling than that described here. These more comprehensive treatments deal explictly with the angular momenta as mathematical operators, and the coupling behaviour then follows from the properties of these operators when added vectorially. Further details can be found in the books listed under Further Reading at the end of this appendix. Here we restrict ourselves to the simple vector picture and use it to emphasize the physical principles underlying the coupling of angular momenta, focussing on electronic angular momentum. In Appendix G we consider the interaction of electronic and rotational angular momenta.

We will start by focussing on the coupling between angular momenta of electrons in atoms, but attention later will shift to molecules. The magnitude of the orbital angular momentum for a single electron in an atom is given by the expression $\hbar\sqrt{l(l+1)}$, where l is the orbital angular momentum quantum number with allowed values 0, 1, 2, . . . , $(n-1)$ and n is the principal quantum number (see Section 4.1). Similarly, the magnitude of the spin angular momentum vector is given by $\hbar\sqrt{s(s+1)}$, where s is the spin quantum number which, for a single electron, can only take the value 1/2. For an electron with both orbital and spin angular momenta, the total angular momentum vector $\boldsymbol{j}$ will be the sum of the two constituent vectors $\boldsymbol{j} = \boldsymbol{l} + \boldsymbol{s}$, where bold is used to specify vector quantities. The magnitude of this vector is $\hbar\sqrt{j(j+1)}$, where j is the corresponding quantum number, and according to the quantum theory of angular momentum this can only take on the values $j = l+s, l+s-1, \ldots, |l-s|$. Series such as this are called Clebsch–Gordan series and arise when the total angular momentum results from a system composed of any two

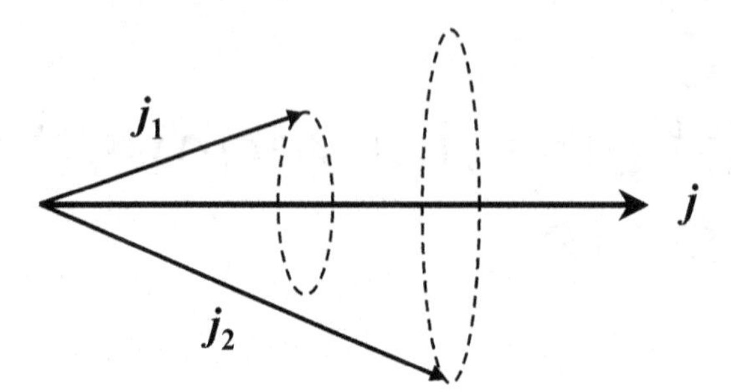

Figure C.1 Vector coupling model of two angular momenta $\boldsymbol{j}_1$ and $\boldsymbol{j}_2$.

sources of quantized angular momentum. In the general case, if two angular momenta, say $\boldsymbol{j}_1$ and $\boldsymbol{j}_2$, can interact then the allowed values for the angular momentum quantum number for the resultant total angular momentum will be given by the Clebsch–Gordan series $j = j_1 + j_2, j_1 + j_2 - 1, \ldots, |j_1 - j_2|$.

C.1 Coupling in the general case: the basics

The orientation of a single angular momentum vector in free space is arbitrary. However, if an electric or magnetic field that interacts with the particle is applied, then a torque is exerted, which forces the angular momentum vector to *precess* around the applied field, as shown in Figure 3.2. If we define the field direction as axis z, then the angular momentum along this axis is constant whereas the angular momenta along the x and y directions are not. The applied field therefore creates an *axis of quantization* and the projection of angular momentum along the axis takes on the values $m\hbar$, where m is an integer or half integer quantum number. In atoms the projection quantum numbers are the familiar m_l and m_s quantum numbers assigned to electrons.

When there are two angular momenta present, any interaction between them creates a torque that forces the individual angular momenta, represented by the vectors $\boldsymbol{j}_1$ and $\boldsymbol{j}_2$, to precess around a common axis. This common axis is the direction of the resultant angular momentum $\boldsymbol{j}\,(= \boldsymbol{j}_1 + \boldsymbol{j}_2)$. The precession of $\boldsymbol{j}_1$ and $\boldsymbol{j}_2$ about $\boldsymbol{j}$ means that the projection of $\boldsymbol{j}_1$ and $\boldsymbol{j}_2$ along $\boldsymbol{j}$ is no longer well defined. In other words, m_1 and m_2 are no longer meaningful quantum numbers. However, the quantum numbers j_1 and j_2 are still good (meaningful) quantum numbers, as are the total angular momentum quantum number j and its projection m. This coupled model is illustrated in Figure C.1. As in the weak coupling case, the possible values of the total angular momentum quantum number j are given by the Clebsch–Gordan series shown earlier.

C.2 Coupling of angular momenta in atoms

In an atom, the spin of an electron can interact with its own orbital angular momentum, with the orbital angular momenta of other electrons, or with the spins of other electrons. The second interaction is very weak and is normally neglected. The first interaction is called spin–orbit coupling. This and the third interaction are invoked to describe the two extreme cases of coupling in atoms, the Russell–Saunders (also called *LS*) and the *jj* schemes.

C.2.1 Russell–Saunders coupling limit

In the Russell–Saunders limit the coupling between the orbital angular momenta is strong. The source of the coupling is the electrostatic interaction between two electrons. Electron spins can also couple together, although spin is a magnetic phenomenon and therefore the coupling is via magnetic fields, which tends to be a weaker effect than electric field coupling. In Russell–Saunders coupling the interaction between the orbital and spin angular momenta of a given electron is assumed to be small compared with the coupling between orbital angular momenta. In this limit the principal torque causes the orbital angular momenta to precess about a common direction, the axis of the total orbital angular momentum of all electrons. The spin angular momenta also couple together through the magnetic interaction of electron spins.

Armed with these assumptions, the vector model described above can be employed to see the effect of coupling on atoms. For illustration, consider an atom with two electrons outside its closed shell. The orbital angular momenta of these electrons are represented by $\boldsymbol{l}_1$ and $\boldsymbol{l}_2$, and the corresponding spin momenta by $\boldsymbol{s}_1$ and $\boldsymbol{s}_2$. The coupling of orbital angular momenta dominates and they form a resultant total orbital angular momentum $\boldsymbol{L} = \boldsymbol{l}_1 + \boldsymbol{l}_2$. In the same way, the spin momenta also couple to form $\boldsymbol{S} = \boldsymbol{s}_1 + \boldsymbol{s}_2$. When this coupling case is valid, the individual orbital angular momenta $\boldsymbol{l}_1$ and $\boldsymbol{l}_2$ precess rapidly around $\boldsymbol{L}$, and $\boldsymbol{s}_1$ and $\boldsymbol{s}_2$ precess rapidly around $\boldsymbol{S}$. $\boldsymbol{L}$ and $\boldsymbol{S}$ can themselves couple with each other, but this coupling, known as spin–orbit coupling, is assumed to be relatively weak. As a result, $\boldsymbol{L}$ and $\boldsymbol{S}$ precess slowly around the resultant total angular momentum $\boldsymbol{J}$ $(= \boldsymbol{L} + \boldsymbol{S})$.

The significance of Russell–Saunders coupling is that a particular electronic state in an atom is well defined by the quantum numbers L and S. The effect of the weak spin–orbit coupling results in closely spaced spin–orbit sub-states, designated by the quantum number J. As detailed in Section 4.1, this gives rise to the familiar $^{2S+1}L_J$ label for electronic states in atoms.

C.2.2 *jj* coupling

The Russell–Saunders scheme describes the electronic states of light atoms rather well but breaks down for heavier atoms, particularly for the lanthanides and actinides. This is due to increasing coupling between the orbital and spin angular momenta of individual electrons to the point where it is no longer negligible, as was assumed in the Russell–Saunders case. In the limit of very strong coupling between orbital and spin angular momenta, the appropriate coupling scheme is known as *jj* coupling. A dominant spin–orbit torque will first couple the spin and orbital momenta of each electron to form resultants, $\boldsymbol{j}_1 = \boldsymbol{l}_1 + \boldsymbol{s}_1$ and $\boldsymbol{j}_2 = \boldsymbol{l}_2 + \boldsymbol{s}_2$. The vectors $\boldsymbol{j}_1$ and $\boldsymbol{j}_2$ interact more weakly, forming the total angular momentum $\boldsymbol{J} = \boldsymbol{j}_1 + \boldsymbol{j}_2$. In *jj* coupling, $\boldsymbol{l}_1$ ($\boldsymbol{l}_2$) and $\boldsymbol{s}_1$ ($\boldsymbol{s}_2$) precess rapidly around $\boldsymbol{j}_1$ ($\boldsymbol{j}_2$), while $\boldsymbol{j}_1$ and $\boldsymbol{j}_2$ precess slowly around their resultant $\boldsymbol{J}$. As a result, only j_1, j_2, J and the projection of

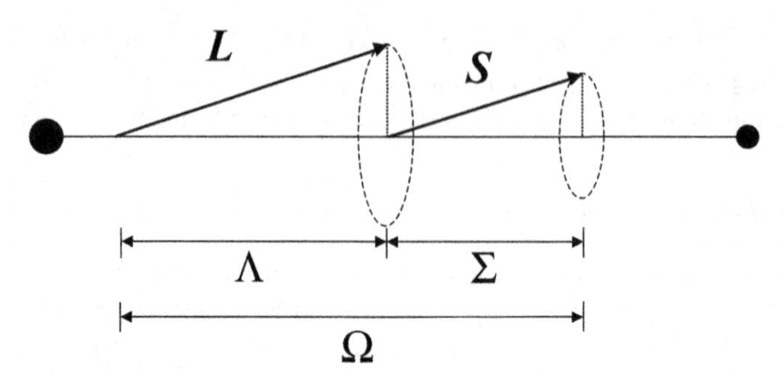

Figure C.2 Illustration of the behaviour of the electronic orbital (***L***) and spin (***S***) angular momenta in a diatomic molecule.

$J (M_J)$ are good quantum numbers, i.e. the quantum numbers L and S in the Russell–Saunders scheme have no significance in *jj* coupling.

Of course it is also possible that the spin–orbit coupling is neither weak enough for Russell–Saunders to be applicable, nor strong enough for true *jj* coupling. Description of this intermediate coupling requires a more mathematical treatment and is not considered here.

C.3 Coupling of electronic angular momenta in linear molecules

The extension of the above ideas to molecules is not difficult. Consider, for example, a linear molecule. As in atoms, electrons in this molecule can also possess angular momenta due to their orbital motion and spin. There is, however, a fundamental difference between atoms and linear molecules in terms of the environment experienced by electrons. If electron–electron interactions are neglected, electrons in atoms are subjected to a spherically symmetric field, whereas in a linear molecule a strong axial electric field exists between the nuclei and the interaction with this field determines the behaviour of the electrons. This field provides a torque on the orbital motion of the electrons and therefore analogous arguments to those employed earlier can also be used to construct a vector model of coupled angular momenta in molecules.

In many-electron molecules, there may be additional torques from the interaction between angular momenta. In the most common case, the strongest torque couples the orbital angular momenta of individual electrons to form a resultant ***L***. Similarly, the spin angular momenta couple together to form a resultant ***S***. If the spin–orbit coupling is relatively weak, then this is clearly the molecular analogue of Russell–Saunders coupling in atoms.

The torque exerted by the electrostatic field along the molecule is important because it causes the orbital angular momentum vector to precess around the internuclear axis, as shown in Figure C.2. This precession is rapid and therefore the corresponding quantum number, L, is not a good quantum number in this limit. However, the projection of ***L*** along the internuclear axis, denoted by the symbol Λ, is a constant of motion and is therefore a good quantum number. To determine the possible values of Λ, the contributions from individual electrons must be considered. With only one unpaired electron, the total orbital angular momentum is the orbital angular momentum of that sole electron, and we can identify a corresponding one-electron orbital angular momentum quantum number $\lambda = 0, 1, 2, 3$, etc., which corresponds to σ, π, δ, and ϕ orbitals, respectively (see Section 4.2.2). Note that the orbitals with non-zero λ are doubly degenerate, which may be viewed as being due to the two possible directions for rotation of the electrons around the internuclear axis, clockwise or anticlockwise. In this sense the projection of orbital angular momentum on the internuclear axis is a signed quantity, but the convention is that λ is always quoted as a positive number. The reader should recognize however that if, for example, $\lambda = 1$ then the actual orbital angular momentum along the internuclear axis is $+\hbar$ or $-\hbar$.

If there is more than one electron, each electron will contribute $\pm\lambda\hbar$ to the orbital angular momentum along the internuclear axis. All filled orbitals will therefore make a zero contribution to Λ since the orbital angular momenta of the electrons in these orbitals cancel. If there are two unpaired electrons, say one in a π orbital and one in a δ orbital, the possible angular momenta are $\pm\hbar$, $\pm2\hbar$, or $\pm3\hbar$. These correspond to $\Lambda = 1, 2$, or 3 and the resulting electronic states are labelled as Π, Δ, and Φ electronic states, respectively. A better and more general way of deriving the possible orbital angular momentum states is by recognizing that the σ, π, δ, etc., labels are actually symmetry labels for the molecular orbitals, i.e. they are irreducible representations of the appropriate linear molecule point group, $D_{\infty h}$ or $C_{\infty v}$. The overall angular momentum must therefore also correspond to one of the irreducible representations of the point group and can be obtained by taking the direct product of the symmetries for each occupied orbital, and taking appropriate care in the application of the Pauli principle (see below). This was the recommended procedure covered in Part I.

The axial electric field in linear molecules does not have a direct effect on the spin angular momenta, since spin is a magnetic phenomenon. However, when $\Lambda \neq 0$ the orbiting motion of the electron(s) generates a magnetic field,[1] which can also cause the total spin angular momentum, $\boldsymbol{S}$, to precess around the internuclear axis. This is none other than spin–orbit coupling, but if the spin–orbit coupling is not as strong as the spin–spin coupling then S remains a good quantum number. As in the case of orbital angular momenta, the projection of $\boldsymbol{S}$ onto the internuclear axis is quantized. The projection quantum number is given the symbol Σ, which is unfortunately the same as the label used to designate electronic states with $\Lambda = 0$. The allowed values of Σ are $-S, -S+1, \ldots, +S$, where S may be integer or half integer depending on the number of unpaired electrons.

As in the case of atoms, spin–orbit coupling leads to spin–orbit sub-states with different energies. In this case the total electronic (orbital + spin) angular momentum is given by the quantum number Ω $(= \Lambda + \Sigma)$. Although Ω is, like Λ, ostensibly a signed quantum number, the accepted convention is to quote the positive value, i.e. $\Omega = |\Lambda + \Sigma|$. The complete label for electronic states in linear molecules is then $^{2S+1}\Lambda_{\Omega}$.

As an example, consider the case of two electrons in two different π molecular orbitals (the choice of different π orbitals avoids difficulties with the Pauli exclusion principle – see

[1] Current flowing in a circular conductor generates a magnetic field perpendicular to the plane of the conductor. We can use the same analogy for an orbiting electron around the internuclear axis to explain how it generates a magnetic field due to its orbital motion.

Appendix E). As $\lambda = 1$ for a π-electron, Λ can be 2 or 0, i.e. Σ or Δ electronic states are possible. The Δ state is doubly degenerate but there are two different Σ states, a Σ^+ and a Σ^- due to the finer interactions of the electrons (a finding that is best seen by evaluating the direct product $\pi \otimes \pi$). The net spin S for the two electrons is 0 or 1, giving rise to the multiplicities 1 and 3. The possible electronic states that may result from the $\pi^1\pi'^1$ configuration are therefore ${}^1\Sigma^+$, ${}^3\Sigma^+$, ${}^1\Sigma^-$, ${}^3\Sigma^-$, ${}^1\Delta$ and ${}^3\Delta$.

C.4 Non-linear molecules

The presence of off-axis nuclei in non-linear molecules usually results in all electronic orbital angular momentum being quenched. The only exceptions to this are high symmetry molecules in spatially degenerate electronic states. A good example is benzene, which in its ground electronic state has the outer electronic configuration . . . $(1a_{2u})^2(1e_{1g})^4$. The resulting electronic state is a ${}^1A_{1g}$ state, in which there is no net orbital or spin angular momentum. If, however, an electron is removed from the HOMO, the resulting ground state of the cation is a ${}^2E_{1g}$ state. E_{1g} is a doubly degenerate representation and so the ground electronic state of the cation does possess orbital angular momentum. The source of this orbital angular momentum is the unimpeded circulation of the unpaired electron in the π system above and below the nuclei in the benzene ring. In the ground state of benzene the net orbital angular momentum is zero because all orbitals are full and therefore the clockwise and counterclockwise contributions cancel. However, in the benzene cation this is no longer the case and spin–orbit coupling splits the resulting ${}^2E_{1g}$ state into two spin–orbit sub-states, which are labelled ${}^2E_{1g(1/2)}$ and ${}^2E_{1g(3/2)}$.

Although orbital angular momentum can exist in non-linear molecules with degenerate electronic states, it is important to recognize that it will still be quenched to a greater or lesser extent. For example in the benzene cation the Jahn–Teller effect, which couples electronic orbital and vibrational motions, acts to quench some of the pure orbital angular momentum.

Further reading

Molecular Quantum Mechanics, 3rd edn., P. W. Atkins and R. S. Friedman, Oxford, Oxford University Press, 1999.

Angular Momentum, R. N. Zare, New York, Wiley, 1988.

Angular Momentum in Quantum Mechanics, A. R. Edmonds, Princeton, Princeton University Press, 1996.

Appendix D
The principles of point group symmetry and group theory

Molecular symmetry is of great importance in the discussion of spectroscopy. It helps to simplify the explanation of complex phenomena, such as molecular vibrations, and is an important aid in the derivation of electronic states and transition selection rules. It also simplifies the application of molecular orbital theory, which is often applied to assign or predict electronic spectra. In many cases, it provides strikingly simple answers to complicated questions.

In its original form, group theory is a rigorous mathematical subject. No attempt will be made here to be rigorous – the aim is simply to summarize the basics as they apply to symmetry, in light of which the spectroscopic applications of the theory can become clearer. Although the concepts introduced here might be valid for any object with symmetry elements, we will apply these only to molecules. This appendix is not intended to be a comprehensive account of point group symmetry and group theory. Instead the intention is to review some of the key principles required for applications in electronic spectroscopy. A newcomer to the subject of symmetry and group theory is first advised to consult an appropriate textbook on this topic, such as one of those listed in the Further Reading at the end of this appendix.

D.1 Symmetry elements and operations

We begin with two fundamental concepts, symmetry *operations* and symmetry *elements*. Symmetry operations are transformations that move the molecule such that it is indistinguishable from its initial position and orientation. For example, the water molecule has mirror image symmetry. An imaginary mirror perpendicular to the molecular plane and passing through the oxygen atom will interchange the two hydrogen atoms, leaving the molecule unchanged in its appearance. This reflection operation is an example of a symmetry operation and is denoted by the symbol σ.

Symmetry elements are geometric objects, such as points, lines or planes. For water, the symmetry element considered so far is a plane of reflection. The water molecule has another mirror plane, this second one being in the plane of the molecule. The two corresponding

Table D.1 *Symmetry elements and symmetry operations*

Symmetry element	Symmetry operation	Symbol
	Identity operation (does nothing)	E
Plane	Reflection through a plane	σ
Axis	Rotation 360°/n around an n-fold axis	C_n
Centre of symmetry (or inversion)	Inversion through a point	i
Improper axis (rotation axis and a perpendicular plane)	Rotation 360°/n about an axis followed by reflection through a plane perpendicular to the axis	S_n

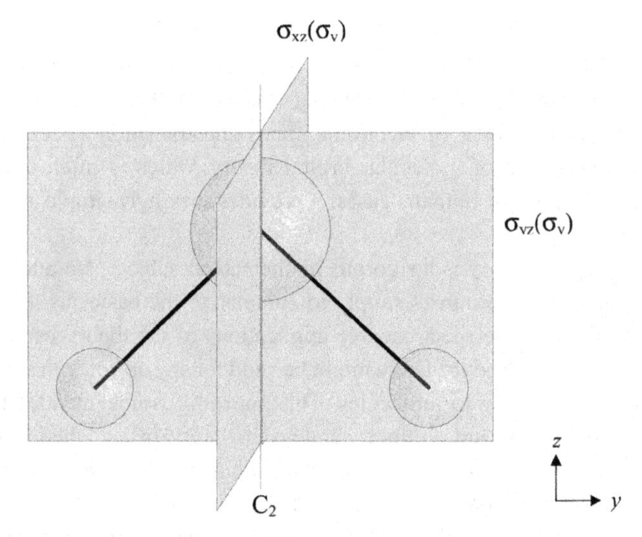

Figure D.1 Symmetry elements of the water molecule.

symmetry operations are distinguished by their subscripts, referring to the chosen coordinate system, as shown in Figure D.1. In addition to mirror planes, the water molecule has a two-fold axis of rotational symmetry that bisects the HOH angle: rotation of the molecule around this axis by 180° leaves the molecule unchanged.

There are five types of symmetry operations that are used for molecules and the corresponding symmetry elements are summarized in Table D.1.

As can be seen from the final column of Table D.1, the symbols used to denote symmetry operations are often accompanied by a subscript. For example, to express that the rotation of the object by 360°/n leaves it indistinguishable, the applied operation is denoted as C_n, e.g. the 180° rotational symmetry of H_2O is denoted as C_2. Another symmetry operation in which rotation plays a part is *improper rotation*. For example, S_6 expresses a six-fold improper rotation that consists of a 60° rotation (360°/6) about an axis followed by reflection in the plane perpendicular to the axis.

Planes of reflection are labelled to indicate their relative orientation in the coordinate system. The reflection in the plane that includes the *principal* axis of rotation (the axis of highest rotational symmetry) is said to be vertical and is labelled as σ_v. Mirror planes perpendicular to this are referred to as horizontal and are denoted by σ_h. In some cases when there is more than one symmetry plane of the same kind (such as the two σ_v planes of water shown in Figure D.1), they are distinguished by subscripts showing which plane they include (e.g. σ_{yz} and σ_{xz}). If an operation is to be performed several times, this is shown as a superscript, e.g. C_3^2 signifies that the C_3 operation is to be carried out twice such that a rotation of 240° takes place.

The identity operation, E, is rather odd in that it corresponds to no net movement of any atom in the molecule. However, this operation is always included because it is important in the mathematics of group theory, as will be seen later. Some operations are equivalent to E. Examples of this are C_3^3 (indicating a 360° rotation), i^2 and σ^2. Improper rotations are less straightforward: S_n^n implies n rotations and n reflections in the plane and this can lead to identity only if n is even.

D.2 Point groups

The collection of symmetry operations applying to molecules of a particular symmetry is called a point group. There are a number of different point groups and the properties of these are collected in *character tables*. Determining which point group the molecule belongs to is the first step in utilizing molecular symmetry.

The diagram below helps in the identification of the most commonly occurring point groups. The classification is achieved by answering a series of simple questions. The first question relates to special groups. These include the two groups for linear molecules, $D_{\infty h}$ and $C_{\infty v}$. These point groups are distinguished by whether or not they have a centre of symmetry (operation i in the above table). Thus, for example, CO_2 has a centre of symmetry (positioned at the carbon atom) and therefore belongs to the $D_{\infty h}$ point group. In contrast, CO does not possess a centre of symmetry and so has $C_{\infty v}$ point group symmetry. Other special groups include tetrahedral (T_d, e.g. the CH_4 molecule), octahedral (O_h, e.g. SF_6), and icosohedral (I_h, e.g. C_{60}).

If a molecule does not belong to any of these special groups, then a series of rules can be used to establish its point group. The starting point is to determine the principal axis of rotation. In the flow chart below, the letter n in the name of the point groups indicates the order of the principal axis. If there is no such axis (other than C_1, which is equivalent to E), there might only be a symmetry plane (this is the case in the C_s group), or an inversion centre (in the C_i group). Molecules with no symmetry other than identity belong to the C_1 group.

If the molecule has an n-fold principal axis, further classification depends on whether or not this is only the consequence of a $2n$-fold improper axis (if the answer is yes the point group is designated as S_{2n}). Molecules belonging to the S_{2n} point groups are rare.

The remaining groups are denoted with the letters C or D. The former have no C_2 axis perpendicular to the principal axis, whereas the latter have such axes. If a σ_h plane exists (i.e. a plane perpendicular to the principal axis), the group is labelled C_{nh} or D_{nh}. Molecules in the C_{nv} and D_{nd} groups have no σ_h plane, only one or more σ_v planes, i.e. containing the principal axis. Here the d subscript arises because the axes of rotational symmetry perpendicular to the principal one do not contain the σ_v planes; these planes dissect the angle between the axes. Such planes are referred to as dihedral and marked as σ_d. If there is no σ_v plane, the point group is called C_n or D_n.

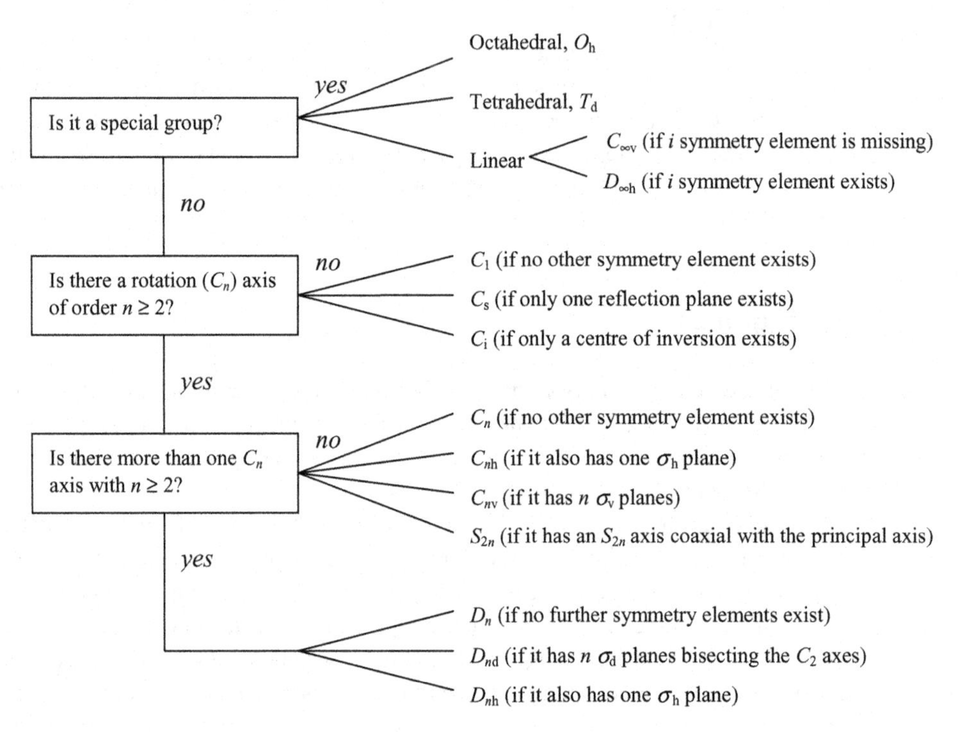

D.3 Classes and multiplication tables

A few properties of point groups are described below, which can be derived from the general properties of mathematical groups.

- Point groups can be characterized by the different applicable symmetry operations possible within the group.
- Multiplication is the subsequent execution of two symmetry operations. The multiplication of symmetry operations is associative (e.g. $(AB)C = A(BC)$) but not necessarily commutative (e.g. $AB \neq BA$ is possible).
- Each symmetry operation has an inverse, such that the operation multiplied with its inverse gives the identity operation.

Table D.2 *Multiplication table for symmetry operations of the C_{3v} point group*

C_{3v}	E	C_3^1	C_3^2	σ_v'	σ_v''	σ_v'''
E	E	C_3^1	C_3^2	σ_v'	σ_v''	σ_v'''
C_3^1	C_3^1	C_3^2	E	σ_v'''	σ_v'	σ_v''
C_3^2	C_3^2	E	C_3^1	σ_v''	σ_v'''	σ_v'
σ_v'	σ_v'	σ_v''	σ_v'''	E	C_3^1	C_3^2
σ_v''	σ_v''	σ_v'''	σ_v'	C_3^2	E	C_3^1
σ_v'''	σ_v'''	σ_v'	σ_v''	C_3^1	C_3^2	E

Table D.3 *Similarity transformations for C_{3v} point group*

Similarity transformation	E	C_3^1	C_3^2	σ_v'	σ_v''	σ_v'''
EXE^{-1}	E	C_3^1	C_3^2	σ_v'	σ_v''	σ_v'''
$(C_3^1)^{-1}XC_3^1$	E	C_3^1	C_3^2	σ_v'''	σ_v'	σ_v''
$(C_3^2)^{-1}XC_3^2$	E	C_3^1	C_3^2	σ_v''	σ_v'''	σ_v'
$(\sigma_v^1)^{-1}X\sigma_v^1$	E	C_3^2	C_3^1	σ_v'	σ_v'''	σ_v''
$(\sigma_v^2)^{-1}X\sigma_v^2$	E	C_3^2	C_3^1	σ_v'''	σ_v''	σ_v'
$(\sigma_v^3)^{-1}X\sigma_v^3$	E	C_3^2	C_3^1	σ_v''	σ_v'	σ_v'''

- Point groups must contain the products of all pairs of elements, the squares of all elements and the reciprocals of all elements.
- The total number of elements is called the *order* of the group.

These concepts will be demonstrated for the C_{3v} point group. A molecule with this point group symmetry is NH_3. It has the following symmetry elements: E (identity operation), C_3^1 (120° rotation about a three-fold axis), C_3^2 (240° rotation about a three-fold axis), σ_v', σ_v'', and σ_v''' (reflection through one of the three equivalent mirror planes, each containing an N–H bond in the case of NH_3). The multiplication table for the point group is shown in Table D.2 (where the column and row heading show the symmetry operations that are multiplied).

The symmetry operations of the point group can be subdivided into classes. If A, B, and X are elements of a group, then the operation XAX^{-1} is called a similarity transformation. If the relationship $XAX^{-1} = B$ holds, A and B are said to be conjugates. A class consists of a complete set of elements that are conjugates of each other.

Looking at the multiplication table of the C_{3v} group, it can be established that the three σ_v operations belong to the same class, as do C_3^1 and C_3^2, because they are connected by similarity transformations (see Table D.3). The identity operation is in a class of its own.

In general the E, i and σ_h operations are always in a class of their own. In contrast, all σ_v operations of a group form a class together, as do the σ_d operations.

D.4 The matrix representation of symmetry operations

All symmetry operations correspond to geometrical transformations and can be represented by matrices. Each such matrix represents a single operation. These matrices obey the same multiplication rules as the symmetry operations. The effect of different symmetry operations on an arbitrary point, represented by its coordinates (x, y, z), will be shown for the water molecule. This will be done by using the matrix representation of the operations.

In the C_{2v} point group for the water molecule, the identity operation can be represented by a unit matrix:

$$\begin{pmatrix} 1 & 0 & 0 \\ 0 & 1 & 0 \\ 0 & 0 & 1 \end{pmatrix} \begin{pmatrix} x \\ y \\ z \end{pmatrix} = \begin{pmatrix} x \\ y \\ z \end{pmatrix} \tag{D.1}$$

Notice that the coordinates of the arbitrary point are expressed as a column matrix.

According to the axis scheme shown in Figure D.1, the C_2 operation reverses the sign of the x and y coordinates and so is equivalent to the 3×3 matrix shown below:

$$\begin{pmatrix} -1 & 0 & 0 \\ 0 & -1 & 0 \\ 0 & 0 & 1 \end{pmatrix} \begin{pmatrix} x \\ y \\ z \end{pmatrix} = \begin{pmatrix} -x \\ -y \\ z \end{pmatrix} \tag{D.2}$$

Reflection through the xz plane reverses the sign of the y coordinate and in matrix notation is equivalent to

$$\begin{pmatrix} 1 & 0 & 0 \\ 0 & -1 & 0 \\ 0 & 0 & 1 \end{pmatrix} \begin{pmatrix} x \\ y \\ z \end{pmatrix} = \begin{pmatrix} x \\ -y \\ z \end{pmatrix} \tag{D.3}$$

The matrix representations are often complicated to deduce. Luckily, as will be seen later, for practical purposes it is unnecessary to derive these representations. It should be noted that these matrices are 3×3 because they were derived for a triatomic molecule. The dimensionality of these transformation matrices depends on the number of atoms in the system. The actual composition of the matrices is also determined by the choice of coordinate system. Hence in any point group, it is possible to devise an infinite number of matrix representations of the symmetry operations.

If a similarity transformation exists that transforms all matrices of the representation into block diagonal form, the initial representation is said to be *reducible*. A block-diagonal matrix has the appearance of a matrix constructed from smaller matrices located along the diagonal. Such a matrix can be illustrated by the following 9×9 matrix that has been block-diagonalized into a 2×2, a 5×5, and another 2×2 matrix:

$$\left[\begin{array}{cc|ccccc|cc} a_{11} & a_{12} & 0 & 0 & 0 & 0 & 0 & 0 & 0 \\ a_{21} & a_{22} & 0 & 0 & 0 & 0 & 0 & 0 & 0 \\ \hline 0 & 0 & a_{33} & a_{34} & a_{35} & a_{36} & a_{37} & 0 & 0 \\ 0 & 0 & a_{43} & a_{44} & a_{45} & a_{46} & a_{47} & 0 & 0 \\ 0 & 0 & a_{53} & a_{54} & a_{55} & a_{56} & a_{57} & 0 & 0 \\ 0 & 0 & a_{63} & a_{64} & a_{65} & a_{66} & a_{67} & 0 & 0 \\ 0 & 0 & a_{73} & a_{74} & a_{75} & a_{76} & a_{77} & 0 & 0 \\ \hline 0 & 0 & 0 & 0 & 0 & 0 & 0 & a_{88} & a_{89} \\ 0 & 0 & 0 & 0 & 0 & 0 & 0 & a_{98} & a_{99} \end{array}\right]$$

If no transformation exists that brings the representation to a block-diagonal form, it is said to be *irreducible*.

Unlike the arbitrary matrix representations above, irreducible representations are unique: they are the simplest representations of the symmetry group. It is, however, often rather difficult to find the appropriate similarity transformation to bring the matrix representation to an irreducible form. Luckily, as will be shown later, it is usually sufficient to use the *characters* of the representation, where the character is defined as the trace of the corresponding matrix (i.e. the sum of its diagonal elements). Dealing with characters is much simpler than dealing with matrix representations, and these can be collected together into tables for general use.

There are five important rules that form the basis of the derivation of character tables. The reader who is only interested in the use of character tables can simply skip these.

(i) The number of irreducible representations of a group equals the number of classes in the group.

(ii) The order of the group, h, is determined by the dimension of its irreducible representations, i.e.

$$h = \sum_i l_i^2 \tag{D.4}$$

where l_i is the dimension of the ith irreducible representation.

(iii) The sum of the squares of the characters in any irreducible representation is equal to the order of the group,

$$h = \sum_R [\chi_i(R)]^2 \tag{D.5}$$

where $\chi_i(R)$ is the character (trace of the matrix) representing the Rth symmetry operation in the ith irreducible representation.

(iv) In any irreducible representation the characters that belong to symmetry operations in the same class are identical.

(v) The following expression holds for the characters of two different irreducible representations (orthogonality relation):

$$\sum_R \chi_i(R)\chi_j(R) = 0, \text{ where } i \neq j \tag{D.6}$$

D.5 Character tables

The properties of point groups can be summarized using character tables. Character tables list the possible symmetry operations for a given point group along with the irreducible representations and their characters. The character tables for the most important point groups can be found at the end of this appendix.

The way character tables are arranged is illustrated below for the C_{3v} point group.

C_{3v}	E	$2C_3$	$3\sigma_v$		
A_1	1	1	1	z	x^2+y^2, z^2
A_2	1	1	−1	R_z	
E	2	−1	0	$(x, y), (R_x, R_y)$	$(x^2-y^2, xy), (xz, yz)$

This character table consists of four sections, separated above by double lines (optional). In the leftmost column, beneath the point group symbol, are the irreducible representations. These are sometimes also called symmetry species, or simply the symmetry. The uppermost irreducible representation is always the totally symmetric one, for which all characters are equal to 1. These characters can be seen to the right of the A_1 label.

Conventionally, the symbols for irreducible representations are determined in the following way. One-dimensional representations are marked with the letters A or B, two-dimensional representations by E, three-dimensional ones usually with the letter T.[1] For one-dimensional representations, the letter A is used when the character for the rotation around the principal axis is +1 (i.e. when it is symmetric for this transformation) and B when this character is −1. The symmetry with respect to the rotation around the axis perpendicular to the principal axis (or in its absence reflection in the σ_v plane) is shown as a subscript, 1 for the symmetric and 2 for the antisymmetric representation. Reflections in the σ_h plane are designated with a prime (symmetric) or double prime (antisymmetric). The subscripts g and u denote the symmetric or antisymmetric nature of the representation with respect to inversion.

[1] Exceptions to this are the linear molecule point groups $D_{\infty h}$ and $C_{\infty v}$, where labels such as σ and π are preferred over A and E. This is discussed again later in the appendix.

The second section of the C_{3v} character table gives the character for each irreducible representation and for each class of symmetry operations. It is useful to note that the character for the identity operation is equal to the dimension of the irreducible representation.

The final two columns provide information about the symmetries of cartesian vectors (x, y, z),[2] products of these vectors, and rotations about the cartesian axes (R_x, R_y, and R_z). This information is useful for determining spectroscopic selection rules. For example, the z coordinate axis in the C_{3v} point group transforms like the totally symmetric (A_1) irreducible representation, because it is unaffected by the operations of the group. R_z also appears on its own and transforms as the A_2 irreducible representation. In contrast to these, x and y (and similarly R_x and R_y) jointly form a representation. This arises because, after the C_3 operation is performed, the resulting vector will contain both x and y components. As a result, x and y are inseparable in this respect, and so they jointly form a representation and transform as the E irreducible representation.

D.6 Reducible representations, direct products, and direct product tables

There are many occasions in spectroscopy when it is necessary to multiply irreducible representations, or in the language of group theory, calculate their *direct products*. The direct product is obtained by multiplying the characters for each symmetry element. The resulting representations are often reducible. It can be proved that the number of times (a_i) an irreducible representation occurs within a reducible one can be determined using the following formula:

$$a_i = \frac{1}{h}\sum_R \chi_{\text{red}}(R)\chi_i(R) \tag{D.7}$$

where $\chi_{\text{red}}(R)$ is the character of the reducible representation corresponding to operation R, and $\chi_i(R)$ is the character of the irreducible representation. The summation is over all symmetry operations and h is the order of the group.

This rule can be illustrated by determining the direct product of the E species with itself within the C_{3v} point group, i.e. $E \otimes E$. As the characters for the E species are 2, -1, and 0, the characters of the direct product will be $\Gamma = 4$, 1 and 0. This is a reducible representation that can be decomposed to irreducible representations using the formula above, yielding

$$a_{A_1} = \tfrac{1}{6}[1(1)(4) + 2(1)(1) + 3(1)(0)] = 1$$

$$a_{A_2} = \tfrac{1}{6}[1(1)(4) + 2(1)(1) + 3(-1)(0)] = 1$$

$$a_E = \tfrac{1}{6}[1(2)(4) + 2(-1)(1) + 3(0)(0)] = 1$$

[2] The symmetry of a cartesian vector is the same as the symmetry of the corresponding cartesian axis. For example, the x axis has both positive and negative regions and any rotation about this axis will leave these unmoved. On the other hand, a C_2 rotation about an axis perpendicular to the x axis and passing through the origin will transform x into $-x$, and vice versa. In other words, the x axis in this instance will be antisymmetric with respect to C_2. Thus symmetry operations can be applied to cartesian vectors in a manner identical to their application to molecules.

Table D.4 *Direct product table for point groups* C_2, C_{2v}, C_{2h}, C_3, C_{3v}, C_{3h}, D_3, D_{3h}, D_{3d}, C_6, C_{6v}, C_{6h}, D_6, S_6, D_{6h}

	A_1	A_2	B_1	B_2	E_1	E_2
A_1	A_1	A_2	B_1	B_2	E_1	E_2
A_2		A_1	B_2	B_1	E_1	E_2
B_1			A_1	A_2	E_2	E_1
B_2				A_1	E_2	E_1
E_1					$A_1 + [A_2] + E_2$	$B_1 + B_2 + E_1$
E_2						$A_1 + [A_2] + E_2$

Table D.5 *Direct product table for point groups* T, T_h, T_d, O, O_h

	A_1	A_2	E	T_1	T_2
A_1	A_1	A_2	E	T_1	T_2
A_2		A_1	E	T_2	T_1
E			$A_1 + [A_2] + E$	$T_1 + T_2$	$T_1 + T_2$
T_1				$A_1 + E + [T_1] + T_2$	$A_2 + E + T_1 + T_2$
T_2					$A_1 + E + [T_1] + T_2$

Table D.6 *Direct product table for point groups* $C_{\infty v}$ *and* $D_{\infty h}$

	Σ^+	Σ^-	Π	Δ
Σ^+	Σ^+	Σ^-	Π	Δ
Σ^-		Σ^+	Π	Δ
Π			$\Sigma^+ + [\Sigma^-] + \Delta$	$\Pi + \Phi$
Δ				$\Sigma^+ + [\Sigma^-] + \Gamma$

Note that the number of symmetry operations in each class needs to be considered, and these are the first numbers in each term inside the square brackets. The result is that the direct product $E \otimes E$ can be reduced to $A_1 + A_2 + E$. That this finding is correct can be confirmed by adding up the characters of the three irreducible representations, which will yield the original reducible representation.

Fortunately, it is not necessary to use (D.7) every time direct products of irreducible representations are required. Instead, direct product tables are available, which allow the task to be carried out quickly and easily. Direct product tables often prove themselves to be just as useful in spectroscopy as character tables, and it is important to be comfortable with their use. Three direct product tables are, D.1, D.2, and D.3, are shown above, covering a wide range of point groups.

Interpretation and use of the direct product tables requires a little care. First, notice that each table applies to a number of different point groups. In some cases the irreducible representations in the table do not correspond exactly to those of one of the listed point groups. For example, the four irreducible representations of the C_{2h} point group are A_g, B_g, A_u, and B_u. None of these appears in Table D.4, and yet this is the direct product table that is supposed to apply to the C_{2h} point group. To find the direct products we do the following. First, if the irreducible representations of the point group have no numerical subscripts, the corresponding subscripts in the direct product table are ignored. Second, if the irreducible representations have u or g subscripts, or they have a $'$ or $''$ superscript, the following additional product rules apply:

- For g and u subscripts: $\Gamma_g \otimes \Gamma_g = \Gamma_g$, $\Gamma_g \otimes \Gamma_u = \Gamma_u$, $\Gamma_u \otimes \Gamma_g = \Gamma_u$, and $\Gamma_u \otimes \Gamma_u = \Gamma_g$
- For $'$ and $''$ superscripts: $\Gamma' \otimes \Gamma' = \Gamma'$, $\Gamma' \otimes \Gamma'' = \Gamma''$, $\Gamma'' \otimes \Gamma' = \Gamma''$, and $\Gamma'' \otimes \Gamma'' = \Gamma'$

Thus, for example, if the direct product $B_g \otimes A_u$ is required for the C_{2h} point group, the above rules show that $B_g \otimes A_u = B_u$. As another example, the $E_1 \otimes E_1$ direct product in the C_6 point group is found to be $A_1 + [A_2] + E_2$. The significance of the square bracket around A_2 will be seen later.

It is sometimes necessary to extend the concept of direct products to a higher number of terms. As an example, a triple direct product can be calculated by taking the direct product of any pair of representations, and then using the result to calculate its direct product with the third. This operation is commutative, so the order of multiplication does not matter. Triple direct products are particularly useful in the discussion of spectroscopic selection rules (see Section 7.1.2).

There are certain simple rules regarding direct products that are helpful to remember and which can readily be checked by consulting the direct product tables.

- The direct product of the totally symmetric irreducible representation with a non-totally symmetric representation gives the non-totally symmetric representation (as all characters of the totally symmetric species are 1).
- The direct product of any one-dimensional irreducible representation with itself gives the totally symmetric representation.
- The direct product of a higher-dimensional species with itself will be reducible and always includes the totally symmetric irreducible representation.

D.7 Cyclic and linear groups

The discussion above shows how to interpret, and use, the character tables for most point groups. However, there are two types of groups that are a little more complicated. One of these falls into the category of the so-called cyclic groups. They are called cyclic because all their symmetry elements can be generated from different powers of one of their elements. Cyclic groups can easily be recognized from their character tables, as the characters of two-dimensional species contain a function and its complex conjugate. Examples include the groups C_3, C_5, C_{3h}, and many others. The other category that presents difficulties at first sight is the linear molecule point groups $C_{\infty v}$ and $D_{\infty h}$.

Consider the C_3 point group as an example of a cyclic group. The character table for this group is as follows:

C_3	E	C_3	C_3^2		$\varepsilon = \exp(2\pi i/3)$
A	1	1	1	z, R_z	x^2+y^2, z^2
E	$\begin{Bmatrix}1\\1\end{Bmatrix}$	$\begin{Bmatrix}\varepsilon\\\varepsilon^*\end{Bmatrix}$	$\begin{Bmatrix}\varepsilon^*\\\varepsilon\end{Bmatrix}$	$(x, y)(R_x, R_y)$	(x^2-y^2, xy) (yz, xz)

In this table, the symbol ε stands for the quantity $\exp(2\pi i/3)$, where $i = \sqrt{-1}$, and ε^* is the complex conjugate of ε, i.e. $\exp(-2\pi i/3)$. It can be shown that, for the purposes of many physical applications, the two rows belonging to the E representation can be added, so that the resulting row only contains real numbers. When this is done the following table is obtained:

C_3	E	C_3	C_3^2		
A	1	1	1	z, R_z	x^2+y^2, z^2
E	2	$2\cos 2\pi/3$	$2\cos 2\pi/3$	$(x, y), (R_x, R_y)$	$(x^2-y^2, xy), (xz, yz)$

which can be used like any other character table. As an example, we can try to reduce the direct product $E \otimes E$. As the characters for the E representation are 2, $2\cos 2\pi/3$, and $2\cos 2\pi/3$, the characters of the direct product will be $\Gamma = 4, 4\cos^2 2\pi/3$, and $4\cos^2 2\pi/3$. It can be shown by applying well-known trigonometric relationships, namely $\sin^2 x + \cos^2 x = 1$ and $\cos 2x = \cos^2 x - \sin^2 x$, that the characters of the direct product are equal to $\Gamma = 4$, $2 + 2\cos 2\pi/3$, and $2 + 2\cos 2\pi/3$. It is easy to see that this is simply the sum of three species, i.e. $E \otimes E = 2A + E$.

The character tables for linear molecules, $C_{\infty v}$ and $D_{\infty h}$, are also somewhat peculiar at first sight. These two groups differ in the existence of the centre of symmetry as a symmetry element. As an example, the character table for the point group $C_{\infty v}$ is shown below.

$C_{\infty v}$	E	$2C_\infty^\phi$	...	$\infty\sigma_v$		
$A_1 \equiv \Sigma^+$	1	1	...	1	z	x^2+y^2, z^2
$A_2 \equiv \Sigma^-$	1	1	...	-1	R_z	
$E_1 \equiv \Pi$	2	$2\cos\phi$	...	0	$(x, y), (R_x, R_y)$	(xz, yz)
$E_2 \equiv \Delta$	2	$2\cos 2\phi$	...	0		(x^2-y^2, xy)
$E_3 \equiv \Phi$	2	$2\cos 3\phi$	...	0		
...	...	...	...	...		

First, there is an infinite number of classes because rotation about any angle ϕ about the C_∞ axis is a symmetry operation and each of these C_∞^ϕ elements belongs to a different class. Similarly, there is an infinite number of σ_v planes. The consequence of an infinite number of symmetry elements is that there is also an infinite number of irreducible representations. The labelling of these is often slightly confusing. On the one hand, they are sometimes named according to the conventions described above, i.e. A_1, A_2, E_1, E_2, etc. More usually they are labelled according to the convention introduced in electronic structure theory to describe electronic states of linear molecules, namely Σ, Π, Δ, Φ, etc. As described in Sections 4.2.2 and 4.2.3, in electronic states these labels correspond to different values of the angular momentum quantum number Λ.

The direct products of irreducible representations in linear groups can be calculated in a similar manner to other point groups. Taking $\Sigma^- \otimes \Pi$ as an example, the characters of the direct product are $\Gamma = 2, 2\cos\phi, \ldots, 0$, i.e. $\Sigma^- \otimes \Pi = \Pi$. Trigonometric relationships need to be invoked for the direct products of two- or higher-dimensional representations. For example, the characters of the direct product $\Pi \otimes \Pi$ are $\Gamma = 4, 4\cos^2\phi \ldots, 0$. Using the above relationships it can be shown that $4\cos^2\phi = 2 + 2\cos 2\phi$, and hence $\Pi \otimes \Pi = \Delta + \Sigma^+ + \Sigma^-$. In practice such manipulations are not necessary and direct products can be obtained simply by inspecting Table D.6.

D.8 Symmetrized and antisymmetrized products

In the description of spectroscopic states it is sometimes necessary to invoke the symmetrized and antisymmetrized product of two functions, instead of simply taking their product. For functions f_i and f_j, the symmetrized product is $\frac{1}{2}(f_if_j + f_jf_i)$, whereas the antisymmetrized product is $\frac{1}{2}(f_if_j - f_jf_i)$. It can be proved that both of these products are reducible representations of the point group. In many examples, the antisymmetrized product simply vanishes.

Symmetrized and antisymmetrized products have special importance when the electronic state is derived for two electrons. The resulting electronic state can be obtained from the direct product of the symmetry species of the molecular orbitals. Careful consideration of the Pauli principle is required if the electrons reside in degenerate orbitals and this is a topic considered in more detail in the next appendix. In direct product tables antisymmetrized direct products are displayed in square brackets.

Further reading

Good introductory accounts of symmetry and point group theory in chemical and spectroscopic applications can be found in the following books:

Group Theory and Chemistry, D. M. Bishop, New York, Dover, 1993.
Molecular Symmetry and Group Theory, R. L. Carter, New York, Wiley, 1998.
Chemical Applications of Group Theory, F. A. Cotton, New York, Wiley, 1990.
Molecular Symmetry and Group Theory: A Programmed Introduction to Chemical Applications, A. Vincent, Chichester, Wiley, 2001.

More advanced aspects, most notably consideration of flexible molecules, which cannot be treated adequately by point group theory, can be found in the following books:

Molecular Symmetry and Spectroscopy, P. R. Bunker and P. Jensen, Ottawa, NRC Press, 1998.
Symmetry, Structure and Spectroscopy of Atoms and Molecules, W. J. Harter, New York, Wiley, 1993.

Selected character tables

C_1	E
A	1

C_s	E	σ_h		
A'	1	1	x, y, R_z	x^2, y^2, z^2, xy
A''	1	−1	z, R_x, R_y	yz, xz

C_i	E	i		
A_g	1	1	R_x, R_y, R_z	$x^2, y^2, z^2, xy, xz, yz$
A_u	1	−1	x, y, z	

C_2	E	C_2		
A	1	1	z, R_z	x^2, y^2, z^2, xy
B	1	−1	x, y, R_x, R_y	yz, xz

C_3	E	C_3	C_3^2		$\varepsilon = \exp(2\pi i/3)$
A	1	1	1	z, R_z	$x^2 + y^2, z^2$
E	$\begin{Bmatrix} 1 \\ 1 \end{Bmatrix}$	ε ε^*	ε ε^*	$(x, y), (R_x, R_y)$	$(x^2 - y^2, xy)$ (yz, xz)

C_4	E	C_4	C_2	C_4^3		
A	1	1	1	1	z, R_z	$x^2 + y^2, z^2$
B	1	−1	1	−1		$x^2 - y^2, xy$
E	$\begin{Bmatrix} 1 \\ 1 \end{Bmatrix}$	i −1	−1 −1	−i i	(x, y), (R_x, R_y)	(yz, xz)

C_6	E	C_6	C_3	C_2	C_3^2	C_6^5		$\varepsilon = \exp(2\pi i/6)$
A	1	1	1	1	1	1	z, R_z	$x^2 + y^2, z^2$
B	1	−1	1	−1	1	−1		
E_1	$\begin{Bmatrix} 1 \\ 1 \end{Bmatrix}$	ε ε^*	$-\varepsilon^*$ $-\varepsilon$	−1 −1	$-\varepsilon$ $-\varepsilon^*$	ε^* ε	(x, y), (R_x, R_y)	(xz, yz)
E_2	$\begin{Bmatrix} 1 \\ 1 \end{Bmatrix}$	$-\varepsilon^*$ $-\varepsilon$	$-\varepsilon$ $-\varepsilon^*$	1 1	$-\varepsilon^*$ $-\varepsilon$	$-\varepsilon$ $-\varepsilon^*$		$x^2 - y^2, xy$

D_2	E	$C_2(z)$	$C_2(y)$	$C_2(x)$		
A	1	1	1	1		x^2, y^2, z^2
B_1	1	1	−1	−1	z, R_z	xy
B_2	1	−1	1	−1	y, R_y	xz
B_3	1	−1	−1	1	x, R_x	yz

C_{2v}	E	$C_2(z)$	$\sigma_v(xz)$	$\sigma_v(yz)$		
A_1	1	1	1	1	z	x^2, y^2, z^2
A_2	1	1	−1	−1	R_z	xy
B_1	1	−1	1	−1	x, R_y	xz
B_2	1	−1	−1	1	y, R_x	yz

C_{3v}	E	$2C_3(z)$	$3\sigma_v$		
A_1	1	1	1	z	$x^2 + y^2, z^2$
A_2	1	1	−1	R_z	
E	2	−1	0	$(x, y), (R_x, R_y)$	$(x^2 - y^2, xy), (xz, yz)$

C_{4v}	E	$2C_4$	C_2	$2\sigma_v$	$2\sigma_d$		
A_1	1	1	1	1	1	z	$x^2 + y^2, z^2$
A_2	1	1	1	−1	−1	R_z	
B_1	1	−1	1	1	−1		$x^2 - y^2$
B_2	1	−1	1	−1	1		xy
E	2	0	−2	0	0	$(x, y), (R_x, R_y)$	(xz, yz)

C_{2h}	E	C_2	i	σ_h		
A_g	1	1	1	1	R_z	x^2, y^2, z^2, xy
B_g	1	−1	1	−1	R_x, R_y	xz, yz
A_u	1	1	−1	−1	z	
B_u	1	−1	−1	1	x, y	

D_{2h}	E	$C_2(z)$	$C_2(y)$	$C_2(x)$	i	$\sigma(xy)$	$\sigma(xz)$	$\sigma(yz)$		
A_g	1	1	1	1	1	1	1	1		x^2, y^2, z^2
B_{1g}	1	1	−1	−1	1	1	−1	−1	R_z	xy
B_{2g}	1	−1	1	−1	1	−1	1	−1	R_y	xz
B_{3g}	1	−1	−1	1	1	−1	−1	1	R_x	yz
A_u	1	1	1	1	−1	−1	−1	−1		
B_{1u}	1	1	−1	−1	−1	−1	1	1	z	
B_{2u}	1	−1	1	−1	−1	1	−1	1	y	
B_{3u}	1	−1	−1	1	−1	1	1	−1	x	

$\mathbf{D_{3h}}$	E	$2C_3$	$3C_2$	σ_h	$2S_3$	$3\sigma_v$		
A_1'	1	1	1	1	1	1		x^2+y^2, z^2
A_2'	1	1	−1	1	1	−1	$R_z, (x, y)$	(x^2-y^2, xy)
E'	2	−1	0	2	−1	0		
A_1''	1	1	1	−1	−1	−1	$z, (R_x, R_y)$	(xz, yz)
A_2''	1	1	−1	−1	−1	1		
E''	2	−1	0	−2	1	0		

$\mathbf{D_{6h}}$	E	$2C_6$	$2C_3$	C_2	$3C_2'$	C_2''	i	$2S_3$	$2S_6$	σ_h	$3\sigma_d$	$3\sigma_v$		
A_{1g}	1	1	1	1	1	1	1	1	1	1	1	1		x^2+y^2, z^2
A_{2g}	1	1	1	1	−1	−1	1	1	1	1	−1	−1	R_z	
B_{1g}	1	−1	1	−1	1	−1	1	−1	1	−1	1	−1		
B_{2g}	1	−1	1	−1	−1	1	1	−1	1	−1	−1	1		
E_{1g}	2	1	−1	−2	0	0	2	1	−1	−2	0	0	(R_x, R_y)	(xz, yz)
E_{2g}	2	−1	−1	2	0	0	2	−1	−1	2	0	0		(x^2-y^2, xy)
A_{1u}	1	1	1	1	1	1	−1	−1	−1	−1	−1	−1		
A_{2u}	1	1	1	1	−1	−1	−1	−1	−1	−1	1	1	z	
B_{1u}	1	−1	1	−1	1	−1	−1	1	−1	1	−1	1		
B_{2u}	1	−1	1	−1	−1	1	−1	1	−1	1	1	−1		
E_{1u}	2	1	−1	−2	0	0	−2	−1	1	2	0	0	(x, y)	
E_{2u}	2	−1	−1	2	0	0	−2	1	1	−2	0	0		

$\mathbf{D_{2d}}$	E	$2S_4$	C_2	$2C_2'$	$2\sigma_d$		
A_1	1	1	1	1	1		x^2+y^2, z^2
A_2	1	1	1	−1	−1	R_z	
B_1	1	−1	1	1	−1		x^2-y^2
B_2	1	−1	1	−1	1	z	xy
E	2	0	−2	0	0	$(x, y), (R_x, R_y)$	(xz, yz)

$\mathbf{T_d}$	E	$8C_3$	$3C_2$	$6S_4$	$6\sigma_d$		
A_1	1	1	1	1	1		$x^2+y^2+z^2$
A_2	1	1	1	−1	−1		
E	2	−1	2	0	0		$(2z^2-x^2-y^2, x^2-y^2)$
T_1	3	0	−1	1	−1	(R_x, R_y, R_z)	
T_2	3	0	−1	−1	1	(x, y, z)	(xy, xz, yz)

$\mathbf{C_{\infty v}}$	E	$2C_\infty^\phi$	...	$\infty\sigma_v$		
$A_1 \equiv \Sigma^+$	1	1	...	1	z	x^2+y^2, z^2
$A_2 \equiv \Sigma^-$	1	1	...	−1	R_z	
$E_1 \equiv \Pi$	2	$2\cos\phi$	...	0	$(x, y), (R_x, R_y)$	(xz, yz)
$E_2 \equiv \Delta$	2	$2\cos 2\phi$	...	0		(x^2-y^2, xy)
$E_3 \equiv \Phi$	2	$2\cos 3\phi$	...	0		
...	...	...	...	...		

$\mathbf{D}_{\infty h}$	E	$2C_{\infty}^{\phi}$	...	$\infty\sigma_v$	i	$2S_{\infty}^{\phi}$	...	∞C_2		
Σ_g^+	1	1	...	1	1	1	...	1		x^2+y^2, z^2
Σ_g^-	1	1	...	−1	1	1	...	−1	R_z	
Π_g	2	$2\cos\phi$	...	0	2	$-2\cos\phi$	...	0	(R_x, R_y)	(xz, yz)
Δ_g	2	$2\cos 2\phi$	...	0	2	$2\cos 2\phi$	...	0		(x^2-y^2, xy)
...	...	...	...	...	...	...	...	...		
Σ_u^+	1	1	...	1	−1	−1	...	−1	z	
Σ_u^-	1	1	...	−1	−1	−1	...	1		
Π_u	2	$2\cos\phi$	...	0	−2	$2\cos\phi$	...	0	(x, y)	
Δ_u	2	$2\cos 2\phi$	...	0	−2	$-2\cos 2\phi$	...	0		
...	...	...	...	...	...	...	...	...		

Appendix E More on electronic configurations and electronic states: degenerate orbitals and the Pauli principle

The Pauli exclusion principle states that no two electrons in an atom or molecule can share entirely the same set of quantum numbers. This requirement follows from the nature of electronic wavefunctions, which must be antisymmetric with respect to the exchange of any identical electrons. This has an impact in the determination of the electronic states possible from a given electronic configuration.

E.1 Atoms

Consider, for example, the carbon atom, which has a ground electronic configuration $1s^2 2s^2 2p^2$. Suppose that one of the $2p$ electrons is excited to a $3p$ orbital. To determine the electronic states that are possible from this configuration, the process described in Section 4.1 can be followed. The $1s$ and $2s$ orbitals are full and so we can focus on the p electrons only. The possible values of the total orbital angular momentum quantum number L are 2, 1 or 0. Similarly, the total spin quantum number must be 1 or 0 and so the possible electronic states that result from the $1s^2 2s^2 2p^1 3p^1$ configuration are ^{3}D, ^{1}D, ^{3}P, ^{1}P, ^{3}S, and ^{1}S. It is therefore initially tempting to propose that electronic states of the same spatial and spin symmetry arise from the ground electronic configuration. Such an assumption would be wrong because it ignores the Pauli principle.

In contrast to the excited configuration considered above, in the ground configuration of the carbon atom the two p electrons have the same principal quantum numbers. To satisfy the Pauli principle, we must therefore avoid those electronic states of the carbon atom in which the two p electrons possess exactly the same values for the remaining quantum numbers. This means that the electrons cannot be in a $2p$ orbital with the same m_l and m_s quantum numbers. The acceptable arrangements of the electrons within the three $2p$ orbitals are summarized in Table E.1. Notice that in contrast to the excited configuration $1s^2 2s^2 2p^1 3p^1$, only three electronic states (^{3}P, ^{1}D, and ^{1}S) are possible from the configuration $1s^2 2s^2 2p^2$.

Table E.1 *Possible arrangement of electrons for a $2p^2$ configuration*

$m_l = -1$	$m_l = 0$	$m_l = +1$	Electronic state
	↑	↑	
	↓	↑	
	↓	↓	
↑		↑	
↓		↑	^{3}P
↓		↓	
↑	↑		
↓	↑		
↓	↓		
		↑↓	
	↑	↓	
↑		↓	^{1}D
↑	↓		
↑↓			
	↑↓		^{1}S

Constructing a table of electron arrangements amongst orbitals such as that shown in Table E.1 is clearly a cumbersome process. A neater way of arriving at the same conclusions follows from the symmetries of the orbital and spin wavefunctions with respect to the exchange of identical particles. Taking spin first, the spin wavefunction for a singlet state is

$$\Psi_s = \frac{1}{\sqrt{2}}[\alpha(1)\beta(2) - \alpha(2)\beta(1)] \tag{E.1}$$

Triplet states have three possible wavefunctions due to the three-fold degeneracy of unit angular momentum states (cf. p orbitals in atoms), which are given by

$$\Psi_t^{(+1)} = \alpha(1)\alpha(2) \tag{E.2}$$

$$\Psi_t^{(0)} = \frac{1}{\sqrt{2}}[\alpha(1)\beta(2) + \beta(1)\alpha(2)] \tag{E.3}$$

$$\Psi_t^{(-1)} = \beta(1)\beta(2) \tag{E.4}$$

In the above expressions the labels 1 and 2 refer to the two electrons and α and β refer, respectively, to spin up ($m_s = +\frac{1}{2}$) and spin down ($m_s = -\frac{1}{2}$) wavefunctions for the individual electrons. The superscripts on the wavefunctions on the left-hand side of each equation refer to the spin projection quantum number, M_S, which can have the values 1, 0 or -1 for the case where $S = 1$.

Interchange of the two electrons corresponds to switching the locations of the 1 and 2 labels in parentheses. For the singlet state, this changes the sign of the wavefunction so the spin singlet wavefunction is antisymmetric with respect to exchange of identical electrons.

In contrast, all three triplet wavefunctions remain unchanged on switching the electron indices and so triplet wavefunctions are symmetric with respect to electron exchange.

The symmetry of the spatial (orbital angular momentum) part of the electronic wavefunction with respect to electron exchange can also be determined straightforwardly, but we shall avoid the details. In effect, the desired wavefunction is a linear combination of the wavefunctions for the individual $2p$ orbitals in much the same way as the spin wavefunctions can be expressed as linear combinations of the individual spin up and spin down wavefunctions, α and β. The key result is that electronic states with L even are symmetric with respect to electron exchange, whereas those with L odd are antisymmetric. Thus L even states can only combine with the singlet spin function in order to satisfy the Pauli principle, and so we deduce that the only possible singlet states are ^{1}D and ^{1}S. Similarly, only one triplet state can be formed, ^{3}P.

E.2 Molecules

Exactly the same ideas apply to molecules. In molecules, as in atoms, equivalent orbitals are degenerate orbitals and these only arise for molecules that possess relatively high symmetries. Consider, for example, a molecule with C_{6v} symmetry and all orbitals filled except the outer pair, which have e_1 symmetry. Now suppose that there are two electrons in the e_1 orbitals. This is clearly a case where the Pauli principle needs to be taken into account. As we have seen elsewhere (Section 4.2), the possible spatial symmetries of the overall electronic state can be obtained by taking the direct product of the symmetries of the individual orbitals, $e_1 \otimes e_1$. The result can be obtained from Table D.4 in Appendix D, and is $A_1 + [A_2] + E_2$.

The square brackets around the A_2 representation are employed to show that this corresponds to an antisymmetrized product (see Section D.8), which means that the A_2 spatial wavefunction arising from the orbital configuration $(e_1)^2$ is antisymmetric with respect to electron exchange. Consequently, only a triplet spin state is possible for this spatial symmetry. In contrast the A_1 and E_2 spatial wavefunctions are symmetric with respect to electron exchange and can only combine with a singlet spin state. Thus we deduce that the possible electronic states arising from the $(e_1)^2$ configuration are $^3A_2$, $^1E_2$, and $^1A_1$.

Appendix F
Nuclear spin statistics

Some atomic nuclei possess spin angular momentum, and this can couple with other angular momenta in a molecule, notably the overall rotational angular momentum, and with the net electron spin (if any), to cause additional structure in a spectrum. This additional structure is known as *hyperfine* structure. Hyperfine splittings are normally very small and are only resolved in very high resolution spectroscopy. However, the effect of nuclei on molecular spectra can also be observed in lower resolution experiments through the phenomenon known as *nuclear spin statistics*. This manifests itself as an alternation of intensities in the rotational structure for molecules with a rotational symmetry C_2 or higher. Examples were met in the Case Studies described in Chapters 16, 21, and 28.

A general expression for the total wavefunction of the molecule was given by equation (7.11). In reality, the total wavefunction also includes one more term, the wavefunction due to nuclear spin, ψ_{ns}:

$$\Psi(\boldsymbol{r},\ \boldsymbol{R}) = \psi_e(\boldsymbol{r},\ \boldsymbol{R}_e).\psi_v(\boldsymbol{R}).\psi_r(\boldsymbol{R}).\psi_{ns} \tag{F.1}$$

For purposes of this discussion, nuclei with half-integer spins (such nuclei are called fermions because they obey Fermi–Dirac statistics) must be differentiated from those with integer spins (called bosons because they can be described using Bose–Einstein statistics). The generalized Pauli principle states that the total wavefunction of the system must be antisymmetric with respect to the exchange of two identical fermions but symmetric for the exchange of identical bosons.

To establish the symmetry of the overall molecular wavefunction with respect to exchange of identical nuclei, it is necessary to consider the effect of nuclear exchange on each term in equation (F.1). We will simplify things somewhat by focussing on homonuclear diatomic molecules (this discussion would be irrelevant for heteronuclear diatomics since they do not possess identical nuclei). Dealing with the electronic wavefunction first, the symmetry with respect to nuclear exchange depends on the symmetry of the electronic wavefunction. For a totally symmetric ($^1\Sigma_g^+$) electronic state, the electronic wavefunction is totally symmetric with respect to nuclear exchange. However, for other electronic states the wavefunction may be antisymmetric, e.g. $^1\Sigma_u^+$ or $^3\Sigma_g^-$. We will concentrate on the totally symmetric case but the arguments below will differ for the antisymmetric electronic states. In a diatomic molecule the vibrational wavefunction is always totally symmetric with respect to the exchange of nuclei since the wavefunction depends only on the separation of the nuclei, and this is unchanged by a permutation of the nuclei. Note that in polyatomic molecules the vibrational wavefunction is not always totally symmetric with respect to the exchange of identical nuclei.

F.1 Fermionic nuclei

If the identical nuclei are fermions, the overall molecular wavefunction must be antisymmetric with respect to nuclear exchange. In a diatomic molecule in a totally symmetric electronic state only the rotational and nuclear spin states need to be considered to determine the symmetry of the overall wavefunction. In this book we have not discussed the explicit form of the rotational wavefunctions of molecules. However, it can be shown that for diatomic molecules the symmetry of ψ_r for the interchange of identical nuclei is $(-1)^J$ where J is the rotational quantum number. Thus rotational levels with even J are symmetric and those with odd J are antisymmetric. Consequently, for the product $\psi_r\psi_{ns}$ to be antisymmetric, a symmetric ψ_{ns} must be associated with a rotational level having odd J, whereas an antisymmetric ψ_{ns} combines with even J levels. It can be shown that, for different nuclear spins, the number of symmetric and antisymmetric nuclear spin states is given by the following formulae:

$$g_n^{symm} = (2I + 1)(I + 1) \tag{F.2}$$

$$g_n^{antisymm} = (2I + 1)I \tag{F.3}$$

The nuclear spins I of selected nuclei are given in Table F.1. For a nuclear spin of $I = ½$, as found for example in each nucleus in H_2, there are four possible nuclear spin wavefunctions, three of which are symmetric and one which is antisymmetric, i.e. there are three times as many symmetric as antisymmetric states (cf. the spin wavefunctions for two electrons shown in the previous appendix). These are known as *ortho* and *para* states, respectively. The ortho states are associated with odd J values, whereas the para states are associated with even J. Transitions originating from these states will have corresponding differences in their intensities due to the 3:1 alternation in statistical weights.

F.2 Bosonic nuclei

For nuclei with integer spins, the total wavefunction must be symmetric with respect to exchange of identical nuclei. If for example $I = 1$, there are six symmetric and three antisymmetric nuclear spin wavefunctions. The symmetric nuclear spin wavefunctions combine with even J states and will have approximately twice the population of odd J states. As above, these differences in population will be reflected in the intensities of transitions originating in these states.

If we consider a molecule with two identical nuclei possessing zero spin, such as in the $^{12}C_2$ molecule, antisymmetric nuclear spin states will be missing. The ground electronic

Table F.1 *Nuclear spin quantum numbers for some selected nuclei*

Nucleus	I
^{1}H	½
^{2}H (D)	1
^{3}H (T)	½
^{12}C	0
^{13}C	½
^{14}C	0
^{14}N	1
^{15}N	½
^{16}O	0
^{19}F	½
^{31}P	½
^{32}S	0
^{35}Cl	3⁄2
^{37}Cl	3⁄2
^{79}Br	3⁄2
^{81}Br	3⁄2
^{127}I	5⁄2

state of C_2 is $^1\Sigma_g^+$ and so is symmetric with respect to nuclear exchange. Since ^{12}C nuclei are bosons, we have the seemingly bizarre but true situation that the molecule can only exist in rotational energy levels with even *J*. In terms of spectroscopy, this will mean that every other rotational line in the spectrum will be missing. The linear triatomic molecule C_3 also behaves in this manner and the role of nuclear spin statistics in interpreting the rotational structure of this molecule was discussed in Chapter 16.

Appendix G Coupling of angular momenta: Hund's coupling cases

The discussion of angular momentum coupling in Appendix C focussed on electronic (orbital and spin) angular momenta. Other types of angular momenta may be present in molecules and their coupling to electronic angular momenta can have an important impact in spectroscopy. In this appendix rotational angular momentum is added to the pot and its interaction with electronic angular momenta is considered. The discussion is restricted to linear molecules, and several limiting cases, known as Hund's coupling cases, are briefly described.

G.1 Hund's case (a)

Hund's case (a) coupling builds upon the orbital + spin coupling already described in Appendix C. The orbital angular momenta in a molecule are assumed to be coupled to the internuclear axis by an electrostatic interaction and spin–orbit coupling leads to the spin angular momenta also precessing around the same axis. However, the spin–orbit coupling is not too strong to blur the distinction between orbital and spin angular momenta. Rotation in a linear molecule leads to rotational angular momentum and yields a vector $\boldsymbol{R}$ that is oriented perpendicular to the internuclear axis, as shown in Figure G.1.

In Hund's case (a) it is assumed that the interaction between the electronic and rotational angular momenta is weak, and hence the former (the orbital angular momentum $\boldsymbol{L}$ and the spin angular momentum $\boldsymbol{S}$) continue to precess rapidly around the internuclear axis with projections whose sum is equal to Ω $(= \Lambda + \Sigma)$. The total angular momentum $\boldsymbol{J}$, electronic + rotational, is the vector sum of $\boldsymbol{R}$ and $\boldsymbol{\Omega}$. The vectors $\boldsymbol{R}$ and $\boldsymbol{\Omega}$ precess about vector $\boldsymbol{J}$.

In Hund's case (a) the quantum numbers J, Ω, Λ, S, and Σ are all well defined. We could also add a quantum number to define the rotational angular momentum but this would be redundant if we already know J and the electronic angular momentum quantum numbers. Since Ω is the quantum number representing the projection of J on the internuclear axis, the minimum possible value of J is Ω. The allowed values of J are therefore Ω, $\Omega + 1$, $\Omega + 2$, $\Omega + 3$, etc. If the number of unpaired electrons is odd then Ω will be a half-integer quantum number and therefore J also has half-integer values only.

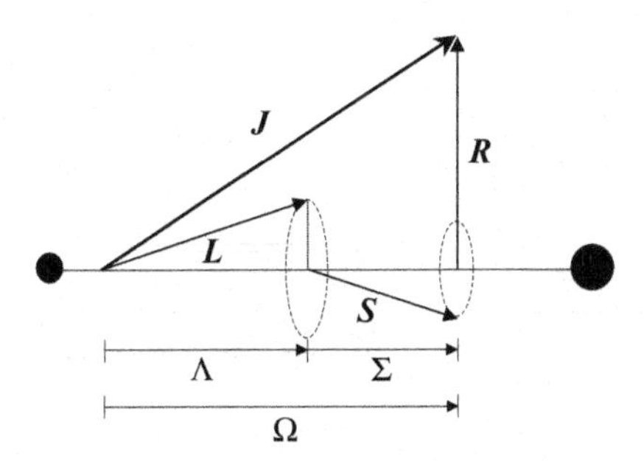

Figure G.1 Illustration of Hund's case (a) coupling. The strong axial electric field along the internuclear axis causes the total electron orbital (***L***) and spin (***S***) angular momenta to precess rapidly about the internuclear axis. The components of these angular momenta along the internuclear axis are well defined, giving quantum number Ω, and this couples with the rotational angular momentum of the molecule (***R***) to form a resultant, ***J***.

The rotational energy levels of Hund's case (a) molecules can be derived by analogy with symmetric top rotational energy level formulae. The angular momentum about the internuclear axis, denoted by quantum number Ω, is equivalent to the projection of rotational angular momentum, K, in a prolate symmetric top (see equation 6.15), and so we can write

$$E_{J,\Omega} = BJ(J+1) + (A-B)\Omega^2 \tag{G.1}$$

A is inversely related to the moment of inertia of the electrons and by definition is therefore very large. The $A\Omega^2$ term can in fact be ignored since it is a purely electronic term that contributes equally to all rotational energy levels, leaving the expression

$$E_{J,\Omega} = B[J(J+1) - \Omega^2] \tag{G.2}$$

Molecules showing Hund's case (a) behaviour possess orbital angular momentum. The rotational energy levels of a molecule in a $^3\Pi$ state are shown in Figure G.2 as an illustration. In this example three spin–orbit sub-states arise whose separation depends on the magnitude of the spin–orbit coupling. Notice that the lowest rotational level in each sub-state has the value Ω for that sub-state.

The basis of Hund's case (a) coupling is that the orbital and spin angular momenta remain firmly coupled to the internuclear axis even when the molecule rotates. This is a good approximation but in practice the rotation does induce some uncoupling and this grows in magnitude as the speed of rotation increases, i.e. as J increases. This uncoupling removes the two-fold degeneracy in Λ and is therefore known as Λ-*doubling*. This splitting of each rotational level is shown in Figure G.2, but is exaggerated and in practice the effect of Λ-doubling can only be resolved in high resolution experiments. Notice that the two components for a given J can be distinguished by an additional symmetry label, the *parity* of the energy level (±). This refers to the symmetry with respect to inversion (switching coordinates (x, y, z) to $(-x,-y,-z)$) of all particles in a laboratory-fixed axis system, i.e. one not attached to the molecule. We shall not consider this any further except to say that it is helpful in the determination of transition selection rules (for example see

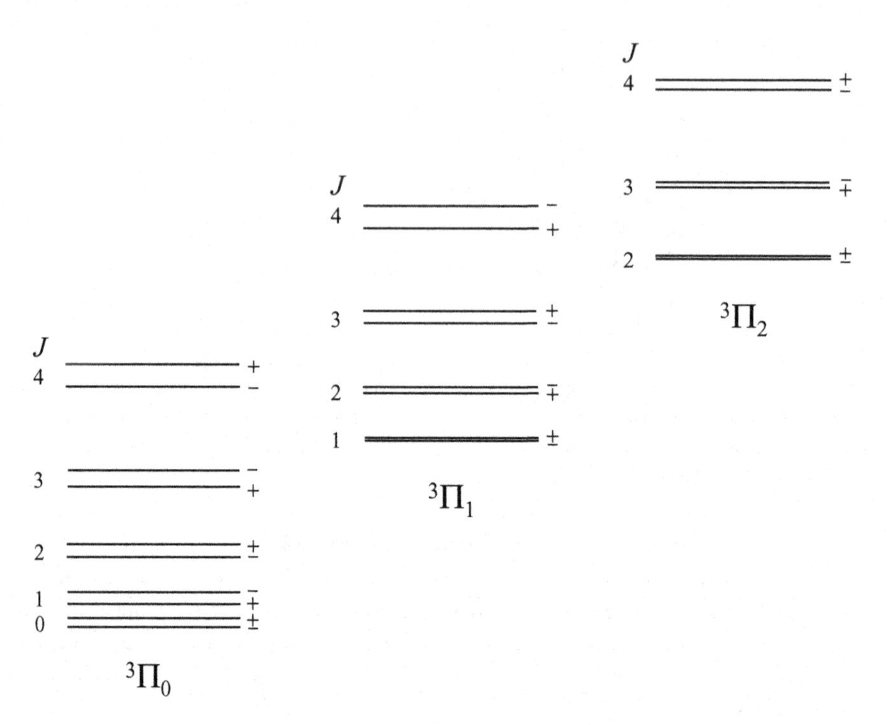

Figure G.2 Rotational energy levels of a molecule in a $^3\Pi$ electronic state satisfying Hund's case (a) coupling. Spin–orbit coupling splits the $^3\Pi$ state into the spin–orbit components $^3\Pi_0$, $^3\Pi_1$, and $^3\Pi_2$, where the subscript refers to the quantum number Ω. Each rotational level within a particular spin–orbit component is split into a doublet due to Λ-doubling, but the size of this effect is much exaggerated in this diagram.

Chapter 24). More details can be found in the books listed in the Further Reading section at the end of this appendix.

G.2 Hund's case (b)

The premise of Hund's case (b) is that the spin–orbit coupling is no longer strong enough to tie the precession of $\boldsymbol{S}$ to the internuclear axis. This most commonly occurs when $\Lambda = 0$, but it is also known in molecules with $\Lambda \neq 0$ under certain conditions (see below). Assuming $\Lambda = 0$, only the spin and rotational angular momenta remain and these couple together and precess around the resultant $\boldsymbol{J}$. More generally, we have the situation shown in Figure G.3, where the possibility of a non-zero $\boldsymbol{\Lambda}$ has been included. The precession of the orbital angular momentum around the internuclear axis remains rapid and the total angular momentum excluding electron spin, designated as vector $\boldsymbol{N}$, is then given by $\boldsymbol{R} + \boldsymbol{\Lambda}$. A weak interaction then occurs between $\boldsymbol{N}$ and $\boldsymbol{S}$ and these vectors precess slowly about the total angular momentum vector $\boldsymbol{J}$.

The quantum numbers used to define Hund's case (b) states are J, N, Λ, and S. Notice that Ω is no longer a good quantum number in the Hund's case (b) limit, since precession of the electron spin is no longer tied to the internuclear axis. If $\Lambda = 0$ then the lowest value

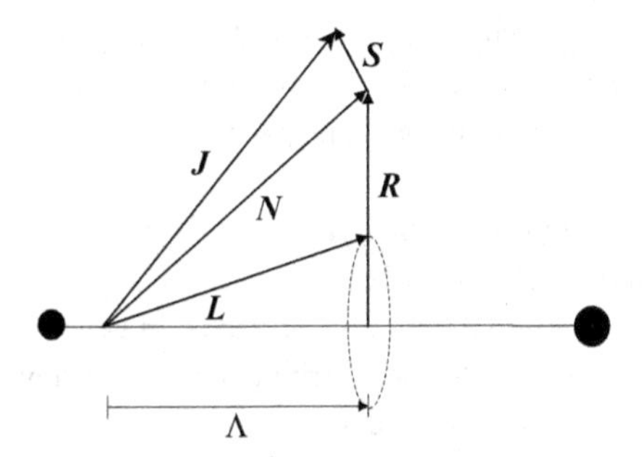

Figure G.3 Illustration of Hund's case (b) coupling. In Hund's case (b) spin–orbit coupling is no longer strong enough to couple $\boldsymbol{S}$ to the internuclear axis. However, $\boldsymbol{L}$ (if non-zero) is still coupled to the internuclear axis and together with the rotational angular momentum $\boldsymbol{R}$ this forms a resultant $\boldsymbol{N}$. The total angular momentum $\boldsymbol{J}$ is obtained from the vector addition $\boldsymbol{N} + \boldsymbol{S}$.

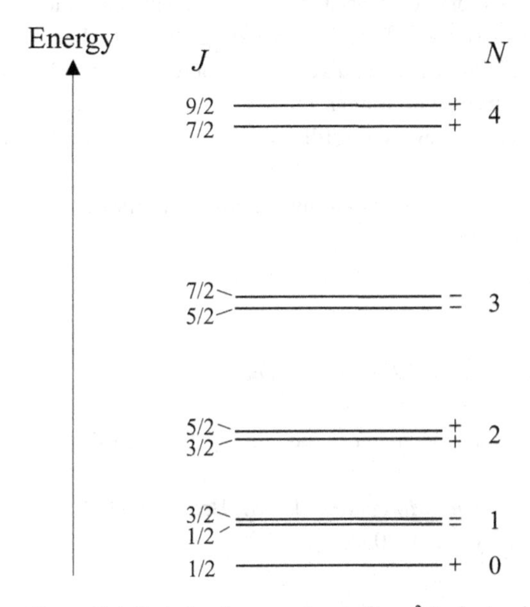

Figure G.4 Rotational energy levels for a $^2\Sigma$ electronic state. The interaction between the spin and rotational angular momenta gives rise to a spin–rotation splitting for each rotational energy level (except the lowest level). The labels + or – for each level refer to the parity (see text for more details).

of N is zero and therefore the allowed values of N are the same as for the rotational energy levels of a closed-shell linear molecule, i.e. 0, 1, 2, 3, etc. J has allowed values $N+S, N+S-1, N+S-2, \ldots, |N-S|$, and therefore J will be an integer if there is an even number of unpaired electrons and half-integer for an odd number of unpaired electrons.

The rotational energy levels for Hund's case (b) are similar to those of closed-shell molecules. However, the effect of interaction between the rotational motion and spin cannot be entirely neglected. This spin–rotation coupling is small but observable in high resolution experiments because it gives rise to a splitting of each rotational level except for the lowest. For example, the rotational energy levels of a molecule in a $^2\Sigma$ state are shown in Figure G.4. Each rotational energy level is split into a doublet and the splitting increases as the rotational energy increases. In fact the splitting can be shown to be $\gamma(N+1/2)$ where γ is a quantity known as the spin–rotation coupling constant. Notice that the two components of a spin–rotation doublet have the same parity, in contrast to the opposite parities for the components of a Λ-doublet.

Finally, we note that a molecule may switch from satisfying Hund's case (a) to Hund's case (b) behaviour if it is sufficiently rotationally excited. This can occur when rotation is fast enough to uncouple $\boldsymbol{S}$ from precession around the internuclear axis. In general, Hund's case (a) coupling applies when $A \gg BJ$, where A is the spin–orbit coupling constant for the electronic state. Transition towards case (b) behaviour occurs when $A \approx BJ$.

G.3 Other Hund's coupling cases

Hund's cases (a) and (b) are satisfactory for describing the rotational energy levels of the great majority of linear molecules. However, three other coupling cases have been proposed. The most commonly encountered is probably Hund's case (c), where the spin–orbit coupling is now sufficiently large that Λ and S are no longer defined, but Ω is still a good quantum number. The resulting rotational energy levels are still given by the energy level expression (G.1).

Further details of Hund's coupling cases, including the less well-known cases (d) and (e), can be found in the books listed below.

Further reading

Molecular Spectra and Molecular Structure. I. Spectra of Diatomic Molecules, G. Herzberg, Malabar, Florida, Krieger Publishing, 1989.

The Spectra and Dynamics of Diatomic Molecules, 2nd edn., H. Lefebvre-Brion and R. W. Field, Academic Press, 2004.

Rotational Spectroscopy of Diatomic Molecules, J. M. Brown and A. Carrington, Cambridge, Cambridge University Press, 2003.

Appendix H
Computational simulation and analysis of rotational structure

Except for the very simplest cases, the analysis of rotational structure in the spectra of molecules is nowadays carried out using computer simulation. The essence of this approach can be divided into three parts: (i) the calculation of the rotational energy levels of a molecule using known or estimated spectroscopic constants; (ii) the calculation of the relative intensities of rotational lines; (iii) the adjustment of the spectroscopic constants to give a simulated spectrum that matches experiment. Each of these is briefly considered below.

H.1 Calculating rotational energy levels

The starting point for simulating any spectrum is to calculate the energies of the levels involved in the spectroscopic transitions. Once these have been obtained, transition energies are then simply the difference in energy between the appropriate pairs of levels involved in the transitions.

For closed-shell linear molecules the calculation of rotational energy levels is trivial, since the energies are given by equation (6.4) in the rigid rotor limit, while in the more realistic non-rigid case the expression

$$E_J = BJ(J+1) - D_J J^2(J+1)^2 \tag{H.1}$$

usually suffices. In equation (H.1) B and J have their usual meaning and D_J is known as the centrifugal distortion constant, which allows for the fact that bonds tend to lengthen as the molecule rotates faster and faster. The values of B and D_J will be different for different electronic and/or vibrational states but once their values are known, or are estimated, then the energies for specific rotational transitions can be calculated. For rotational structure in electronic transitions the contribution from electronic and vibrational changes is a constant quantity that can simply be added to all transitions within the rotational envelope.

In more complicated examples it may no longer be possible to write down the rotational energies in a closed form such as that shown in equation (H.1). This is found to be the case for open-shell molecules (free radicals) and also for asymmetric tops. To illustrate why this happens and how it can be tackled, we choose the asymmetric top as an example.

The general form for the classical kinetic energy of a rotating molecule was given in equation (6.11), which can be recast in the following form:

$$E = AR_a^2 + BR_b^2 + CR_c^2 \tag{H.2}$$

R_a, R_b, and R_c represent the rotational angular momentum about principal axes a, b, and c and A, B, and C are the rotational constants of the molecule. As described in Section 6.2.4, in quantum mechanics the classical angular momenta are replaced by *operators* whose properties can be used to predict the resulting quantized energy levels. The operator form of (H.2) looks exactly the same, but instead of the energy on the left-hand side we now have the so-called Hamiltonian, H_{rot}, which is a mathematical operator, i.e.

$$H_{\text{rot}} = AR_a^2 + BR_b^2 + CR_c^2 \tag{H.3}$$

In symmetric and spherical tops the Hamiltonian in equation (H.3) can be simplified through the use of symmetry and used to calculate very simple expressions for the rotational energy levels, as was seen in Sections 6.2.4 and 6.2.5. Unfortunately, the lower symmetry in asymmetric tops makes it impossible to derive a simple and general formula for their energy levels.

The alternative approach for determining the rotational energies of asymmetric tops is a numerical procedure that involves three key steps.

(i) It is assumed that the wavefunction for rotational motion can treated as a superposition of symmetric top-like wavefunctions. The technical way of describing this is to say that the asymmetric rotor wavefunction is expanded in a basis of symmetric rotor wavefunctions, and this expansion is exact if sufficient symmetric top wavefunctions (the so-called *basis functions*) are employed. To grasp this idea, you may find it helpful to draw an analogy with the expansion of molecular orbitals in terms of atomic orbitals. This is the LCAO expansion of MOs and the atomic orbitals form a basis set for describing the MOs.

(ii) The next step is to express the Hamiltonian in a form such that it can be used to operate on the chosen basis functions to deliver useful results. In the case of an asymmetric rotor the Hamiltonian in (H.3) can be rewritten as

$$H = \alpha \boldsymbol{R}^2 + \beta R_c^2 + \gamma(R_+^2 + R_-^2) \tag{H.4}$$

where α, β, and γ are simple functions of the rotational constants A, B, and C but whose detailed forms we do not need to consider here. $\boldsymbol{R}$ is the total rotational angular momentum operator and the operators R_+ and R_- are functions of R_a and R_b with useful properties specified below.

(iii) In the limit that the asymmetric rotor behaves like a symmetric top the third term in (H.4) is zero and the rotational energy levels can be obtained immediately from the resulting Hamiltonian. However, in a real asymmetric top the final term cannot be ignored and as a result the energy cannot be obtained directly from (H.4). Instead a Hamiltonian *matrix* is constructed where the elements of this matrix are obtained by letting the Hamiltonian operate on the basis functions chosen in step 1. R_+ and R_- are key here because they connect basis functions (equivalent to symmetric rotor states) which differ in K by ± 2, where K is the projection quantum number in the symmetric rotor limit. In other words R_+ and R_- are *raising* and *lowering* operators which mix together character from different K levels in the pure symmetric rotor. The energy levels of the asymmetric rotor can then be obtained from the Hamiltonian matrix by a process known as matrix diagonalization. This is a laborious procedure for all but the simplest of matrices but which is well suited for computer calculations.

It is important to recognize that values for the rotational constants A, B, and C must be chosen beforehand in order for the above procedure to work, i.e. we do not obtain general expressions for the rotational energies but specific values given the chosen spectroscopic constants.

Matrix diagonalization is also used to calculate the rotational energy levels in other systems, e.g. open-shell linear molecules. It is a common procedure and lies at the heart of most rotational structure analysis programs. Further details about the basis functions and rotational Hamiltonians used can be found in the Further Reading section at the end of this appendix.

H.2 Calculating transition intensities

For absorption transitions the relative transition intensities[1] are the products of two factors, the transition line strength and a Boltzmann term that describes the relative population of the lower level involved in the transition at a given temperature. The transition line strength is a quantity that depends on the rotational wavefunctions in the upper and lower states and is obtained from the transition dipole moment (see Section 7.2) evaluated over the rotational basis functions. Once the rotational energy levels have been determined, evaluation of transition intensities is a relatively rapid process. Forbidden transitions will obviously give a zero relative intensity.

The processes described in this section and H.1 can be used to simulate the rotational structure in a spectrum. Rather than generate a stick spectrum, it is more useful to associate a linewidth with each transition in the simulation to match that seen in the experiment. This generates a more realistic looking spectrum which is easier to compare with experiment. Examples are shown in Chapters 22, 24, and 28.

H.3 Determining spectroscopic constants

So far we have considered in outline how a spectrum can be generated assuming values for the relevant spectroscopic constants. However, more usually the aim is the reverse process in which spectroscopic constants of a molecule are to be determined from a spectrum. Clearly one could make a guess at the constants, simulate the spectrum, and then visually compare it

1 We are not interested here in the absolute transition intensities. These depend on the experimental arrangement as well as the properties of the molecules under investigation.

with the experimental spectrum. If the agreement is good, then one could reasonably assume that the constants employed are fairly close to the true values. However, if the agreement between the simulation and experiment is poor, it may take an awfully long time to determine the rotational and other constants by arbitrary adjustments followed by visual comparison with experiment. What is required is a more systematic and faster procedure for carrying out essentially the same process. The approach that is employed involves *least-squares fitting*.

The reader will be familiar with the least-squares fitting of straight lines in graphs. This is the process (also known as linear regression) that finds the best straight line through experimental data by minimizing the sum of the squares of the differences $y_{i(\mathrm{line})} - y_{i(\mathrm{expt})}$ for each value of x. Unfortunately, this simple least-squares procedure is not applicable to rotational analyses because the energy levels, and therefore the transition energies, depend non-linearly on the spectroscopic constants. This makes the fitting procedure more complicated and solutions can only be found by an iterative process. Nevertheless, standard computational procedures are well known for carrying out non-linear least-squares fits and can be incorporated into computer programs for spectral analysis [1]. The fitting process involves minimizing, in a least-squares sense, the difference between the rotational line positions in the simulation versus experiment by adjusting the spectroscopic constants.

Many programs have been written for simulating and fitting rotationally resolved spectra. Three examples that are widely used can be followed up from References [2]–[4]. It is important to recognize that many programs are written with specific situations in mind. An example is the AsyrotWin program (Reference [4]), which is designed for simulating closed-shell asymmetric rotors, i.e. it will not deal with open-shell asymmetric tops. Obviously anyone wishing to make use of such a program must first establish that it can deal with their particular problem. These programs should not be thought of as 'black boxes' since they usually require substantial user input. The user must decide on the model to be used, the starting estimates for spectroscopic constants, and the specific lines in the experimental spectrum that will be used in the fit. Furthermore, each line chosen in the experimental data must be associated with a particular transition in the simulated spectrum. If the initial estimates of the spectroscopic constants are poor, then the fitting process may converge on a solution that is not the true best fit. The usual way of proceeding is to first try out a few approximate simulations to see if the starting spectroscopic constants yield a simulated spectrum somewhat similar to that observed experimentally. Only when this first stage is satisfactorily achieved is it sensible to attempt a least-squares fit.

Transition intensities are not used in the fitting but comparison of the simulated relative intensities with those observed experimentally can be a useful way of checking whether the fit is realistic or not. The simulated intensities are also the means by which the temperature of the sample can be determined.

References

1. *Numerical Recipes in C++: the Art of Scientific Computing*, 2nd edn., W. H. Press, S. A. Teukolsky, W. T. Vettering, and B. P. Flannery, Cambridge, Cambridge University Press, 2002. A version of this book is also available for FORTRAN and Basic programmers.

2. DSParFit, a computer program for least-squares fitting of the rotational structure in spectra of diatomic molecules. Details can be found at the website http://scienide.uwaterloo.ca/~leroy/dsparfit/.
3. SpecView, a program for simulating rotational structure in electronic spectra. This is able to deal with many different types of rotors with closed or open shells. Further details can be found at the following website: http://molspect.mps.ohio-state.edu/goes/specview.html.
4. AsyrotWin, a program for the analysis of band spectra in closed-shell asymmetric tops. This program is described in the following article: R. H. Judge and D. J. Clouthier, *Comput. Phys. Commun.* **135** (2001) 293.

Further reading

Molecular Rotation Spectra, H. W. Kroto, New York, Dover Publications, 1992.

Angular Momentum, R. N. Zare, New York, Wiley, 1988.

The Spectra and Dynamics of Diatomic Molecules, 2nd edn., H. Lefebvre-Brion and R. W. Field, Academic Press, 2004.

Rotational Spectroscopy of Diatomic Molecules, J. M. Brown and A. Carrington, Cambridge, Cambridge University Press, 2003.

Index

Printed in the United States
By Bookmasters